Electron Paramagnetic Resonance Spectroscopy

Patrick Bertrand

Electron Paramagnetic Resonance Spectroscopy

Fundamentals

UGA Editions

Springer

Patrick Bertrand
Aix-Marseille University
Marseille, France

This translation has been supported by **UGA Éditions**, publishing house of Université Grenoble Alpes and the **région Auvergne-Rhône-Alpes**. https://www.uga-editions.com/.

ISBN 978-3-030-39665-7 ISBN 978-3-030-39663-3 (eBook)
https://doi.org/10.1007/978-3-030-39663-3

Translated, revised and adapted from "La Spectroscopie de Résonance Paramagnétique Electronique, Vol. I: Fondements", P. Bertrand, EDP Sciences, 2010.

Cover illustration: Alice Giraud, from: elements provided by the author.

This Springer imprint is published by the registered company Springer Nature Switzerland AG
The registered company address is: Gewerbestrasse 11, 6330 Cham, Switzerland

Grenoble Sciences

The French version of this book received the "Grenoble Sciences" label. "Grenoble Sciences" directed by Professor Jean Bornarel, was between 1990 and 2017 an expertising and labelling centre for scientific works, with a national accreditation in France. Its purpose was to select the most original high standard projects with the help of anonymous referees, then submit them to reading comittees that interacted with the authors to improve the quality of the manuscripts as long as necessary. Finally, an adequate scientific publisher was entrusted to publish the selected works worldwide.

About this Book

This book is translated, revised and adapted from *La spectroscopie de résonance paramagnétique éléctronique – Fondements* by Patrick Bertrand, EDP Sciences, Grenoble Sciences Series, 2010, ISBN 978-2-7598-0554-9.

The translation from original French version has been performed by: Maighread Gallagher, scientific editor, TWS Editing.

The reading committee of the French version included the following members:

- ▹ E. Belorizky, professor at Grenoble Alpes University
- ▹ J.L. Cantin, lecturer at Pierre & Marie Curie University, Paris 6
- ▹ G. Chouteau, professor at Grenoble Alpes University
- ▹ P. Turek, professor at Strasbourg University

Preface

After long being reserved for physicists capable of developing the equipment and interpreting the spectra, electron paramagnetic resonance (EPR) spectroscopy was found to be a very effective technique for the study of extremely varied phenomena. Examples illustrating the diversity and advantages of its current applications include:

- ▹ Development of new materials for electronics, telecommunications, imaging, lasers, and future quantum computers.
- ▹ Dating of rocks and sediments for mineralogy, paleontology and archaeology, the search for the first traces of life in fossilised organic matter.
- ▹ Analysis of the catalytic mechanism of enzymes and their models, in particular in the context of biotechnologies.
- ▹ Design of traps to study radicals produced by oxidative stress, identification of radical species involved in allergies, elucidation of the role played by metal ions in neurodegenerative diseases.
- ▹ Dosimetric techniques, such as *in vivo* oxymetry or measurement of ionising radiation absorbed during accidental irradiation.
- ▹ Chemistry of the environment. For example, tracing the organic matter dissolved in water or studying the role played by transition ions in the degradation of plant matter.

During conferences and summer schools, frequent requests were made for books dealing with the basics of EPR spectroscopy and its applications, written for beginners. Indeed, students and researchers who wish to acquire the basics of EPR or who want to complete their knowledge will only find books that are often old and voluminous in their libraries. The principles and methods of EPR spectroscopy are generally illustrated in these works by studies performed on single-crystal samples containing molecules with a high symmetry. These studies are very interesting from a theoretical point of view, but they are difficult for beginners to understand and they fail to provide the practical information required to interpret the complex spectra produced by most samples used in current research, in which several types of low symmetry molecules are found with completely random orientation.

For this reason, we considered it timely to produce a "manual" for students and researchers which is accessible to a broad audience, and written to help them interpret the spectra and understand the applications of this technique. This book is organised in two parts:

- ▹ The first volume is devoted to the bases of EPR spectroscopy. Emphasis is placed on understanding the phenomena determining the *shape* and *intensity* of the spectrum. Our approach is different to that of the classical works through the following points:
  - *The progressive nature of the description*. For example, the first two chapters can be read by, and will be of benefit to, any student enrolled in a primary degree course, the word *Hamiltonian* only appears in chapter 3, and zero-field splitting terms are only dealt with in chapter 6. Each chapter finishes with a "points to consider in applications" section and a few simple exercises, the solutions to which are given at the end of each chapter. The exercises are numerical applications, spectrum analyses or demonstrations of results used in the main text. The main terms used in the book and their definitions can be found in the *glossary*.
  - *Learning at several levels*: each chapter includes *complements* which need not be read to understand the main body of the text, but which may be of interest to more demanding readers. They contain details of some calculations, and specific points are developed. *Appendices* of a more general nature can be found at the end of the book.
  - *Novel content*. The formalism of quantum mechanics is considered here as a "toolbox" to analyse the shape and intensity of the EPR spectrum, the justification for this approach can be found in specialised texts. The basic tool, perturbation theory, is presented in a less abstract manner than in the usual formalism. We have insisted on the differences between magnetic interactions to which the unpaired electrons in a centre are subjected and the spin Hamiltonian operators which reproduce their effects on the ground state. The *intensity* of the lines, which is central to the resonance phenomenon, is often treated too briefly in the classical works. We have detailed the calculation resulting in its expression, insisting on the practical advantages of intensity

measurements which could be more frequently exploited in EPR applications. Most of the expressions used are demonstrated as we share the opinion of our colleague Blaise Pascal on this point: "*People are generally better persuaded by the reasons which they have themselves discovered than by those which have come into the mind of others.*" [*Pensées*, 1670]. The work is illustrated with numerous original diagrams and by previously unpublished spectra kindly communicated by colleagues.

▹ The second volume presents examples of applications of EPR spectroscopy in various fields ranging from chemistry, to biology and physics. It was written by specialists and extensively relies on the contents of the first volume, although development of some points are sometimes necessary. It is aimed at both beginners and confirmed researchers. Its objectives are, indeed, multiple:
- To show how the principles developed in the first volume are implemented in real cases.
- To present a collection of methods and techniques that can be exploited in applications other than those for which they were initially developed. For example, a procedure to calibrate Fe^{3+} ions in a kaolinite can be very useful to quantify the signal due to "free iron" in samples of metalloproteins. Similarly, the problems raised by the calculation of the spectrum of a pair of centres coupled by weak dipolar and exchange interactions are always the same, whether dealing with rare earth ions substituted into a crystal or metal centres in a biological macromolecule.
- To illustrate the wide range of current applications of EPR spectroscopy. This specificity merits the emphasis given here, in these days of multidisciplinarity as buzz-word.

Notes on physical quantities and their units

▹ The only mechanical quantity used in this volume is the *angular momentum* which is expressed in joule second [J s]. In practice, it is replaced by a dimensionless "reduced angular momentum" defined in section 1.4.1. The magnetic quantities are the *magnetic field*, which is expressed in tesla [T] and the *magnetic moment* which is expressed in ampère square metre [A m^2] or in joule $tesla^{-1}$ [J T^{-1}]. In some works, the magnetic field is still expressed in gauss [G] (1 G = 10^{-4} T).

▹ The international unit for energy, the joule [J], is not at all appropriate when measuring the small differences between energy levels in atoms and molecules which are exploited by spectroscopic techniques. In EPR spectroscopy, we therefore use:

- The *frequency* ν defined by $\nu = E/h$, where E is the energy and h is Planck's constant ($h = 6.6261 \times 10^{-34}$ J s), expressed in MHz or GHz:

$$1 \text{ MHz} = 6.6261 \times 10^{-28} \text{ J}$$

- The *wave number* defined by $1/\lambda = E/hc$, where c is the speed of light in a vacuum, expressed in cm^{-1}:

$$\begin{aligned} 1 \text{ cm}^{-1} &= 1.9864 \times 10^{-23} \text{ J} \\ &= 2.9979 \times 10^{4} \text{ MHz} \end{aligned}$$

The fundamental physical constants and some formulae for the conversion of units are given on page 1.

Acknowledgements

Several colleagues contributed to this volume by providing figures and spectra or by critically rereading certain chapters:

Carole Baffert, Valérie Belle, Frédéric Biaso, Bénédicte Burlat, Pierre Ceccaldi, Pierre Dorlet, Emilien Etienne, André Fournel, Yves Frapart, Philippe Goldner, Didier Gourier, Stéphane Grimaldi, Christophe Léger, Marlène Martinho, Béatrice Tuccio, Philippe Turek, Hervé Vezin and, in particular, Geneviève Blondin.

I am very grateful to them for the gift of their most precious asset: some of their time.

I also want to thank all the members of the team at UGA Editions, whose professional attitudes and readiness to answer all my questions I greatly appreciated.

Contents

Fundamental constants
Unit conversions

Reference: *Handbook of Chemistry and Physics* (2009–2010) CRC

Fundamental constants

Electron rest mass	$m_e = 9.109382 \times 10^{-31}$ kg
Proton rest mass	$m_p = 1.672622 \times 10^{-27}$ kg
Elementary charge	$e = 1.602176 \times 10^{-19}$ C
Speed of light in a vacuum	$c = 2.997925 \times 10^{8}$ m s^{-1}
Boltzmann constant	$k_B = 1.380650 \times 10^{-23}$ J K^{-1}
Vacuum permeability	$\mu_0 = 4\pi \times 10^{-7}$ H m^{-1}
Bohr radius	$a_0 = 4\pi\varepsilon_0\hbar^2/m_e e^2 = 5.291772 \times 10^{-11}$ m
Avogadro constant	$\mathcal{N}_A = 6.022142 \times 10^{23}$ mol^{-1}
g factor for electrons	$g_e = 2.002319$
Planck's constant	$h = 6.626069 \times 10^{-34}$ J s
	$\hbar = 1.054670 \times 10^{-34}$ J s
Electronic Bohr magneton	$\beta = 9.274009 \times 10^{-24}$ J T^{-1}
	$h/\beta = 7.144773 \times 10^{-11}$ T s
Nuclear Bohr magneton	$\beta_N = 5.050783 \times 10^{-27}$ J T^{-1}
Gas constant	$R = 8.314472$ J mol^{-1} K^{-1}
Faraday constant	$\mathcal{F} = 9.648534 \times 10^{4}$ C mol^{-1}

Unit conversions

$$1 \text{ eV} = 1.602176 \times 10^{-19} \text{ J}$$

$$1 \text{ cm}^{-1} = 1.986445 \times 10^{-23} \text{ J}$$

$$1\ \text{MHz} = 6.626069 \times 10^{-28}\ \text{J}$$

$$1\ \text{cm}^{-1} = 2.997925 \times 10^{4}\ \text{MHz}$$

$$k_\text{B} = 0.6950356\ \text{cm}^{-1}\ \text{K}^{-1}$$

$$= 2.083664 \times 10^{4}\ \text{MHz K}^{-1}$$

Chapter 1

The electron paramagnetic resonance phenomenon

1.1 – What is a spectroscopy experiment?

1.1.1 – Exchange of energy between matter and radiation

In physics, a distinction is made between two entities of very different natures:

- ▹ *Matter* composed of nuclei, atoms and molecules, which are characterised by their energy levels.
- ▹ *Radiation* which is composed of *electromagnetic waves*. Each electromagnetic wave is formed by an electric field $\mathbf{e}_1(t)$ and a magnetic field $\mathbf{b}_1(t)$ which vary in a sinusoidal manner with the same *frequency* ν. These fields are perpendicular to each other and they propagate in a vacuum at a speed, c, equal to 2.9979×10^8 m s^{-1}. An electromagnetic wave *transports energy*, the amount of which is proportional to the square of the amplitude of the electric and magnetic fields.

Matter and radiation can *exchange energy* in discrete amounts known as *photons* equal to $h\nu$, where h is Planck's constant ($h = 6.6261 \times 10^{-34}$ J s) and ν, as above, is the frequency of the wave. These exchanges make the radiation perceptible, either directly or through the use of measurement devices. How we detect them depends largely on the frequency value, which is why various frequency ranges were discovered and named as physics progressed (figure 1.1).

P. Bertrand, *Electron Paramagnetic Resonance Spectroscopy*,
https://doi.org/10.1007/978-3-030-39663-3_1

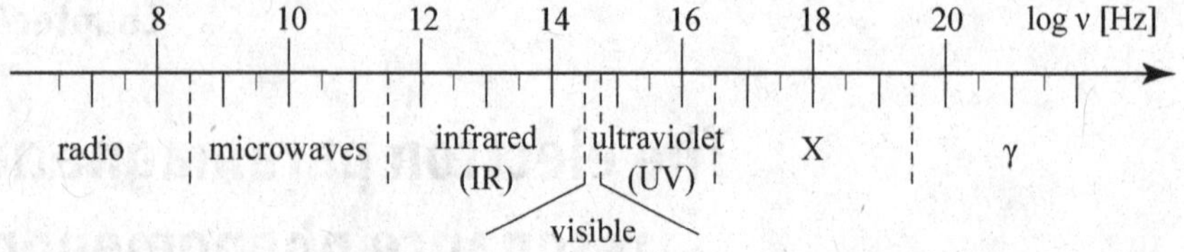

Figure 1.1 – The different frequency ranges of electromagnetic radiation.

Energy exchanges between matter and radiation play a fundamental role in all the processes occurring in the Universe. Charles Fabry clearly stated the reason for this in his work *Les Radiations* published in 1946: *Electromagnetic waves propagate through space independently of all matter; in a vacuum they lose no energy, and no electromagnetic waves can appear. Radiation can neither appear nor disappear in the absence of matter. The two fundamental processes, emission of radiation and its absorption, involve energy exchange between matter and radiation*. Nowadays, these exchanges occupy a growing place in our daily lives, for example in the fields of telecommunication (radio, television, telephone, internet) or health (laser microsurgery, X-ray radiology, radiotherapy and sterilisation of food by ultraviolet and gamma radiation). In a much more basic way, radiation plays an essential role in how we perceive the world: through the absorption of visible radiation by the cells in our retinas which triggers the sequence of processes resulting in *vision*; and through the absorption of infrared radiation by the nerves in our skin, causing us to feel *heat*. Finally, we should recall that life would probably not exist on Earth if it were not for photosynthesis, through which photosynthetic organisms exploit molecules with extraordinary properties, the chlorophylls, to harvest solar radiation.

1.1.2 – Spectroscopic techniques

Energy exchange between radiation and matter only occurs in certain conditions. In particular:

- ▹ Some *interaction* must occur between the electromagnetic field and elementary entities of the matter.
- ▹ The photon's energy $h\nu$ must be equal to the energy difference between two energy levels for these entities.

Long before these conditions became known, numerous experiments relying on energy exchanges were performed and helped advance our knowledge of the nature of radiation and the composition of matter. The first of these experiments was performed in 1669, when Isaac Newton undertook the rational study of the decomposition of white light by a prism. Indeed, it was Newton who used the term "spectrum" to describe the unexpected appearance of coloured lines on a screen. Since the end of the 19th century, we have understood that energy exchanges can be used to study the properties of a sample *at the scale of atoms and molecules*. Spectroscopy experiments consist in causing electromagnetic waves with a well-defined frequency to interact with the components of a sample to "probe" their energy levels. Under appropriate conditions, this interaction induces *transitions* between energy levels, leading to detectable and measurable energy absorption. By varying the frequency ν of the electromagnetic waves, a *spectrum* composed of absorption lines providing information on the components of a sample is obtained:

- ▹ The *shape* of the spectrum reflects the energy level diagram, which depends on the nature and details of the structure of the components.
- ▹ Its *intensity*, which is generally proportional to the number of components present in the sample, can be exploited in assays, titrations or kinetic experiments.

Researchers are generally only interested in selected energy levels of well-defined microscopic entities, and they use a spectrometer "working" in an appropriate frequency range, over which the interaction between the radiation and the components of the sample is optimised. All the available radiation frequency ranges are exploited by different spectroscopic techniques:

- ▹ radiowaves in NMR (Nuclear Magnetic Resonance) to collect information on molecules by splitting the energy levels of their nuclei through application of a magnetic field,
- ▹ microwaves to study the electronic structure of molecules by EPR (Electron Paramagnetic Resonance) thanks to their magnetic properties,
- ▹ infrared radiation is used to probe the rotational and vibrational energy levels in molecules,
- ▹ UV-visible radiation reveals the electronic energy levels of molecules and their vibrational properties thanks to the Raman effect,
- ▹ X-rays allow the energy levels of atoms to be explored,
- ▹ gamma radiation is used in Mössbauer spectroscopy to study the electronic properties of atoms and molecules based on the energy levels of some nuclei.

Any given spectroscopic technique can be used in many different ways. For example, to identify a molecule among a limited number of different molecules, it is sufficient to compare its "spectral signature" to those contained in a "library", which itself is often stored in the memory of experienced scientists. Similarly, titration or kinetic experiments can be performed by measuring *relative variations* in the intensity of the spectrum. In other types of applications, researchers can encounter spectral forms which they have never previously observed nor even seen in the literature. To identify the molecules producing these spectra, they must be able to extract a maximum number of characteristics, possibly through numerical simulation, and to interpret them to discover their structure. Likewise, to determine the number of molecules contained in a sample, it is essential to know how to measure the intensity of the spectrum and compare it to that of a reference sample of known concentration. Finally, the experimental conditions in which the spectrum is recorded may affect its shape and/or intensity. The adjustable experimental parameters are not the same in all spectroscopies, but the *temperature* of the sample, the *frequency* and *intensity* of the radiation with which it exchanges energy always plays a significant role.

Scientists who wish to use spectroscopic techniques for these ends must know all the parameters which determine the shape and intensity of the spectrum.

1.2 – Magnetic spectroscopy techniques

An electromagnetic field can interact with the microscopic entities of matter in two ways:

1. The electric component $\mathbf{e}_1(t)$ interacts with the *electric dipole moments* due to the heterogeneous distribution of charges at the molecular scale. This interaction can induce numerous transitions in atoms and molecules; it is responsible for the phenomena mentioned in section 1.1.1 and is often used in spectroscopic techniques.
2. The magnetic component $\mathbf{b}_1(t)$ interacts with the *magnetic moments* of some so-called *paramagnetic* entities. These are mainly:
    - some *atomic nuclei*, their magnetic moment is due to the protons and neutrons,
    - atoms and molecules with *unpaired electrons*, such as free radicals and transition ion complexes. Hereafter, the entities for which the magnetic moment is of electronic origin will be called *paramagnetic centres*.

The magnetic-type interaction concerns only certain types of matter entities and they are much weaker than electric interactions. In addition, the energy levels between which magnetic interactions induce transitions are often only split *in the presence of a magnetic field*. For these reasons, this interaction only plays a minor role in the energy exchanges between radiation and matter taking place in the world around us. However, it can be exploited to perform *spectroscopy experiments* by placing paramagnetic entities in a magnetic field **B**. As the energy exchange with radiation only takes place when the splitting between energy levels created by **B** is adjusted to that of the energy of the incident photons, "magnetic spectroscopy techniques" are therefore termed "resonant": Nuclear Magnetic Resonance (NMR) for the study of nuclei, Electron Paramagnetic Resonance (EPR) when examining paramagnetic centres. These experiments provide information on the atoms and molecules through the magnetic characteristics of their nuclei or electrons.

Besides NMR and EPR, which are extensively used in laboratories, other magnetic spectroscopy techniques have been developed for specific applications:

- ▹ When the unpaired electrons in a paramagnetic centre interact with a paramagnetic nucleus, the intensity of its EPR spectrum is modified when NMR transitions are induced between some of the nucleus's energy levels. This effect is exploited in ENDOR (Electron Nuclear DOuble Resonance) spectroscopy, which very precisely measures these interactions.
- ▹ Transitions between energy levels of some paramagnetic nuclei can occur without requiring application of a field **B**. These transitions require very high-energy gamma radiation, as used in Mössbauer spectroscopy.

In the following chapters, we will gradually discover the relationships existing between the shape of the EPR spectrum produced and the characteristics of the paramagnetic centres present in the sample. In this chapter, we will start to familiarise ourselves with paramagnetic centres, and we describe the principle of an EPR experiment.

1.3 – Diversity of paramagnetic centres

1.3.1 – Electrons have two magnetic moments

According to classical electromagnetic theory, movement of a point of mass m carrying an electric charge q creates a *magnetic moment* **μ** which can be written

$$\boldsymbol{\mu} = (q/2m)\,\boldsymbol{\sigma} \qquad [1.1]$$

where $\boldsymbol{\sigma}$ is a mechanical quantity linked to movement of the point mass, known as its *angular momentum*. Complement 1 provides its definition and verifies relation [1.1] in the specific case of uniform circular movement.

To explain magnetic phenomena at the atomic scale, it must be postulated that this relation also applies to the elementary particles. In particular, any electron for which the movement is characterised by an angular momentum $\boldsymbol{\sigma}_l$ has a magnetic moment given by

$$\boldsymbol{\mu}_l = -(e/2m_e)\,\boldsymbol{\sigma}_l \qquad [1.2]$$

where e is the elementary charge and m_e is the mass of the electron. To specify that the quantities $\boldsymbol{\sigma}_l$ and $\boldsymbol{\mu}_l$ originate from the *movement* of electrons, they are called the *orbital angular momentum* and the *orbital magnetic moment* of the electron. Like many physical quantities, angular momenta are *quantized* at the microscopic level. Thus, their components can only take certain values which are defined as follows: if $\{x, y, z\}$ is any orthogonal system of axes and $(\sigma_{lx}, \sigma_{ly}, \sigma_{lz})$ are the components of $\boldsymbol{\sigma}_l$ in this reference frame, the quantity $\boldsymbol{\sigma}_l^2 = \sigma_{lx}^2 + \sigma_{ly}^2 + \sigma_{lz}^2$ can only take values of the form

$$\boldsymbol{\sigma}_l^2 = \ell(\ell + 1)\hbar^2 \qquad [1.3]$$

where ℓ is one of the numbers

$$\ell = 0, 1, 2, \ldots \qquad [1.4]$$

The symbol $\hbar$ represents the quantity $h/2\pi$, where h is Planck's constant.

In addition, for a fixed value of ℓ, one of the components of $\boldsymbol{\sigma}_l$, for example σ_{lz}, can only take values of the form

$$\sigma_{lz} = m_\ell\,\hbar \qquad [1.5]$$

The possible values of m_ℓ are

$$m_\ell = -\ell, -\ell + 1, \ldots \ell \qquad [1.6]$$

In addition to the angular momentum $\boldsymbol{\sigma}_l$ linked to its movement, each electron has a *spin angular momentum* $\boldsymbol{\sigma}_s$. Its components obey relations similar to equations [1.3] and [1.5]

$$\boldsymbol{\sigma}_s^2 = s(s + 1)\hbar^2; \quad \sigma_{sz} = m_s\hbar$$

But, while ℓ can take any integer value (equation [1.4]), the number s is an *intrinsic characteristic* of the electron like its mass or charge, known as *its spin*. This spin is equal to $s = ½$, therefore m_s can only take the values $\pm ½$.

A *spin magnetic moment* $\boldsymbol{\mu}_s$ defined by a relation similar to equation [1.2] corresponds to the spin angular momentum $\boldsymbol{\sigma}_s$:

$$\boldsymbol{\mu}_s = -g_e (e/2m_e)\, \boldsymbol{\sigma}_s \qquad [1.7]$$

where $g_e = 2.0023$. The electron's *total* magnetic moment can therefore be written

$$\boldsymbol{\mu} = -(e/2m_e)\,(\boldsymbol{\sigma}_l + g_e\, \boldsymbol{\sigma}_s) \qquad [1.8]$$

Given the close relationship existing between angular momenta and magnetic moments, angular momenta will be found throughout this book. It could be thought that any sample of matter, which necessarily contains a very large number of electrons, has a magnetic moment. However, this is not the case for a number of reasons:

- ▹ A sample of matter is composed of atoms or molecules, and we will see that the restrictions imposed on the electrons in these entities are such that the sum of their magnetic moments is often null.
- ▹ Even when the atoms or molecules have a magnetic moment, their resultant at the level of the sample can vanish due to the effects of thermal agitation.

1.3.2 – Paramagnetic atoms

In the simplest model of the atom, electrons are considered to be *independent*, each being subject to the attraction of the nucleus and the *average* repulsive interaction due to the other electrons. In these conditions, it can be shown that [Ayant and Belorizky, 2000; Cohen-Tannoudji *et al.*, 2015]:

- ▹ The electron's position is *indeterminate*: we can only define the probability density of its presence at any point in space, which is equal to the square of a function known as an *atomic orbital*. Each orbital is identified by a triplet of numbers (n, ℓ, m_ℓ). The number n can take one of the values $\{1, 2, 3, \ldots\}$. For any value of n, ℓ can take one of the n values $\{0, 1, 2, \ldots n-1\}$ and m_ℓ one of the $(2\ell + 1)$ values defined by equation [1.6]. This triplet determines the "shape" of the orbital. For example, it should be remembered that the numbers $\ell = 0, 1, 2, 3$ correspond to *s*, *p*, *d*, *f*-type orbitals.
- ▹ In contrast, when an electron "occupies" the orbital defined by (n, ℓ, m_ℓ), its energy and the possible values of the components of its orbital angular momentum are perfectly *determined*:
    - Its energy is defined by the numbers (n, ℓ): $E_{n,\ell}$. This energy is negative and its absolute value is smaller for larger values of these numbers.

- The possible values of the components of its orbital angular momentum in any Cartesian reference frame $\{x,y,z\}$ are defined by equations [1.5] and [1.6].

We also use the following terminology:

▹ A *shell* is composed of all the orbitals characterised by the same value of n. Their number can be shown to be equal to n^2 (exercise 1.1).

▹ A *subshell* is composed by all the $(2\ell+1)$ orbitals characterised by a given (n,ℓ) pair. All of these orbitals correspond to the same energy $E_{n,\ell}$: we say that the energy level $E_{n,\ell}$ is *degenerate* and its degeneracy is equal to $(2\ell+1)$.

Here, the first shells and subshells of atoms are sorted *in increasing order of energy*:

Shell	Subshell	Orbitals	
$n = 1$	$\ell = 0$	$m_\ell = 0$	(1*s* orbital)
$n = 2$	$\ell = 0$	$m_\ell = 0$	(2*s* orbital)
	$\ell = 1$	$m_\ell = -1, 0, 1$	(three 2*p* orbitals)
$n = 3$	$\ell = 0$	$m_\ell = 0$	(3*s* orbital)
	$\ell = 1$	$m_\ell = -1, 0, 1$	(three 3*p* orbitals)
$n = 4$	$\ell = 0$	$m_\ell = 0$	(4*s* orbital)
$n = 3$	$\ell = 2$	$m_\ell = -2, -1, 0, 1, 2$	(five 3*d* orbitals)

As the electrons are assumed to be independent, an "electronic configuration" for the atom can be obtained by attributing an orbital to each one. We are generally interested in the configuration of lowest energy, known as the *ground* configuration, from which the periodic table of the elements is constructed. To obtain this ground configuration, the orbitals are "filled" starting with the innermost of lowest energy, taking Pauli's exclusion principle into account, which requires that an orbital harbours at most two electrons; for one of these electrons $m_s = -\frac{1}{2}$, while for the other $m_s = +\frac{1}{2}$.

We consider an atom with N electrons in a given (n,ℓ) subshell, and term $\boldsymbol{\mu}(n,\ell)$ the sum of their magnetic moments, as defined by equation [1.8]. The projection of $\boldsymbol{\mu}(n,\ell)$ on any axis z is written

$$\mu_z(n,\ell) = -(e/2m_e)\sum_{i=1}^{N}[\sigma_{lz}(i) + g_e\sigma_{sz}(i)]$$

The sum is performed over all the electrons in the subshell. The value of $\mu_z(n, \ell)$ is therefore equal to

$$\mu_z(n,\ell) = -(e/2m_e) \sum_{i=1}^{N} [m_\ell(i) + g_e m_s(i)]$$

When the subshell is *full*, all of the possible values of m_ℓ are used and their sum is null according to the list [1.6]. The same is true for the sum of the values of m_s. The projection of $\boldsymbol{\mu}(n, \ell)$ is therefore null: the full subshells do not contribute to the magnetic moment of an atom. For an atom to have a magnetic moment, one of its subshells, *a priori* the highest energy subshell, must be *incomplete*.

Examination of the periodic table of the elements reveals that numerous atoms have an incomplete subshell (table 1.1).

Table 1.1 – Atoms with an incomplete subshell.

1*s*: H	
2*s*: Li	
2*p*: B, C, N, O, F	
3*s*: Na	
3*p*: Al, Si, P, S, Cl	
4*s*: K	
3*d*: Sc, Ti, V, Cr*, Mn, Fe, Co, Ni, Cu*	first transition series
4*d*: Y, Zr, Nb*, Mo, Tc, Ru*, Rh*, Ag* (Pd**)	second transition series
4*f*: Ce*, Pr*, Nd*, Pm, Sm, Eu, Gd*, Tb*, Dy, Ho, Er, Tm, Lu* (Yb**)	rare earths
5*d*: La*, Hf, Ta, W, Re, Os, Ir, Pt*, Au* (Hg**)	third transition series
5*f*: Th, Pa, U, Np, Pu, Am, Cm, Bk, Cf, Es, Fm, Md, No, Lr	actinides

From potassium onwards, we have only indicated the elements in transition metal series, rare earths and actinides. Superscript * indicates that the atomic subshells are not "regularly" filled. For example, in the first transition series, the electronic configuration of Cr is (Ar) $4s^1\,3d^5$ and that of Cu is (Ar) $4s^1\,3d^{10}$. Superscript ** indicates that the atom is non-paramagnetic, but that its cation can be paramagnetic. Indeed, transition elements, rare earths and actinides can readily release electrons to produce cations which are often paramagnetic.

The existence of an incomplete subshell is *necessary* but not *sufficient* for an atom to be paramagnetic: in their lowest energy state, numerous atoms listed in table 1.1 are not paramagnetic (appendix 1).

1.3.3 – Paramagnetic molecules

In samples studied by EPR, the elements are generally organised in *molecules*. In molecules, the state of the electrons in the inner atomic shells, which are complete, is practically unchanged. In contrast, that of the "valence electrons" in *molecular orbitals*, which contribute to chemical bonds, is profoundly modified. Each molecular orbital can accept two electrons characterised by opposite values of m_s. Various situations can thus emerge:

- ▷ When a molecule has an *even number* of electrons, the molecular orbitals are generally doubly-occupied and the sum of the values of σ_{sz} is null in the ground state. From very general considerations based on symmetry properties, the sum of values of σ_{lz} can also be shown to be null. In general, these molecules are therefore *non-paramagnetic*. There nevertheless exist two exceptions: some molecules with an even number of electrons have two unpaired electrons in their ground state. These are said to be in a "triplet state", as we will see that they can exist in three different spin states. This is the case with B_2 and O_2 (complement 2) and "biradicals" composed of two coupled entities each having an unpaired electron. Metastable triplet states can be created by photolysis of some organic molecules trapped in a solid at low temperature. In addition, aromatic molecules such as naphtalene, for which the ground state is not paramagnetic, have a triplet excited state which can be generated and studied by EPR by continuously exposing the sample to ultraviolet radiation.
- ▷ In molecules with an *odd number* of electrons, an unpaired electron is necessarily present. These molecules are known as *free radicals*. Examples include small molecules such as $OH^{\bullet}$, $CN^{\bullet}$, $NO^{\bullet}$, $NO_2^{\bullet}$, and numerous organic and inorganic molecules. Free radicals are generally highly reactive, but some have a long enough lifetime to allow their study by EPR spectroscopy.
- ▷ Cations of transition, rare earth and actinide elements form mono or polynuclear coordination complexes with a wide variety of ligands. In these "metal complexes", the cations often retain their paramagnetic characteristics, sometimes with a high number of unpaired electrons. These cations are found with the same types of coordination when they are included as substituted cations in some ionic solids. It should be noted that paramagnetic cations are responsible for "ordered magnetism" phenomena, such as ferromagnetism and antiferromagnetism, in certain solids.

Paramagnetic centres exist in many other forms in *solids*. For example, conduction electrons in metals, ionised impurities in semi-conductors, organic molecular conductors, and a wide variety of point defects.

1.4 – Principle of electron paramagnetic resonance experiments

In this section, we describe an EPR experiment based on a simple example, without attempting to be exhaustive, while nevertheless introducing the main parameters determining the *position* and *intensity* of the resonance signal returned by the spectrometer.

1.4.1 – Reduced angular momenta

We will start by simplifying the notations. Equation [1.8], which gives the magnetic moment of an electron, can be written

$$\boldsymbol{\mu} = -\beta(\mathbf{l} + g_e\,\mathbf{s}) \qquad [1.9]$$

where $\mathbf{l} = \boldsymbol{\sigma}_l/\hbar$ and $\mathbf{s} = \boldsymbol{\sigma}_s/\hbar$ are dimensionless "reduced angular momenta", and β is a universal constant known as the "electronic Bohr magneton" which takes the value

$$\beta = e\hbar/2m_e = 9.2740 \times 10^{-24}\ \mathrm{J\,T^{-1}}$$

Hereafter, the magnetic moments will be expressed using reduced angular momenta, like in equation [1.9]. This equation shows that the electronic magnetic moments are of the same order of magnitude as β.

1.4.2 – Interaction between a paramagnetic centre and a magnetic field

We will consider a paramagnetic centre which has a certain energy in the absence of a magnetic field, and for which the magnetic moment takes the form

$$\boldsymbol{\mu} = -g\beta\mathbf{S} \qquad [1.10]$$

where **S** is an angular momentum complying with the usual rules:

- ▹ the value of $\mathbf{S}^2$ is equal to $S(S + 1)$.
- ▹ the projection S_z of **S** on any axis z can take one of the $(2S + 1)$ values:

$$M_S = -S, -S+1, \ldots S$$

The properties of the angular momentum **S** are therefore entirely determined by the *number* S. In general, S is said to be the "spin" of the paramagnetic centre, even if its value is not always only determined by the spin angular momenta of the electrons, as we will see later.

The g "factor" is a positive dimensionless number characterising the paramagnetic centre, which we wish to *measure*. To do so, the centre must be placed in a *magnetic field* **B**. The interaction energy between **μ** and **B** is given by:

$$H = -\boldsymbol{\mu}\cdot\mathbf{B} = g\beta\mathbf{S}\cdot\mathbf{B}$$

If we choose the z axis in the direction of **B**, H is written:

$$H = g\beta B S_z$$

where B is the modulus of **B**. As the S_z component of the angular momentum is quantized, so is the interaction energy H. Its possible values are:

$$E(M_S) = g\beta B M_S; \; M_S = -S, -S+1, \dots S \qquad [1.11]$$

Interaction of the magnetic moment with the magnetic field **B** therefore produces a pattern with $(2S + 1)$ equidistant energy levels separated by $\Delta E = g\beta B$; the lowest levels correspond to the most negative values of M_S. This pattern is centred on the energy level of the paramagnetic centre in the absence of a magnetic field (figure 1.2), and the interaction with the magnetic field is said to "remove the degeneracy" of this level. This phenomenon is known as the Zeeman effect.

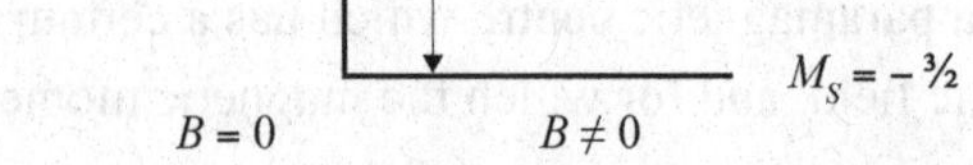

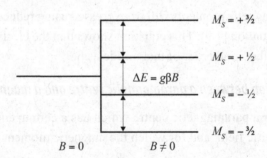

Figure 1.2 – Splitting of energy levels for a paramagnetic centre placed in a magnetic field. The diagram corresponds to $S = 3/2$.

These energy levels will be used to produce the EPR transitions which will allow us to measure the g factor. It should be noted that the splitting of the energy levels produced by the magnetic fields generally available in laboratories is very weak: for the typical value $g = 2.0$, a field of 1 tesla, which is already intense, causes a splitting $\Delta E \approx 2\times10^{-23}$ joule (0.9 cm^{-1}). This value is much smaller than the splitting between vibrational energy levels in molecules.

1.4.3 – EPR transitions

We will now subject the paramagnetic centre placed in the field **B** to an electromagnetic wave of frequency ν. The magnetic component $\mathbf{b}_1(t)$ of this wave takes the form:

$$\mathbf{b}_1(t) = \mathbf{B}_1 \cos 2\pi \nu t$$

In section 3.4, we will see that the interaction between $\mathbf{b}_1(t)$ and the magnetic moment **μ** of the centre produces transitions between the energy levels $E(M_S)$ (equation [1.11]) only if $\mathbf{b}_1(t)$ is *not parallel* to **B**. We will also see that these transitions satisfy certain conditions:

1. They only occur between *adjacent* energy levels in the diagram shown in figure 1.2. This condition can be expressed by saying that the "selection rule" is:

$$\Delta M_S = \pm 1$$

2. The photon energy $h\nu$ of the radiation must be equal to the difference between the energy levels involved in the transitions; this is the "resonance condition". For purely technical reasons, which we will come back to later, the frequency ν of the radiation is constant in an EPR experiment and the value of the magnetic field is varied (figure 1.3a). Transitions take place when B takes the value B_0 such that

$$h\nu = g\beta B_0 \qquad [1.12]$$

These transitions lead to an absorption of electromagnetic energy; in the spectrometer, this absorption generates an *absorption signal* $s(B)$ which is known as a resonance line (figure 1.3b).

When the magnetic moment has the form of equation [1.10], the EPR spectrum will contain a single *resonance line*. Its position B_0 is independent of the *direction* of **B** relative to the paramagnetic centre, since this parameter is not involved in the calculation.

Knowing the values of the frequency ν and of the resonance field B_0, from equation [1.12] we can deduce that of the g factor which characterises the paramagnetic centre. However, we will see that the absorption signal also supplies information on the *number of centres* contained in the sample.

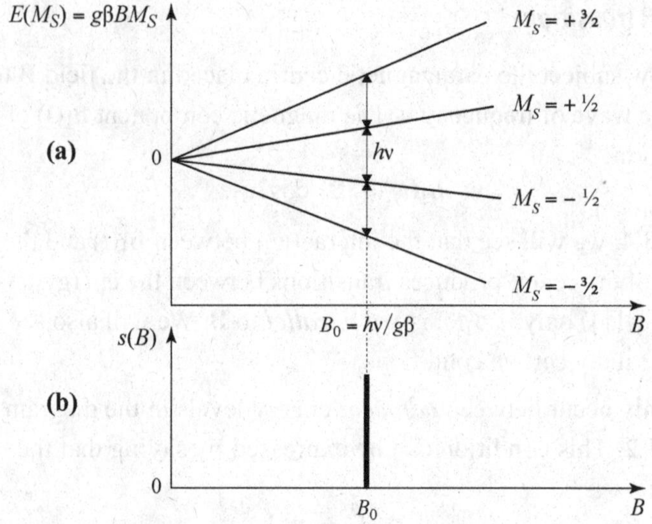

Figure 1.3 – **(a)** Energy levels from figure 1.2 as a function of *B*. **(b)** Formation of the absorption signal at resonance.

1.4.4 – Expression of the absorption signal

A sample containing *N* paramagnetic centres identical to the one described previously is placed in the field **B** and we assume (for simplicity) that the spin *S* is equal to ½. For each centre, the energy levels identified by $M_S = \pm 1/2$ will be split by $\Delta E = g\beta B$. Among all the centres in the sample, some occupy the state characterised by $M_S = -1/2$, while others occupy the state characterised by $M_S = +1/2$. We will use N_- and N_+ to identify the "populations" of the two levels (figure 1.4).

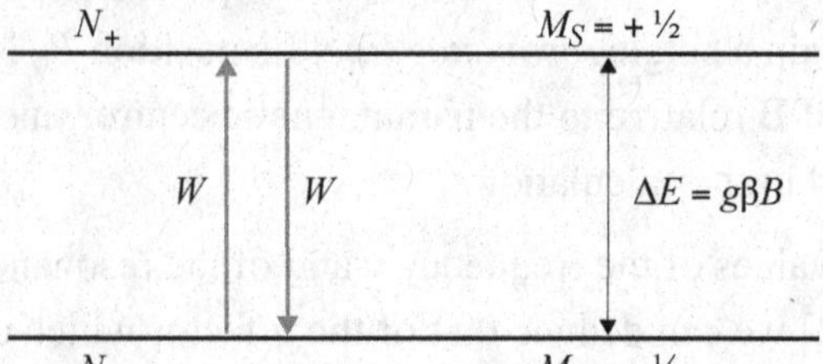

Figure 1.4 – Transitions between energy levels for a paramagnetic centre of spin $S = 1/2$.

Naturally:

$$N_- + N_+ = N \qquad [1.13]$$

If the paramagnetic centres are in *thermal equilibrium* at the temperature T, the populations satisfy the equation:

$$N_+/N_- = \exp(-\Delta E/k_B T) \qquad [1.14]$$

where k_B is the Boltzmann constant. When the sample is subjected to an electromagnetic radiation of frequency ν, transitions take place in each centre when the value of B is such that the resonance condition [1.12] is satisfied. The *rate* of these transitions is characterised by a number W known as the *transition probability per unit time*, which has the following signification (figure 1.4):

- ▹ The number of transitions per second from $M_S = -½$ towards $M_S = +½$ in the sample is equal to $W N_-$. Under the influence of these transitions, the paramagnetic centres *absorb* the power $W N_- \Delta E$.
- ▹ The number of transitions per second from $M_S = +½$ towards $M_S = -½$ is equal to $W N_+$. Under the effect of these transitions, the paramagnetic centres *emit* the power $W N_+ \Delta E$.

The *net power absorbed* at resonance is therefore equal to:

$$P_a = W(N_- - N_+)\,\Delta E \qquad [1.15]$$

From equations [1.13] and [1.14], we can deduce the difference $(N_- - N_+)$:

$$N_- - N_+ = N\frac{1-\exp(-\Delta E/k_B T)}{1+\exp(-\Delta E/k_B T)}$$

which can be written (exercise 1.3):

$$N_- - N_+ = N\tanh(\Delta E/2k_B T) \qquad [1.16]$$

Equation [1.15] thus becomes:

$$P_a = W N h\nu \tanh(h\nu/2k_B T) \qquad [1.17]$$

The spectrometer's detection device is designed such that the *absorption signal* that it delivers is related to P_a by:

$$s = \alpha P_a/(h\nu B_1) \qquad [1.18]$$

where B_1 is the amplitude of the magnetic component $\mathbf{b}_1(t)$ of the radiation and α is a scaling factor which depends on the spectrometer. Using equation [1.17], we obtain:

$$s = \alpha W N \tanh(h\nu/2k_B T)/B_1 \qquad [1.19]$$

The absorption signal is thus proportional to the number N of paramagnetic centres contained in the sample. The other parameters which appear in this expression are determined by the experimental conditions. With the frequencies generally used in EPR (see section 1.5.1), the ratio $(h\nu/2k_BT)$ is small when the temperature exceeds a few kelvins, which allows us to replace $\tanh(h\nu/2k_BT)$ by $(h\nu/2k_BT)$ in equation [1.19]. With a standard spectrometer of frequency $\nu \approx 9$ GHz, the temperature only needs to exceed 1 K for the error introduced to be less than 1.5 %, but higher temperatures are necessary at higher frequencies (exercise 1.5). We can therefore write:

$$s \approx \alpha W N (h\nu/2k_BT)/B_1$$

The signal is proportional to the frequency, and inversely proportional to temperature (Curie's law). We will see in section 5.2 that the transition probability W is proportional to $B_1{}^2$, which implies that s is proportional to B_1.

In the specific case that we have just dealt with, the transition probability per second W is only different from zero when $B = B_0$. This can be expressed mathematically by writing $W = w\delta(B - B_0)$, where $\delta(B - B_0)$ is an *infinitely narrow* function centred on B_0, which verifies:

$$\int_0^{+\infty} \delta(B - B_0)\mathrm{d}B = 1$$

This very specific function is known as "Dirac's function" (see complement 1 in chapter 9). With these notations, equation [1.19] becomes:

$$s(B) = \alpha\, w\, N\,[\tanh(h\nu/2k_BT)/B_1]\,\delta(B - B_0) \qquad [1.20]$$

In practice the EPR lines are not infinitely narrow, and we will see in section 5.2 that the Dirac function must be replaced by a function $f(B - B_0)$ describing their *shape*.

1.5 – Basic EPR spectrometry instrumentation

1.5.1 – A few orders of magnitude

The experiment which we just described requires a magnetic field and a source of radiation. According to equation [1.12], resonance occurs when the frequency ν of the radiation and the value B of the field verify the relation:

$$\nu/B = g\beta/h$$

The g values of paramagnetic centres are generally greater than 1, often being close to 2. For the typical value $g = 2.0$ the $g\beta/h$ ratio is equal to 28 GHz T^{-1}. We saw in the previous section that the absorption signal increases with the frequency ν, and therefore with the magnetic field B. With magnetic fields of the order of 1 tesla, which are available with electromagnets, the frequency of the radiation must be of the order of a few tens of GHz – in the "microwave" range of electromagnetic radiation (see figure 1.1). In this range of magnetic fields and frequencies, it is technically easier to vary the value of B than to adjust the frequency of the radiation. To satisfy the resonance condition, we therefore operate at a fixed frequency and sweep the magnetic field.

To gather more information on some paramagnetic centres, their spectra are sometimes recorded at different frequencies. The names of the microwave bands used in EPR are a carryover from radar technology. For "standard" spectrometers, working at X-band ($\nu \approx 9$ GHz), $h\nu$ is equal to 0.3 cm^{-1} and the field B can vary between 0 and around 0.5 T. Some commercial spectrometers also operate at L ($\nu \approx 1$ GHz), S ($\nu \approx 4$ GHz), Q ($\nu \approx 35$ GHz) and W ($\nu \approx 90$ GHz) bands. So-called high-field spectrometers, with frequencies up to 350 GHz for a magnetic field of 12 T, have been built in some specialised laboratories.

1.5.2 – Simplified description of an EPR spectrometer

On the block diagram in figure 1.5, the elements necessary for the experiment described in section 1.4 are shown:

▹ An electromagnet creates a magnetic field **B** with a well defined *direction*. The field sweep is performed by varying the intensity of the electric current running in the coils. Some problems with measurement of B are addressed in chapter 9.

▹ A generator supplies the microwave radiation. Currently, semiconductor-based devices known as "Gunn diodes" are used. The radiation frequency can be varied over a narrow range to adjust it to that of the resonance cavity (see next section). Its value is indicated by a frequency meter. The radiation, of which the power P can be adjusted between 1 µW and around 200 mW, propagates inside a waveguide.

▹ The sample is placed in a resonance cavity. In most spectroscopic techniques, the power absorbed by a sample is determined by comparing the power of the incident radiation to that of the transmitted radiation. To improve sensitivity, a different approach is used in EPR: the sample is placed in a cavity where the radiation creates a standing wave regime such that the magnetic component $\mathbf{b}_1(t)$ is *perpendicular* to **B** and its amplitude B_1 is *maximal* at the centre of the cavity. This amplitude, which is proportional to $\sqrt{P}$, is around 10^{-4} T for P = 200 mW. The cavity is designed to ensure that the electric component $\mathbf{e}_1(t)$ of the radiation is null at its centre. This design avoids interaction between this electric component and the electric dipoles in the sample.

▷ The radiation reflected by the cavity at resonance is sent towards a detection device which returns the "absorption signal" $s(B)$ (equation [1.18]). The detector is a semiconductor (diode) for which the current I_D depends on the electric component E of the electromagnetic field. For the absorption signal to be proportional to the power absorbed by the sample at resonance, the diode must function in the *linear* portion of the $I_D(E)$ characteristics, around $I_D{}^0$ = 200 μA (inset in figure 1.5). The polarisation current $I_D{}^0$ is created by the radiation which arrives directly *via* the reference arm (figure 1.5). Its power ("bias") can be adjusted as can its dephasing relative to the radiation reflected by the cavity.

To improve the sensitivity of the spectrometer, a detection technique based on *modulation* of the magnetic field is generally used. In these conditions, the signal produced by the detector is not proportional to $s(B)$, but to its *derivative* relative to the field $\mathrm{d}s/\mathrm{d}B$ (complement 3).

▷ Cryogenic equipment can be used to control and monitor the sample temperature.

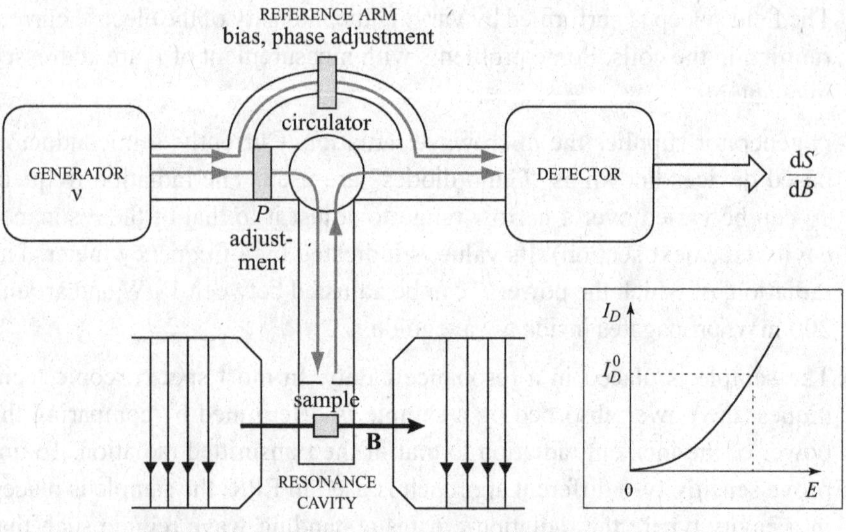

Figure 1.5 – Block diagram for an EPR spectrometer.
The radiation propagates as indicated thanks to the circulator. The inset shows the $I_D(E)$ characteristics for the detection diode. At resonance, the variation of I_D around $I_D{}^0$ is proportional to the power absorbed by the sample.

1.5.3 – Tuning the spectrometer

Before each recording, the spectrometer must be tuned to correctly detect the absorption signal. The main stages in this process are as follows:

- ▷ The frequency of the generator is adjusted to the resonance frequency for the cavity. When the spectrometer is placed in *TUNE* mode (Bruker terminology), the screen displays the power reflected by the cavity as a function of the radiation frequency. When this frequency is changed manually, a very marked narrow minimum is observed at the resonance frequency for the cavity; at this frequency, the power absorbed by the cavity is maximal.
- ▷ "Bias" and phase can be adjusted in *OPERATE* mode. The frequency of the generator is then maintained equal to that of the cavity by the "automatic frequency control" and the manual adjustment made previously can be refined by minimising the error signal (*lock offset*) controlling the regulation. The next step consists in setting the polarisation current for the diode to I_D^0 = 200 µA by adjusting the "bias", after strongly reducing the power P of the incident radiation reaching the cavity so as to eliminate reflected radiation (figure 1.5). To adjust the phase, the reflected radiation is re-established by increasing P. The diode current then results from interference between the reflected radiation and the radiation arriving through the reference arm, and the phase can be adjusted to achieve maximal current.
- ▷ Coupling between the waveguide and the cavity must be such that the cavity reflects no radiation as long as the resonance condition is not satisfied. For this "critical coupling", all of the power P of the incident radiation is absorbed and converted to heat in the walls of the cavity. We first adjust the coupling iris at low power, and fine-tune it by progressively increasing P. When appropriately adjusted, the current retains the value I_D^0 created by the bias whatever the power P.

1.6 – Points to consider in applications

1.6.1 – Electronic and nuclear paramagnetism

Significant differences between the two categories of paramagnetic entities deserve mention. In section 2.2.1, we will see that the magnetic moments of electronic origin are three orders of magnitude greater than their nuclear homologues. Indeed, they are responsible for the magnetic phenomena that can be observed at the macroscopic level, such as magnetisation processes. In addition, paramagnetic centres are atoms and molecules; naturally they are much more diverse than paramagnetic nuclei, although they are much less common than the ubiquitous protons.

1.6.2 – Importance of paramagnetic centres

Numerous molecules have unpaired electrons, which gives them properties that can be exploited in a wide variety of systems:

- Because of their unpaired electron, organic free radicals are generally highly reactive, and they play the role of reaction intermediates in numerous chemical reactions. For example, the radicals produced by oxidative stress, UV radiation or ionising radiation can induce sequences of reactions leading to alterations of DNA in living organisms.
- Inorganic radicals are often unstable, but when they are created in a mineral matrix they can remain trapped for long periods. This process allows their use as markers to date minerals or fossilised organic matter.
- The presence of two unpaired electrons gives molecules in an excited triplet state a certain stability which is observed through the phenomenon of *phosphorescence*.
- Cations in transition ion complexes can easily switch ligands and exchange electrons. They are often good catalysts, and are found in metallo-enzymes where they play the role of active site and oxido-reduction centres, as well as in some proteins which transport specific ligands. When they are too reactive, metal cations can become toxic and be responsible for serious public health problems.
- The paramagnetic centres present in some materials can be used to characterise them. For example, defects in amorphous silica used to produce optical fibres for telecommunications, ionised impurities in semiconductors, new magnetic materials, superconductors.

1.6.3 – Continuous wave EPR and pulsed EPR

In the experiment presented in section 1.4 and in the spectrometer described in section 1.5, the radiation used is a sinusoidal electromagnetic wave of frequency ν. This type of experiment is known as "continuous wave EPR spectrometry", and is the only type dealt with in this volume. EPR can also be performed by applying a sequence of microwave pulses to the sample. By carefully selecting the parameters of this sequence, information which would be difficult, or even often impossible, to obtain by continuous wave EPR can be extracted. Examples of this type of information include the spin-lattice T_1 and spin-spin T_2 relaxation times for paramagnetic centres (see chapter 5) and hyperfine (see chapter 2) or intercentre interactions (see chapter 7) which are too weak to produce resolved structures on the spectrum. The applications of pulsed EPR techniques are growing rapidly [Schweiger and Jeschke, 2001].

1.6.4 – Some observations

▹ Researchers performing magnetic spectroscopy can alter the splitting between energy levels of the molecules they are studying by adjusting the value of the magnetic field. This is particularly the case with EPR, where the spectrum is recorded by scanning the magnetic field. The information contained in the spectrum therefore is not related to the value of this splitting, which is equal to the quantum $h\nu$ of the spectrometer at resonance, but to how it *depends on B*. In the simple example provided in section 1.4, the splitting between two levels is proportional to B, and the information on the paramagnetic centre comes from the number g which determines the proportionality coefficient.

▹ In this introductory chapter, we presented the principle of EPR spectrometry by considering a very simple case. Even in this specific case, we had to apply an *ad hoc* procedure to determine the energy levels for a centre placed in a field **B**, and we used an arbitrary selection rule to determine the allowed transitions. The justification for this approach will appear naturally in the following chapters.

Complement 1 – Magnetic moment created by a rotating point charge

The action of an electric current on a magnetic needle first revealed the close relationship existing between electric and magnetic phenomena (experiment performed by Oersted, 1820). This relationship takes a particularly simple form when the electric circuit is a loop through which a current of intensity i circulates. The general laws of electromagnetism show that this loop creates a *magnetic moment* $\boldsymbol{\mu} = A\, i\, \mathbf{n}$, where A is the area of the loop and $\mathbf{n}$ is a unit vector perpendicular to it. This expression reminds us that magnetic moments are expressed in A m^2, which can also be written J T^{-1} (figure 1.6a). We will now consider a point charge q rotating in a circle at constant speed V (figure 1.6b). By definition, the intensity of the electric current created by its movement is the charge which passes through any point in the trajectory in one second. It is therefore equal to:

$$i = q\,N$$

N is the number of laps completed by the point mass over a second, which is equal to:

$$N = V/(2\pi R)$$

where R is the radius of the trajectory. The *magnetic moment* created by this movement can therefore be written:

$$\boldsymbol{\mu} = (qRV/2)\,\mathbf{n}$$

Moreover, the *angular momentum* associated with the *movement* of the point mass m is equal to:

$$\boldsymbol{\sigma} = (mVR)\,\mathbf{n}$$

It is deduced that $\boldsymbol{\mu}$ and $\boldsymbol{\sigma}$ are linked by:

$$\boldsymbol{\mu} = (q/2m)\,\boldsymbol{\sigma}$$

This relation between magnetic moment and angular momentum, which we have verified here in a specific case, results from the general laws of electromagnetism and applies to any motion of a point charge.

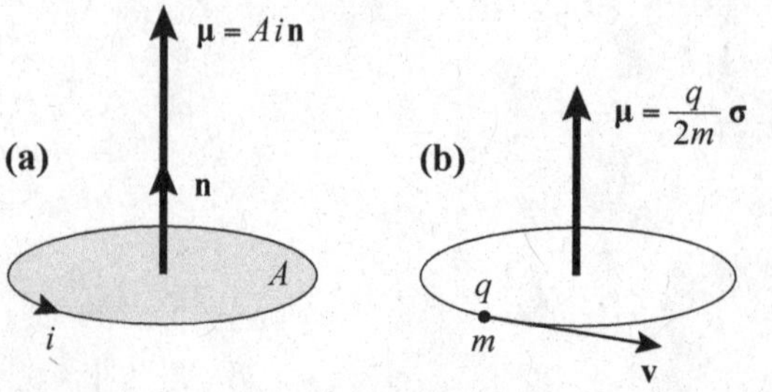

Figure 1.6 – Magnetic moment created by **(a)** a current i circulating in a loop and **(b)** a revolving point of mass m and charge q.

Complement 2 – Why are B_2 and O_2 paramagnetic molecules?

For a homonuclear diatomic molecule, the energy level diagram for the molecular orbitals is the following [Gray, 1965]:

$$
\begin{array}{c}
\underline{\quad\quad}\ \sigma_z^*\ \underline{\quad\quad} \\[2pt]
\underline{\ }\ \pi_x^*\ \underline{\quad}\ \pi_y^*\ \underline{\ } \\[6pt]
\underline{\quad\ }2p\ \underline{\quad\ }\qquad\qquad \underline{\quad\ }2p\ \underline{\quad\ } \\[6pt]
\underline{\quad\quad}\ \sigma_z^b\ \underline{\quad\quad} \\[2pt]
\underline{\ }\ \pi_x^b\ \underline{\quad}\ \pi_y^b\ \underline{\ } \\[6pt]
\underline{\quad\quad}\ \sigma_s^*\ \underline{\quad\quad} \\[8pt]
\underline{\quad\ }2s\ \underline{\quad\ }\qquad\qquad \underline{\quad\ }2s\ \underline{\quad\ } \\[8pt]
\underline{\quad\quad}\ \sigma_s^b\ \underline{\quad\quad}
\end{array}
$$

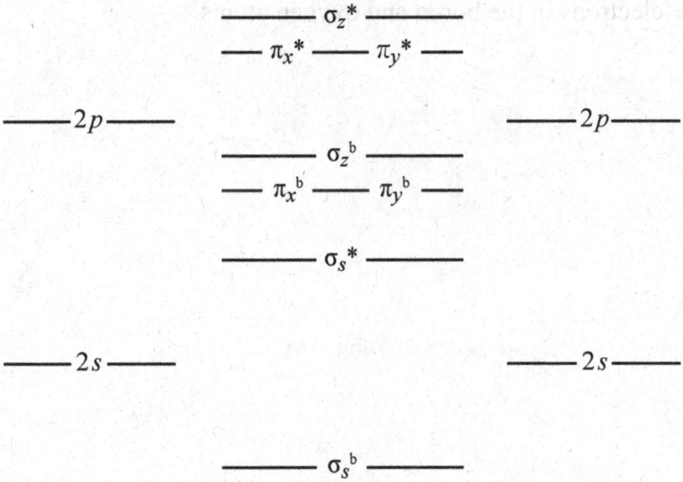

The inter-nuclear axis is labelled z and the superscripts b and * designate the bonding and anti-bonding orbitals, respectively. As the directions x and y are equivalent in diatomic molecules, the orbitals (π_x^b, π_y^b) on the one hand and (π_x^*, π_y^*) on the other hand have the same energy. They are termed "degenerate".

▷ The ground configuration of the boron atom is $2s^2\,2p^1$. That of B_2 can be obtained by attributing the molecular orbitals to the six valence electrons as follows:

$$(\sigma_s^b)^2\,(\sigma_s^*)^2\,(\pi_x^b)^1\,(\pi_y^b)^1$$

Attributing different orbitals to the last two electrons minimises the influence of their Coulomb repulsion. The calculation indicates, and this is experimentally confirmed, that the energy of the molecule is minimal when the total spin is equal to 1 ("triplet" state): the ground state for molecule B_2 is therefore paramagnetic.

▷ As the ground configuration of the oxygen atom is $2s^2\,2p^4$, that of O_2 is obtained by attributing the twelve valence electrons as follows:

$$(\sigma_s^b)^2\,(\sigma_s^*)^2\,(\sigma_z^b)^2\,(\pi_{x,\,y}^b)^4\,(\pi_x^*)^1\,(\pi_y^*)^1$$

In this molecule, the $\sigma_z{}^b$ orbital is more stable than the $\pi_{x,y}{}^b$ orbitals. As in the case of B_2, the ground state is a triplet state.

The paramagnetism of the B_2 and O_2 molecules is therefore due to the degeneracy of the π orbitals as a result of the molecule's symmetry, and to the number of valence electrons in the boron and oxygen atoms.

Complement 3 – How magnetic field modulation affects the signal detected

To increase the signal-to-noise ratio, the field **B** created by the electromagnet is superimposed with a small sinusoidal magnetic field $\mathbf{b}_m(t)$. This field, which is produced by small coils placed on the sides of the cavity, is *parallel* to **B**. It takes the form

$$b_m(t) = \frac{B_m}{2}\cos 2\pi\nu_m t$$

where B_m is the "peak-to-peak" amplitude and ν_m is the modulation frequency, which is generally equal to 100 kHz. If B is the field created by the electromagnet, the total magnetic field applied to the sample is:

$$B + \frac{B_m}{2}\cos 2\pi\nu_m t$$

The absorption signal returned by the detector, which is $s(B)$ in the absence of modulation, becomes $s\left(B + \frac{B_m}{2}\cos 2\pi\nu_m t\right)$. If B_m is small, we can write (figure 1.7a):

$$s\left(B + \frac{B_m}{2}\cos 2\pi\nu_m t\right) \approx s(B) + s'(B)\frac{B_m}{2}\cos 2\pi\nu_m t \qquad [1]$$

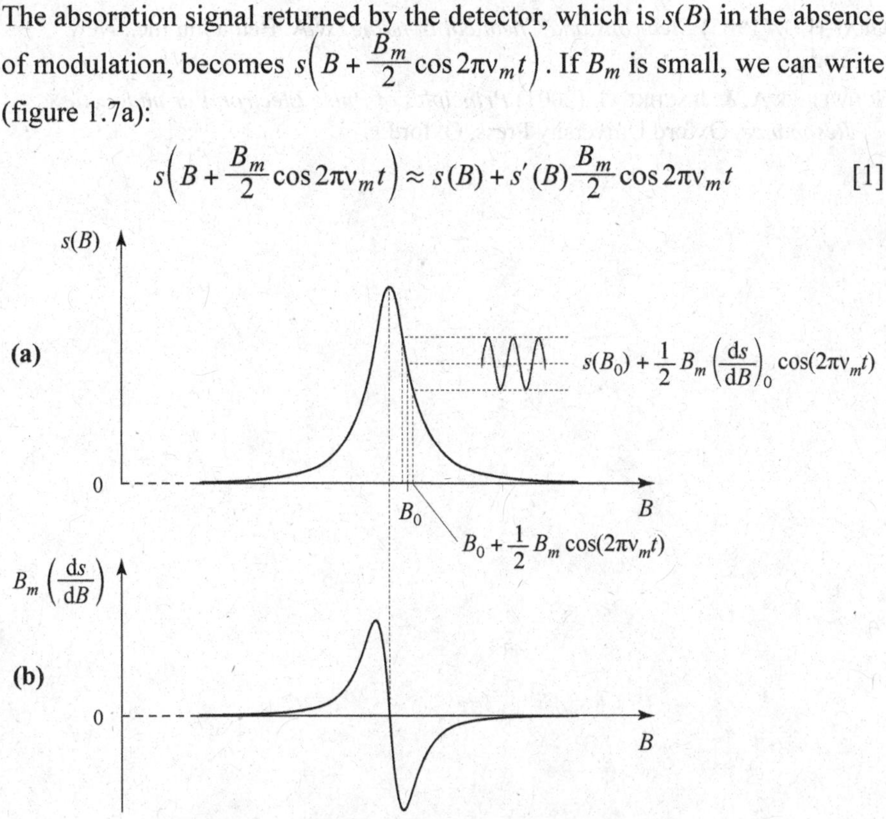

Figure 1.7 – **(a)** Effect of modulation of the magnetic field. **(b)** Shape of the derivative of the absorption signal.

In this expression, information on the absorption signal is present in two forms:

▹ the absorption signal $s(B)$ in the first term.

▹ its derivative $s'(B)$ in the second.

A procedure known as "synchronous detection" can be used to extract the amplitude of the second term with an excellent signal-to-noise ratio. This second term is proportional to $B_m\, s'(B)$ (figure 1.7b) (see section 9.2.2).

Some spectrometers also provide the "second harmonic" proportional to $s''(B)$, which arises from the next term in the development [1].

References

AYANT Y. & BELORIZKY E. (2000) *Cours de Mécanique Quantique*, Dunod, Paris.

COHEN-TANNOUDJI C., DIU B. & LALOË F. (2015) *Quantum Mechanics*, Wiley-VcH, New York.

GRAY H.B. (1965) *Electrons and Chemical Bonding*, W.A. Benjamin Inc., New York.

SCHWEIGER A. & JESCHKE G. (2001) *Principles of Pulse Electron Paramagnetic Resonance*, Oxford University Press, Oxford.

Exercises

1.1. Show that the number of orbitals in an atomic shell identified by n is equal to n^2.

1.2. According to equation [1.9], the magnetic moment of the atoms is of the same order of magnitude as the Bohr magneton. Compare its value to that of the magnetic moment created by a loop of current of radius 1 cm, through which a 1 mA current circulates (see complement 1).

1.3. From equations [1.13] and [1.14], derive expression [1.16] for the population difference $(N_- - N_+)$.

1.4. Calculate the value of the resonance field for a paramagnetic centre for which the g factor is equal to 2.00, at X-band (ν = 9.4 GHz), at Q-band (ν = 35 GHz) and at ν = 350 GHz.

1.5. Above which temperature does the temperature dependence given by Curie's law differ by less than 2 % from that predicted by equation [1.19], at Q-band (ν = 35 GHz) and at ν = 350 GHz?

1.6. As indicated by equation [1.19], the absorption signal is proportional to the number N of paramagnetic centres contained in the sample. However, in some solid samples, this number must sometimes be reduced by substituting non-paramagnetic entities for paramagnetic centres. What do you think is the aim of this "magnetic dilution"?

Answers to exercises

1.1. For a defined value of n, ℓ can take the values $\{0, 1, 2, \ldots n-1\}$ and there will be $(2\ell + 1)$ orbitals for each value of ℓ. The shell thus contains $\sum_{\ell=0}^{n-1}(2\ell + 1)$ orbitals. We can show that this sum is equal to n^2 (which is true for $n = 1$) by mathematical induction. Consider a shell characterised by the number $(n + 1)$. The number of orbitals is $\sum_{\ell=0}^{n}(2\ell + 1)$, i.e., $n^2 + (2n + 1) = (n + 1)^2$.

1.2. $\mu = 3.14 \times 10^{-7}\ \mathrm{J\,T^{-1}}$

1.3. Equations [1.13] and [1.14] produce the following:

$$N_- = \frac{N}{1 + \exp(-\Delta E/k_B T)}; \quad N_+ = \frac{N \exp(-\Delta E/k_B T)}{1 + \exp(-\Delta E/k_B T)}$$

From these values, we deduce the expression for $(N_- - N_+)$. To reveal a hyperbolic tangent, we factorise $\exp(-\Delta E/2k_B T)$ in the numerator and in the denominator.

1.4. $B = 335.8$ mT for $\nu = 9.4 \times 10^9$ Hz;
$B = 1.250$ T for $\nu = 35$ GHz;
$B = 12.50$ T for $\nu = 350$ GHz.

1.5. The difference between x and $\tanh x$ is less than 2 % for $x \leq 0.25$. Therefore, the quantities $h\nu/2k_B T$ and $\tanh(h\nu/2k_B T)$ differ by less than 2 % when $T \geq 2h\nu/k_B$, i.e., $T \geq 3.4$ K for $\nu = 35$ GHz and $T \geq 34$ K for $\nu = 350$ GHz.

1.6. When the paramagnetic centres are too concentrated, their *interactions* cause splitting of the resonance lines (chapter 7) which is revealed by spectral broadening.

Chapter 2

Hyperfine structure of a spectrum in the isotropic regime

2.1 – The various origins of spectral features in EPR

In the previous chapter, we considered a paramagnetic centre of spin S and magnetic moment

$$\boldsymbol{\mu} = -g\beta\mathbf{S} \qquad [2.1]$$

which interacts with a magnetic field **B**. This interaction produces $(2S+1)$ equidistant energy levels defined by:

$$E(M_S) = g\beta B M_S \qquad M_S = -S, -S+1, \ldots S \qquad [2.2]$$

and the resulting spectrum is reduced to a single line centred at $B_0 = h\nu/g\beta$ (figure 1.3).

Luckily, real spectra are not reduced to this spectral form which would provide very little information. They generally contain features of very varied origins, which can be classified as follows:

◇ *Features related to the nature of the paramagnetic centre*

▹ The orientation of **B** relative to the paramagnetic centre has no influence on its interaction with the magnetic moment defined by equation [2.1], and this magnetic moment is therefore termed *isotropic*. But in subsequent chapters we will see that the magnetic moment of molecules is generally *anisotropic* and that the *g number* is replaced by a $\tilde{\mathbf{g}}$ *matrix*. The interaction with **B** creates a set of equidistant energy levels, the splitting between which depends on the *direction* of the field relative to the molecule (section 3.3.2). In these conditions, the shape of the EPR spectrum is determined by how the molecules are organised in the sample (section 4.3).

P. Bertrand, *Electron Paramagnetic Resonance Spectroscopy*,
https://doi.org/10.1007/978-3-030-39663-3_2

▹ In general, the lowest energy levels of paramagnetic centres of spin greater than ½ are already split in the absence of a magnetic field. When these centres are placed in a magnetic field, their energy levels are *not equidistant*, resulting in completely different positions of the resonance lines and consequently producing a spectrum with a different profile (chapters 6 and 8).

◇ *Features emerging due to interactions with other paramagnetic entities*

▹ The unpaired electrons in a paramagnetic centre can interact with the magnetic moments of one or several *nuclei*. These interactions produce a very specific pattern on the spectrum which is known as a *hyperfine structure* (chapters 2 and 4).

▹ Interactions between two paramagnetic centres modify their EPR spectra. When these interactions are weak, the resonance lines split. But when they are strong, like in the case of biradicals and polynuclear complexes of transition metal ions, the EPR spectrum is completely different from what would be obtained by superposing the spectra from each centre. These situations will be studied in detail in chapter 7.

In the following chapters, we will successively address the factors altering the features of the EPR spectrum. It turns out that the hyperfine structure can be quite easily interpreted in a specific but frequent situation, which gives rise to very extensive applications. This situation is the subject of this chapter.

2.2 – Hyperfine interactions

2.2.1 – The nuclear magnetic moment

Like electrons, the nucleons (protons and neutrons) composing the nuclei have an orbital angular momentum **l** linked to their *movement* (the *number* ℓ can take the values $\ell = 0, 1, 2, \ldots$) and a spin angular momentum **s** with $s = ½$. These angular momenta are associated with magnetic moments $\boldsymbol{\mu}_\ell$ and $\boldsymbol{\mu}_s$ for which the expressions are compared to those already given for the electron in table 2.1.

Table 2.1 – Expression of the orbital and spin magnetic moments for particles of matter.

electron	$\boldsymbol{\mu}_\ell = -\beta\,\mathbf{l}$	$\boldsymbol{\mu}_s = -g_e\,\beta\,\mathbf{s}$
proton	$\boldsymbol{\mu}_\ell = +\beta_N\,\mathbf{l}$	$\boldsymbol{\mu}_s = g_p\,\beta_N\,\mathbf{s}$
neutron	$\boldsymbol{\mu}_\ell = \mathbf{0}$	$\boldsymbol{\mu}_s = g_n\,\beta_N\,\mathbf{s}$

β_N is the *nuclear* Bohr *magneton* defined by

$$\beta_N = e\hbar/2M_p = 5.0508 \times 10^{-27}\ \mathrm{J\,T^{-1}}$$

where M_p is the mass of the *proton*. As this mass is around 1800-fold larger than that of the electron, nuclear magnetic moments are three orders of magnitude smaller than their electronic homologues. The orbital magnetic moments conform to what would be predicted from equation [1.1]. The orbital magnetic moment of the proton is similar to that of the electron except for a sign change due to the proton's positive charge; the orbital magnetic moment of the neutron (which has no charge) is null.

The numbers g_e, g_p and g_n which characterise the spin magnetic moments of the three particles are equal to [Lide, 2010]:

$$g_e = 2.0023 \qquad g_p = 5.5857 \qquad g_n = -3.8261$$

The total angular momentum **I** of the nucleus results from the sum of the angular momenta of all its nucleons and the value of I for the ground state depends on whether the number of nucleons is odd or even:

- ▷ When the number is odd, the ground state is paramagnetic with I half-integer.
- ▷ When the number is even, I is an integer or zero.

When the nucleus is paramagnetic, its magnetic moment can be written:

$$\boldsymbol{\mu} = g_N\,\beta_N\,\mathbf{I} \qquad [2.3]$$

Some works use the "magnetogyric ratio" defined by $\gamma_N = g_N\beta_N/\hbar$. The angular momentum **I** complies with the usual rules:

- ▷ the value of $\mathbf{I}^2$ is equal to $I(I + 1)$.
- ▷ the projection I_Z of **I** on any axis Z can take one of the $(2I + 1)$ values:

$$M_I = -I, -I+1, \ldots\ I$$

I is commonly called the "spin" of the nucleus. In table 2.2 we have gathered the data relating to some paramagnetic nuclei likely to interact with the unpaired electrons of paramagnetic centres.

The number g_N depends on the number of protons and neutrons present in the nucleus, and it is negative when the contribution of the neutrons exceeds that of the protons. The data in table 2.2 show that it can vary by more than an order of magnitude in absolute value, which has a direct impact on the strength of the hyperfine interaction. All other things being equal, this interaction is, for example, much stronger with ^{1}H and ^{19}F nuclei than with ^{14}N or chloride nuclei.

In the case of transition ions, we note that the value of g_N is much weaker for the ^{57}Fe isotope and molybdenum isotopes than for the nuclei of vanadium, manganese, cobalt and copper. In section 8.1.2, a similar table will be presented for rare earth nuclei.

Nuclei of spin greater than ½ also have a "quadrupole moment" which interacts with the asymmetric electric field created by the unpaired electrons. In certain cases, this interaction produces visible effects on the EPR spectrum.

Table 2.2 – Natural abundance and characteristics of some paramagnetic nuclei [Lide, 2010].

Nucleus	Natural abundance %	I	g_N
^{1}H	99.99	½	5.5857
^{2}H	0.015	1	0.8574
^{13}C	1.07	½	1.4048
^{14}N	99.63	1	0.4038
^{15}N	0.37	½	−0.5664
^{17}O	0.038	5/2	−0.7575
^{19}F	100	½	5.2578
^{31}P	100	½	2.2632
^{35}Cl	75.76	3/2	0.5479
^{37}Cl	24.24	3/2	0.4561
^{51}V	99.75	7/2	1.4711
^{55}Mn	100	5/2	1.3875
^{57}Fe	2.12	½	0.1812
^{59}Co	100	7/2	1.3228
^{63}Cu	69.2	3/2	1.4849
^{65}Cu	30.8	3/2	1.5878
^{95}Mo	15.8	5/2	−0.3657
^{97}Mo	9.60	5/2	−0.3734

2.2.2 – Hyperfine interactions between unpaired electrons and nuclei

◇ *In an atom or a free ion*

In an atom or a free ion, all directions are equivalent. The magnetic moment is therefore *isotropic* and its expression is similar to equation [2.1]. Interaction between the unpaired electrons and the nucleus of spin I is also *isotropic*, and can be written:

$$H_{hyperfine} = A\,\mathbf{S}\cdot\mathbf{I} \quad [2.4]$$

The hyperfine constant A is a sum of two contributions:

$$A = A_s + A_{dip}$$

- The term A_s is given by the following expression [Atherton, 1993]:

  $$A_s = (2\mu_0/3)\,(g_e\beta g_N\beta_N/2S)\,\rho(0) \quad [2.5]$$

  The constant $\mu_0 = 4\pi \times 10^{-7}$ H m^{-1} (H is the henry) is the vacuum permeability. $\rho(0)$ is the *spin density* at the nucleus, i.e., the difference between the probability densities for electrons with $m_s = ½$ and $m_s = -½$. Its dimension is therefore [length]$^{-3}$. This spin density is necessarily derived from the electrons in the *s orbitals*, as these are the only ones with a non-null probability density at the nucleus.
- The term A_{dip} is due to "dipolar" interactions between the magnetic moments of the electrons and the magnetic moment of the nucleus.

▹ A can easily be calculated when there exists a single unpaired electron in an s orbital. Indeed, $\rho(0)$ is then the probability density for this electron at the nucleus (A_s is said to be due to the "contact" or Fermi mechanism) which can readily be deduced from the expression for this orbital (exercise 2.1), and the dipolar contribution is null due to the spherical symmetry of the s orbitals (appendix 3).

▹ Numerous atoms and ions have one or more unpaired electrons in p, d or f orbitals, for which the probability density at the nucleus is *null*. However, their electrostatic interactions with the electrons in the s orbitals have different effects depending on whether they are characterised by $m_s = +½$ or $m_s = -½$. Because of this dissymmetry, the spatial distributions of these s electrons are slightly different, and a net spin density appears at the nucleus. This mechanism is known as "core polarisation", where "core" stands for all the s orbitals. This mechanism is responsible for the A_s term in transition, rare earth and actinide ions. Theoretical calculations give, for example, $A_s = -37.7$ MHz for $^{57}Fe^{2+}$, $A_s = -34.6$ MHz for $^{57}Fe^{3+}$, and

$A_s = -293$ MHz for Mn^{2+} [Freeman and Watson, 1965]. In these atoms and ions, the dipolar contribution is non-null and can be calculated (appendix 4 and section 8.1.2).

◇ *In a molecule*

In a paramagnetic molecule, the unpaired electrons occupy *molecular orbitals* which are delocalised over several atoms. At any point in space they create a spin density (appendix 5) and the *spin population* corresponds to the fraction of this density found in an orbital of a given atom. When an atom has a spin population, its unpaired electrons can interact with a nearby paramagnetic nucleus in several ways:

1. If it is the nucleus of the atom with the spin population, the contact or core polarisation mechanisms described previously give rise to a term $A_s \mathbf{S}\cdot\mathbf{I}$ similar to that in equation [2.4]. If the nucleus is that of an atom covalently bound to the atom with a spin population, a hyperfine interaction can occur through a mechanism known as "spin polarisation". For example, in radicals of conjugated molecules, a positive spin population ρ_C of the $2p_z$ orbital of the carbon atom in a C–H fragment "polarises" the electrons in the σ(CH) bond such that the probability density of the ($m_s = -½$) electron at the proton is greater than that of the ($m_s = +½$) electron. A_s can be shown to be proportional to ρ_C [McConnell and Chesnut, 1958; Atherton, 1993]. This is the McConnell relation:
$$A_s^{\mathrm{H}} = Q\,\rho_C$$
The proportionality coefficient Q (negative) can be calculated or empirically determined, and thus ρ_C can be deduced from the value of A_s^{H} measured on the EPR spectrum (exercise 2.2).
2. In molecules, dipolar interactions between the magnetic moments of unpaired electrons and of nuclei produce *anisotropic* hyperfine interactions (section 4.4.1).

2.2.3 – The isotropic regime

In chapter 4 we will examine how the anisotropy of the magnetic moment and of the hyperfine interactions affects the shape of the EPR spectrum. But a specific situation exists where these effects are not visible on the spectrum, which considerably simplifies its interpretation. In liquid solution, paramagnetic molecules exposed to shocks from solvent molecules become endowed with rotational Brownian motion, sometimes termed "rotational diffusion". When this random movement is such that all the orientations relative to **B** are rapidly explored with the same probability, an *averaging effect* occurs which causes all the anisotropic effects to disappear from the spectrum. In this "isotropic regime", the electron magnetic moment and the hyperfine interaction with the nuclei are described by equations [2.1] and [2.4], where the numbers g and A are replaced by mean values, denoted g_{iso} and A_{iso}. The averaging effect only occurs if the rotational motion is "fast enough"; in section 4.5 we will see what this actually means. We will also find the expression of the g_{iso} and A_{iso} parameters, but we can already indicate that A_{iso} can often be identified to the parameter A_s which, as we have seen, is proportional to the spin density at the nucleus. In particular, this is the case with *free radicals*. In this type of molecule, the unpaired electron occupies a molecular orbital delocalised over several atoms. This electron can thus interact with a number of paramagnetic nuclei such as ^{1}H, ^{13}C, ^{14}N, ^{19}F or ^{31}P. These interactions create a *hyperfine structure* which is clearly visible on the EPR spectrum, and relatively easy to interpret in the isotropic regime. This feature is determined by how the spin density is distributed over the different atoms in the molecule and therefore constitutes a veritable spectroscopic signature of the radical.

In the remainder of this chapter, we will focus on the EPR spectrum produced by a sample containing paramagnetic centres exposed to hyperfine interactions in the isotropic regime.

2.3 – EPR spectrum for a centre which interacts with a single nucleus in the isotropic regime

2.3.1 – Expression of the energy levels for the centre

Consider a paramagnetic centre of spin S placed in a field **B**. In the isotropic regime, interaction of its unpaired electrons with **B** can be written $g_{iso}\beta\mathbf{B}\cdot\mathbf{S}$. We assume that these electrons also interact with a nucleus of moment $g_N\beta_N\mathbf{I}$, such that the sum of the interactions to which the {paramagnetic centre – nucleus} system is exposed can be written:

$$H = g_{iso}\beta\mathbf{B}\cdot\mathbf{S} + A_{iso}\mathbf{S}\cdot\mathbf{I}$$

There also exists a term $-g_N \beta_N \mathbf{B} \cdot \mathbf{I}$ corresponding to the interaction between the nuclear magnetic moment and the magnetic field, but since the nuclear Bohr magneton is very small, it can be neglected when we are interested in the EPR spectrum. To express the scalar products, we use an orthonormal reference frame $\{X, Y, Z\}$ where Z is parallel to $\mathbf{B}$, and we denote B the modulus of $\mathbf{B}$. H then becomes:

$$H = g_{iso} \beta B S_Z + A_{iso} (S_X I_X + S_Y I_Y + S_Z I_Z)$$

Generally, the hyperfine constant A_{iso} is much smaller than $g_{iso} \beta B$. In section 3.5, it will be seen that the terms not containing S_Z have a negligible effect on the energy levels, which makes it possible to simplify H as follows:

$$H = g_{iso} \beta B S_Z + A_{iso} S_Z I_Z \qquad [2.6]$$

The possible values of S_Z are $M_S = -S, -S+1, \ldots S$, and those of I_Z are $M_I = -I, -I+1, \ldots I$, the possible energy values are thus as follows:

$$E(M_S, M_I) = g_{iso} \beta B M_S + A_{iso} M_S M_I \qquad [2.7]$$

There therefore exist $(2S+1)(2I+1)$ energy levels identified by the (M_S, M_I) pairs.

We will see that *the shape* of the EPR spectrum can be predicted from equation [2.7]. Before treating the general case, we will start with a few simple examples.

2.3.2 – EPR spectrum for $S = ½$ and $I = ½$

In this case, four different (M_S, M_I) pairs exist. Equation [2.7] leads to the following energy levels:

$$E(½, ½) = g_{iso} \beta B/2 + A_{iso}/4$$
$$E(½, -½) = g_{iso} \beta B/2 - A_{iso}/4$$
$$E(-½, -½) = -g_{iso} \beta B/2 + A_{iso}/4$$
$$E(-½, ½) = -g_{iso} \beta B/2 - A_{iso}/4$$

They are represented in figure 2.1 for a fixed value of B and in figure 2.2a as a function of B, assuming a *positive* A_{iso}. In section 3.5, we demonstrate that the interaction between the magnetic moment $\boldsymbol{\mu}$ of the centre and the magnetic component of the radiation can induce transitions between energy levels which obey the selection rule:

$$\Delta M_I = 0 \quad \text{and} \quad \Delta M_S = \pm 1$$

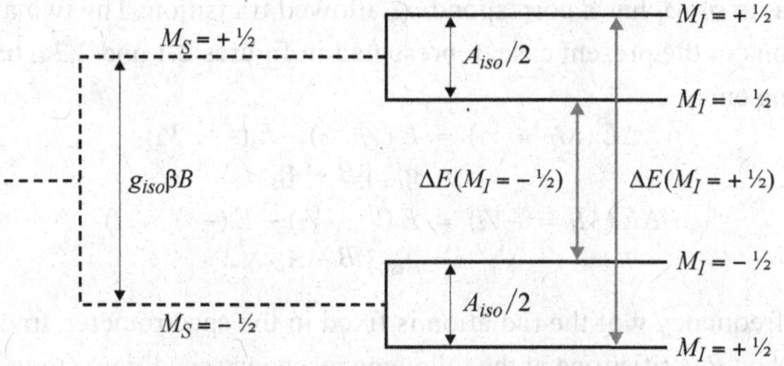

Figure 2.1 – Energy levels for a paramagnetic centre with spin $S = ½$ placed in a field **B**, which interacts with a nucleus of spin $I = ½$. The values of M_I identifying the energy levels are given for $A_{iso} > 0$, and the gray arrows indicate the allowed transitions.

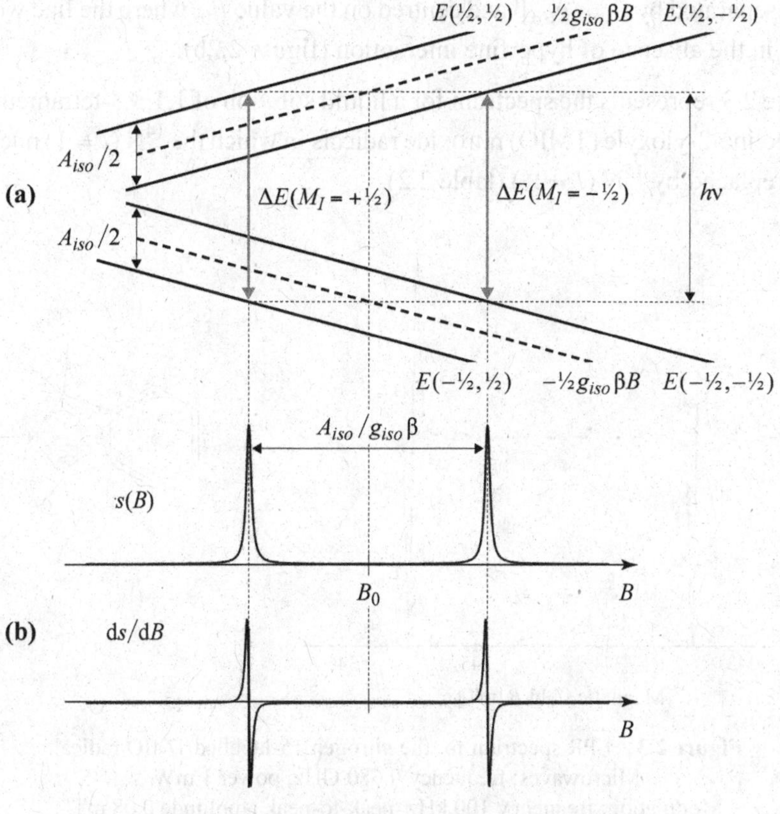

Figure 2.2 – **(a)** Energy levels from figure 2.1 as a function of B. **(b)** Shape of the absorption signal and of its derivative.

Each value of M_I has a corresponding allowed transition. The two allowed transitions in the present case, represented in figures 2.1 and 2.2a, have the following energy:

$$\Delta E\,(M_I = \tfrac{1}{2}) = E\,(\tfrac{1}{2}, \tfrac{1}{2}) - E\,(-\tfrac{1}{2}, \tfrac{1}{2}) = g_{iso}\,\beta B + A_{iso}/2$$

$$\Delta E\,(M_I = -\tfrac{1}{2}) = E\,(\tfrac{1}{2}, -\tfrac{1}{2}) - E\,(-\tfrac{1}{2}, -\tfrac{1}{2}) = g_{iso}\,\beta B - A_{iso}/2$$

As the frequency ν of the radiation is fixed in the spectrometer, transitions occur when B verifies one of the following resonance conditions (figure 2.2a):

$$\Delta E\,(M_I = \tfrac{1}{2}) = h\nu\,;\; B(M_I = \tfrac{1}{2}) = B_0 - A_{iso}/2g_{iso}\,\beta$$

$$\Delta E\,(M_I = -\tfrac{1}{2}) = h\nu\,;\; B(M_I = -\tfrac{1}{2}) = B_0 + A_{iso}/2g_{iso}\,\beta$$

where we have set $B_0 = h\nu/g_{iso}\,\beta$. These transitions produce two hyperfine lines separated by $A_{iso}/g_{iso}\,\beta$ and centred on the value B_0, where the line would form in the absence of hyperfine interaction (figure 2.2b).

Figure 2.3 represents the spectrum for a liquid solution of 1,1,3,3-tetramethylisoindoline-2-yloxyle (TMIO) nitroxide radicals in which the ^{14}N (I = 1) nucleus was replaced by ^{15}N (I = ½) (table 2.2).

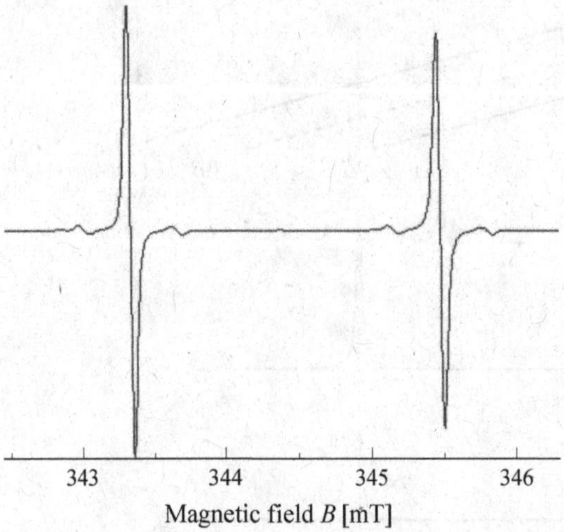

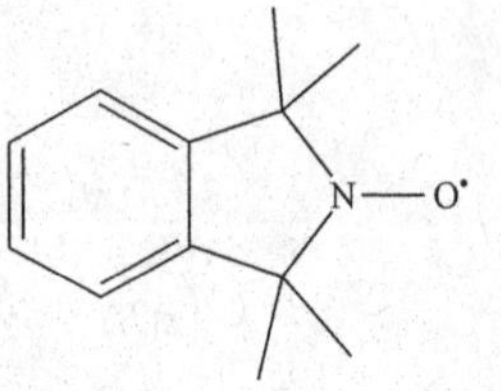

Figure 2.3 – EPR spectrum for the nitrogen 15-labelled TMIO radical. Microwaves: frequency 9.680 GHz, power 1 mW. Modulation: frequency 100 kHz, peak-to-peak amplitude 0.08 mT.

In these molecules, the spin density is mainly localised on the N and O atoms. Polarisation of the N–O bond varies with the solvent's dielectric constant such that the spin density on the nitrogen atom increases with the polar nature of the solvent [Jeunet *et al.*, 1986]. The spectrum is dominated by a "doublet" due to the interaction between the unpaired electron and the ^{15}N nucleus (exercise 2.3). Low-intensity lines are also present on either side of the lines of the doublet. These lines are the result of hyperfine interactions with ^{13}C ($I = ½$) nuclei, present in around 1 % of molecules (table 2.2), and indicate that the spin density on the carbon atoms is non-null.

Note – In figures 2.1 and 2.2a, the energy levels are identified by the numbers M_I, assuming a *positive* A_{iso}. Equation [2.7] shows that we can simultaneously change the sign of M_I and A_{iso} without modifying the energy-level diagram (figure 2.2a) or the EPR spectrum (figure 2.2b).

In general, the *sign* of the M_I values corresponding to the various hyperfine lines cannot be determined and we can only deduce the *absolute value* of the hyperfine constants from the spectrum. To determine their sign, detailed studies must be performed in solid media, or alternative magnetic spectroscopy techniques must be used, such as triple resonance (electron, nucleus, nucleus) or Mössbauer spectroscopy when the nucleus is ^{57}Fe (however, see section 5.4.3). Throughout this chapter, for simplicity, we have assumed that the hyperfine constants are *positive*.

2.3.3 – EPR spectrum for $S = ½$ and $I = 1$

In this case, there exist six (M_S, M_I) pairs. By proceeding as above, we can determine the energies for the three allowed transitions:

$$\Delta E\,(M_I = 1) = E(½, 1) - E(-½, 1) = g_{iso}\beta B + A_{iso}$$

$$\Delta E\,(M_I = 0) = E(½, 0) - E(-½, 0) = g_{iso}\beta B$$

$$\Delta E\,(M_I = -1) = E(½, -1) - E(-½, -1) = g_{iso}\beta B - A_{iso}$$

These transitions are represented in figure 2.4 for a fixed B. Transitions occur when B verifies one of the following resonance conditions:

$$\Delta E\,(M_I = 1) = h\nu\,; \quad B(M_I = 1) = B_0 - A_{iso}/g_{iso}\beta$$

$$\Delta E\,(M_I = 0) = h\nu\,; \quad B(M_I = 0) = B_0$$

$$\Delta E\,(M_I = -1) = h\nu\,; \quad B(M_I = -1) = B_0 + A_{iso}/g_{iso}\beta$$

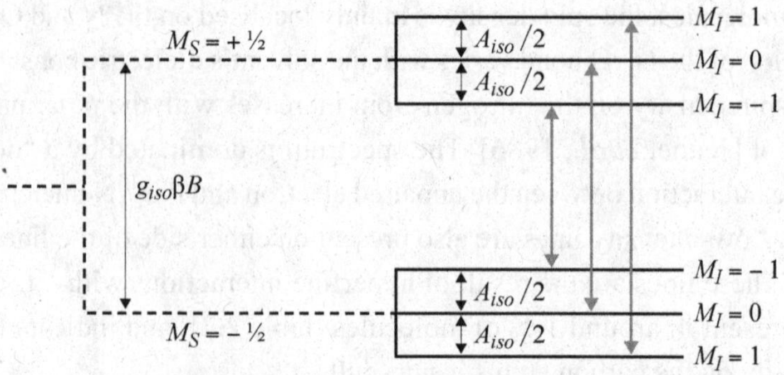

Figure 2.4 – Energy levels for a paramagnetic centre of spin $S = ½$ placed in a field **B**, and interacting with a nucleus of spin $I = 1$. The values of M_I, which determine the energy levels, are given for $A_{iso} > 0$.

The spectrum is composed of a pattern of 3 lines separated by $A_{iso}/g_{iso}\beta$, centred at B_0 (figure 2.5).

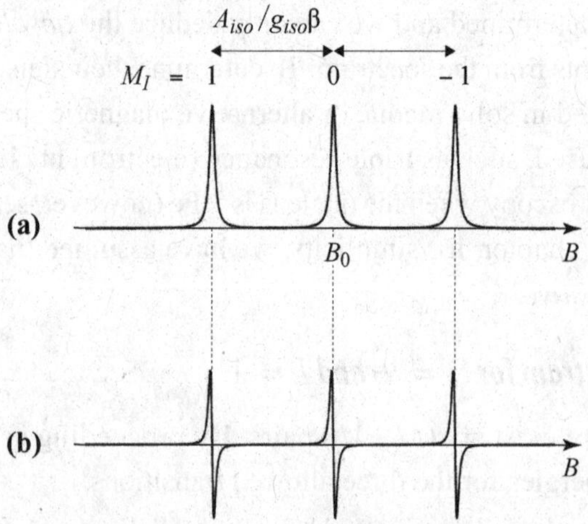

Figure 2.5 – EPR spectrum corresponding to the energy-level diagram shown in figure 2.4. Attribution of lines for $A_{iso} > 0$. **(a)** Absorption signal and **(b)** its derivative.

The spectrum for a liquid solution of 1-oxyl-2,2,5,5-tetramethyl-δ^3-pyrroline-3-methyl-methanethiosulfonate (MTSL) nitroxide radicals in which the unpaired electron interacts with a ^{14}N nucleus, is represented in figure 2.6a. Each line of the "triplet" is surrounded by two low-intensity lines due to interaction with ^{13}C nuclei, as in figure 2.3.

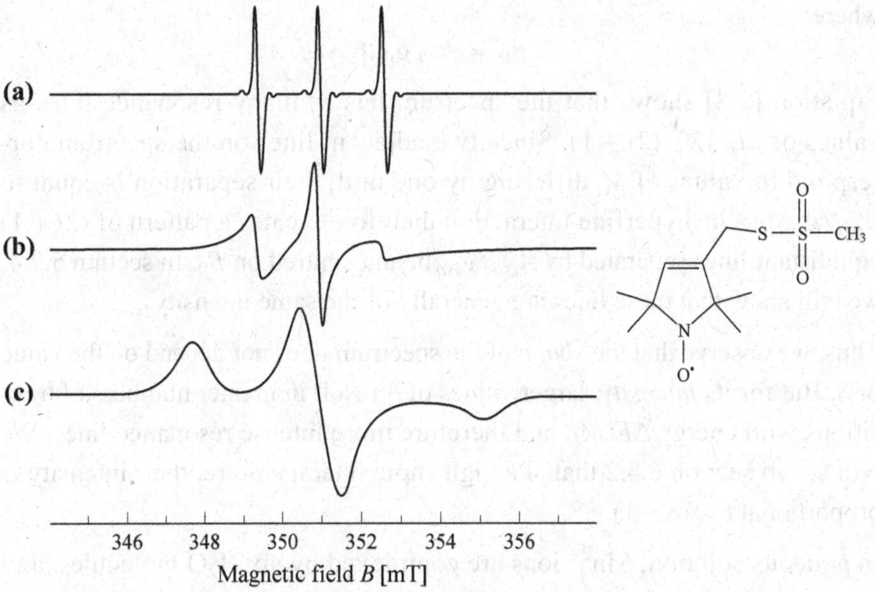

Figure 2.6 – Spectrum for a nitroxide radical (MTSL) recorded at X-band in various mobility conditions: **(a)** in liquid solution at room temperature; **(b)** attached to an enzyme (human pancreatic lipase) at room temperature; **(c)** in frozen solution, T = 100 K. Microwaves: frequency 9.850 GHz, power **(a)**, **(b)** 10 mW **(c)** 0.4 mW. Modulation: frequency 100 kHz, peak-to-peak amplitude **(a)** 0.1 mT **(b)** 0.3 mT **(c)** 0.6 mT.

2.3.4 – General case

Up to now, we have considered a centre of spin $S = ½$ for which the unpaired electron interacts with a nucleus of spin $I = ½$ or $I = 1$. We will now examine the general case where the unpaired electrons of a centre of spin S interact with a nucleus of spin I, where S and I take *any values*. When this centre is placed in a magnetic field, its energy levels are given by equation [2.7]. Because of the selection rule $\Delta M_I = 0, \Delta M_S = \pm 1$, transitions from energy level $E(M_S, M_I) = g_{iso}\beta B M_S + A_{iso} M_S M_I$ can only take place towards the following two levels:

$$E(M_S+1, M_I) = g_{iso}\beta B(M_S+1) + A_{iso}(M_S+1)\,M_I$$
$$E(M_S-1, M_I) = g_{iso}\beta B(M_S-1) + A_{iso}(M_S-1)\,M_I$$

These two transitions have the same energy $\Delta E(M_I) = g_{iso}\beta B + A_{iso} M_I$, and the resonance fields are given by:

$$B(M_I) = B_0 - (A_{iso}/g_{iso}\beta)\,M_I \qquad [2.8]$$

where:

$$B_0 = h\nu / g_{iso}\beta$$

Equation [2.8] shows that the spectrum has as many resonance lines as values of M_I, i.e., $(2I + 1)$. Since two adjacent lines on the spectrum correspond to values of M_I differing by one unit, their separation is equal to $A_{iso} / g_{iso}\beta$. The hyperfine interaction therefore creates a pattern of $(2I + 1)$ equidistant lines separated by $A_{iso} / g_{iso}\beta$, and centred on B_0. In section 5.2.4, we will show that these lines are generally of the same intensity.

Thus, we observe that the *shape* of the spectrum does not depend on the value of S. But for its *intensity*, larger values of S result in greater numbers of transitions with energy $\Delta E(M_I)$ and therefore more intense resonance lines. We will see in section 6.4.2 that at a high enough temperature, their intensity is proportional to $S(S + 1)$.

In aqueous solution, Mn^{2+} ions are complexed by six H_2O molecules, and their spin is $S = 5/2$. At room temperature, the spectrum shown in figure 2.7 is observed, the characteristic shape of which is due to hyperfine interaction with the ^{55}Mn nucleus of spin $I = 5/2$ (table 2.2) (see exercise 2.5).

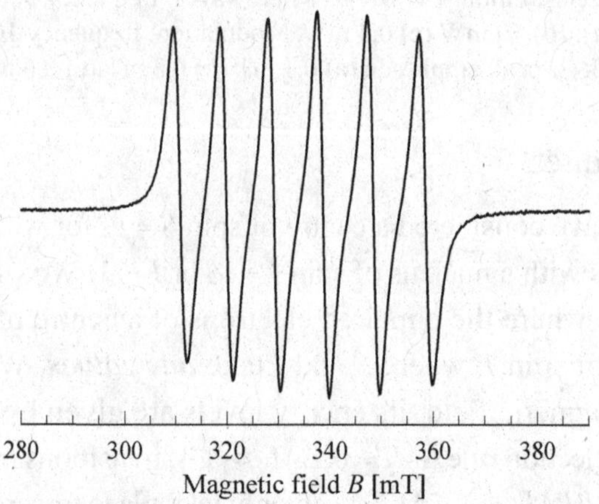

Figure 2.7 – EPR spectrum for an aqueous solution of Mn^{2+} ions at room temperature. Microwaves: frequency 9.3872 GHz, power 10 mW. Modulation: frequency 100 kHz, peak-to-peak amplitude 1 mT.

2.4 – EPR spectrum for a centre which interacts with several nuclei in the isotropic regime

2.4.1 – Hyperfine interactions with several equivalent nuclei

Nuclei are said to be *equivalent* from the point of view of hyperfine interactions when they have the same spin and they are characterised by the same hyperfine constant A_{iso}.

For example, this is the case with an organic radical for which the unpaired electron interacts in the same way with three protons from a CH_3 group, or for a transition ion complex such as $Cu(NH_3)_4{}^{2+}$ where the unpaired electron from the Cu^{2+} ion interacts in the same way with the four ^{14}N nuclei from the ligands.

We will first examine the case of a centre of spin S, its unpaired electrons interact with *two equivalent nuclei* of angular momentum $\mathbf{I}_1$ and $\mathbf{I}_2$, with $I_1 = I_2 = I$. Equation [2.6] then becomes:

$$H = g_{iso}\beta B\, S_Z + A_{iso}\, S_Z I_{1Z} + A_{iso}\, S_Z I_{2Z}$$

If we denote M_1 and M_2 the respective values of I_{1Z} and I_{2Z}, the possible energy values are given by:

$$E(M_S, M_1, M_2) = g_{iso}\beta B\, M_S + A_{iso} M_S\,(M_1 + M_2)$$

The selection rule becomes:

$$\Delta M_S = \pm 1;\ \Delta M_1 = 0;\ \Delta M_2 = 0$$

The transitions allowed from level $E(M_S, M_1, M_2)$ to levels $E(M_S + 1, M_1, M_2)$ and $E(M_S - 1, M_1, M_2)$ have the same energy:

$$\Delta E(M_1, M_2) = g_{iso}\beta B + A_{iso}(M_1 + M_2) \qquad [2.9]$$

For each value of $\Delta E(M_1, M_2)$ there is a corresponding resonance line for which the position can be determined from $\Delta E(M_1, M_2) = h\nu$. Since M_1 and M_2 can take any of the $(2I + 1)$ values $(-I, -I + 1, \ldots, I)$, there exist $(2I + 1)^2$ pairs (M_1, M_2). Among these pairs, some produce the same $(M_1 + M_2)$ value, and their lines are added. To see how this looks, it is convenient to represent equation [2.9] on a diagram (figure 2.8). Starting from $g_{iso}\beta B$, we first add the term $A_{iso} M_1$ considering all possible values of M_1, then we add $A_{iso} M_2$ taking all values of M_2 into consideration.

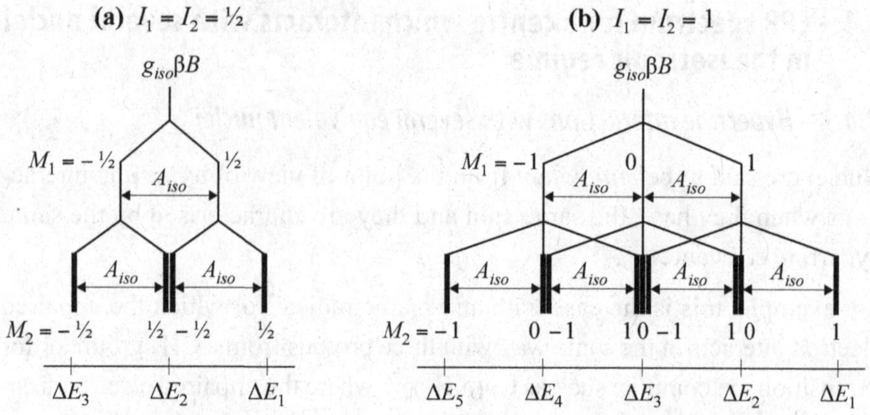

Figure 2.8 – Construction of the quantity $\Delta E(M_1,M_2)$ defined by equation [2.9] for **(a)** $I_1 = I_2 = ½$ and **(b)** $I_1 = I_2 = 1$.

▷ For $I_1 = I_2 = ½$, the 4 pairs (M_1, M_2) give the following ΔE and resonance fields (figure 2.8a):

$$(½, ½):\ \Delta E_1 = g_{iso}\beta B + A_{iso}\,;\quad B_1 = B_0 - A_{iso}/g_{iso}\beta$$
$$(½, -½), (-½, ½):\ \Delta E_2 = g_{iso}\beta B\,;\quad B_2 = B_0$$
$$(-½, -½):\ \Delta E_3 = g_{iso}\beta B - A_{iso}\,;\quad B_3 = B_0 + A_{iso}/g_{iso}\beta$$

As ΔE_2 is obtained twice, the intensity of the central line will be doubled (figure 2.9). The *positions* of the 3 lines are the same as in the case of a single nucleus of spin $I = 1$ (figure 2.5), but their *relative intensities* take the proportions (1 : 2 : 1).

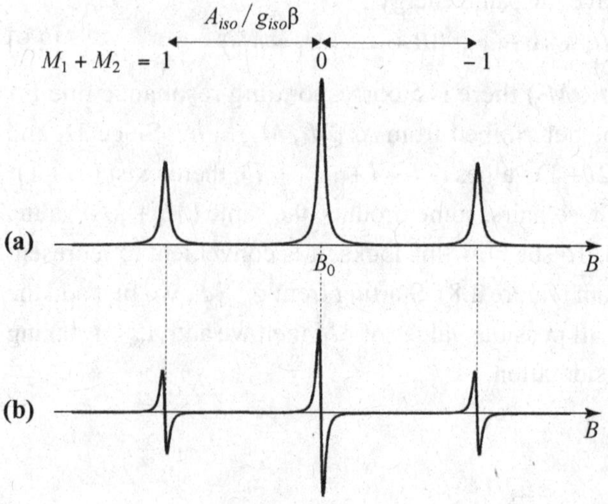

Figure 2.9 – EPR spectrum corresponding to figure 2.8a.
(a) Absorption signal and **(b)** its derivative.

Figure 2.10 shows the EPR spectrum for a liquid solution of the 2,6-di-*tert*-butyl-4-(4-oxo-4H-chromen-2-yl) phenyloxyl radical, for which the hyperfine structure is produced by interaction with the two meta protons in the phenoxyl group.

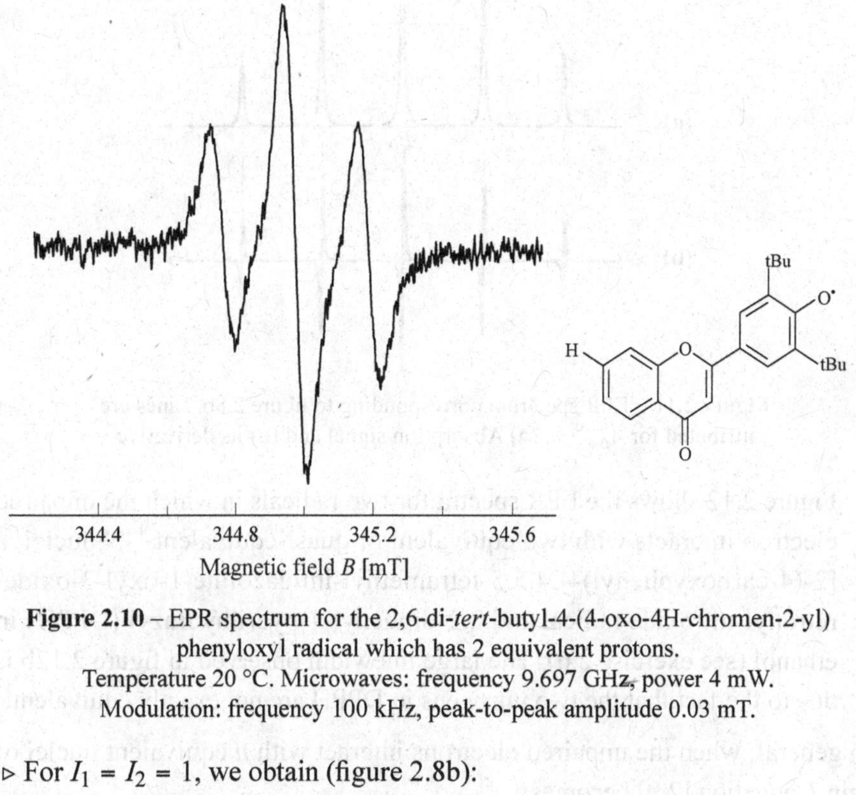

Figure 2.10 – EPR spectrum for the 2,6-di-*tert*-butyl-4-(4-oxo-4H-chromen-2-yl) phenyloxyl radical which has 2 equivalent protons. Temperature 20 °C. Microwaves: frequency 9.697 GHz, power 4 mW. Modulation: frequency 100 kHz, peak-to-peak amplitude 0.03 mT.

▷ For $I_1 = I_2 = 1$, we obtain (figure 2.8b):

$$\begin{aligned}
\Delta E_1 &= g_{iso}\beta B + 2A_{iso}\,; & B_1 &= B_0 - 2A_{iso}/g_{iso}\beta \\
\Delta E_2 &= g_{iso}\beta B + A_{iso}\,; & B_2 &= B_0 - A_{iso}/g_{iso}\beta \\
\Delta E_3 &= g_{iso}\beta B\,; & B_3 &= B_0 \\
\Delta E_4 &= g_{iso}\beta B - A_{iso}\,; & B_4 &= B_0 + A_{iso}/g_{iso}\beta \\
\Delta E_5 &= g_{iso}\beta B - 2A_{iso}\,; & B_5 &= B_0 + 2A_{iso}/g_{iso}\beta
\end{aligned}$$

The positions of the 5 lines are the same as in the case of a single nucleus of spin $I = 2$, but their relative intensities vary in the proportions (1:2:3:2:1) (figure 2.11).

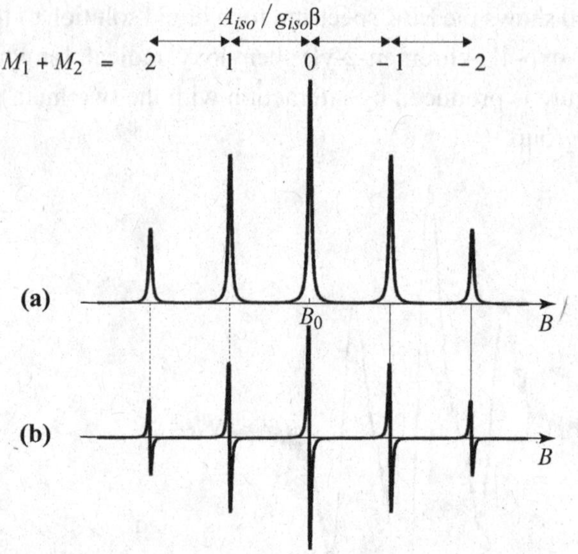

Figure 2.11 – EPR spectrum corresponding to figure 2.8b. Lines are attributed for $A_{iso} > 0$. **(a)** Absorption signal and **(b)** its derivative.

Figure 2.12 shows the EPR spectra for two radicals in which the unpaired electron interacts with two equivalent or quasi-equivalent ^{14}N nuclei: a [2-(4-carboxyphenyl)-4,4,5,5-tetramethyl-imidazoline-1-oxyl-3-oxide] nitronyl-nitroxide radical and 2,2-diphenyl-1-pycrilhydrazyl (DPPH) in ethanol (see exercise 2.6). The large linewidth observed in figure 2.12b is due to the fact that the two nitrogens in DPPH are not exactly equivalent.

In general, when the unpaired electrons interact with n equivalent nuclei of spin I, equation [2.9] becomes:

$$\Delta E(M_1, M_2, \ldots, M_n) = g_{iso}\beta B + A_{iso}(M_1 + M_2 + \ldots + M_n) \quad [2.10]$$

Since all the M_k can take the values $(-I, -I+1, \ldots, I)$, there exist $(2I+1)^n$ multiplets $(M_1, M_2, \ldots, M_n)$.

In equation [2.10], the sum $(M_1 + M_2 + \ldots + M_n)$ can take the values $-nM_I, -nM_I+1, \ldots, +nM_I$. The *position* of the lines is therefore identical to what would be observed if the electrons interacted with a single nucleus of spin nI, and the number of hyperfine lines is equal to $(2nI + 1)$.

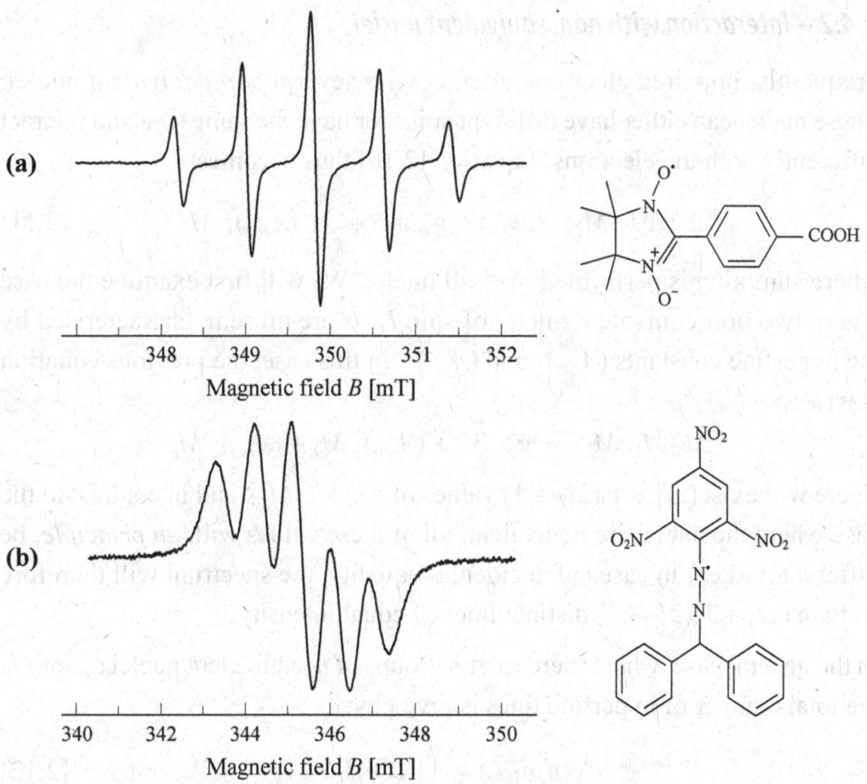

Figure 2.12 – Spectra for two radicals where the unpaired electron interacts with two ^{14}N nuclei.
(a) A nitronyl-nitroxide-type radical with equivalent nitrogen atoms.
Microwaves: frequency 9.825 GHz, power 10 mW.
Modulation: frequency 100 kHz, amplitude 0.1 mT.
(b) DPPH (4 mM) in ethanol at 20 °C. The nitrogen atoms are quasi-equivalent.
Microwaves: frequency 9.698 GHz, power 5 mW.
Modulation: frequency 100 kHz, amplitude 0.1 mT.

To determine their relative intensities, a diagram similar to that shown in figure 2.8 could theoretically be used, but in practice, counting the multiplets ($M_1, M_2, \dots, M_n$) that lead to a given value of ($M_1 + M_2 + \dots + M_n$) is a well-known mathematical problem. Thanks to a simple algorithm, we can represent the result in the form of a “Pascal’s triangle” for each value of I (complement 2).

As in the case where the paramagnetic centre interacts with a single nucleus, the *shape* of the spectrum does not depend on the value of S, rather the *intensity* of the lines increases with S.

2.4.2 – Interaction with non-equivalent nuclei

Frequently, unpaired electrons interact with several *non-equivalent* nuclei. These nuclei can either have different spins, or have the same spin and interact differently with the electrons. Equation [2.10] thus becomes:

$$\Delta E(M_1, M_2, \ldots, M_n) = g_{iso}\,\beta B + \sum_{k=1}^{n} (A_{iso})_k\, M_k \qquad [2.11]$$

where summing is performed over all nuclei. We will first examine the case where two non-equivalent nuclei of spin I_1, I_2 are present, characterised by the hyperfine constants $(A_{iso})_1$ and $(A_{iso})_2$. In this case, the previous equation is written:

$$\Delta E(M_1, M_2) = g_{iso}\,\beta B + (A_{iso})_1\, M_1 + (A_{iso})_2\, M_2$$

There will exist $(2I_1 + 1)(2I_2 + 1)$ values of $\Delta E(M_1, M_2)$, and in contrast to the case where the nuclei are equivalent, all of these values will, *on principle*, be different. Except in cases of accidental equality, the spectrum will therefore contain $(2I_1 + 1)(2I_2 + 1)$ distinct lines of equal intensity.

In the general case, where there exist p groups of n_i equivalent nuclei of spin I_i, the total number of hyperfine lines is given by:

$$N(p, n_i, I_i) = \prod_{i=1}^{p} (2n_i I_i + 1) \qquad [2.12]$$

and the spectrum extends over a field range equal to (exercise 2.7):

$$\Delta B = \sum_{i=1}^{p} 2n_i I_i \frac{(A_{iso})_i}{g_{iso}\,\beta} \qquad [2.13]$$

If the total number of nuclei, n, is small and if the values of $(A_{iso})_k$ are sufficiently different, we can still use a diagram to determine the relative intensities of the lines. When n is too large, we will use numerical simulation instead.

Interpretation of the hyperfine structure of complex spectra is facilitated by *isotopic substitution* experiments, in which some nuclei are replaced by isotopes with different magnetic characteristics.

For example, suppose that the unpaired electrons in a molecule interact with several non-equivalent protons. If we selectively replace one of these protons by a deuteron (^2H nucleus), the corresponding nuclear spin shifts from ½ to 1 and its hyperfine coupling constant, which is proportional to the factor g_N (equation [2.5]), is reduced around 6.5-fold (table 2.2). This modifies the hyperfine pattern, making it possible to attribute a hyperfine constant to the proton which underwent substitution. An example of this type of study can be found in [El Assan *et al.*, 2006].

2.5 – Important points for applications

2.5.1 – Importance of hyperfine interactions

Although hyperfine interactions are usually much weaker than interactions between unpaired electrons and the field **B**, the features they create on the EPR spectrum are very visible in the isotropic regime and they therefore constitute a very useful spectroscopic signature for the identification of paramagnetic molecules. Indeed, by analysing these features it is possible to determine the g_{iso} parameter and to characterise the different paramagnetic nuclei interacting with the electrons: number of equivalent nuclei, values of their spin and of the A_{iso} parameter.

2.5.2 – Free radicals

For numerous radicals in solution at room temperature, their half-life is long enough to allow their study by continuous wave EPR. Spectra of transient radicals can also be studied in *stopped flow* devices or by using time-resolved EPR techniques. In these radicals, the unpaired electron interacts with nuclei of spin $I = ½$ such as ^{1}H, ^{19}F, ^{31}P and even ^{13}C (detectable despite its low natural abundance) (table 2.2), or $I = 1$ such as ^{14}N. In this type of molecule, the A_{iso} parameter can be identified with the A_s term which is proportional to the spin density at the nucleus. Thus, paramagnetic nuclei constitute veritable *probes of the local spin population* (exercises 2.4 and 2.6). The study of free radicals, which are mainly identified by analysing the hyperfine pattern of the EPR spectrum in the isotropic regime, constitutes a very important field of applications of EPR spectroscopy.

In appendix 5, the precise definition of spin density in a paramagnetic molecule can be found, along with a simple example showing how the hyperfine constants measured on the EPR spectrum can be used to model the electronic structure of a radical. However, it should be remembered that the EPR spectrum recorded in the isotropic regime does not allow us to determine the *sign* of the hyperfine constants, nor as a consequence that of the spin density.

2.5.3 – Transition ion complexes

For transition ion complexes, the number g_{iso} provides information on the nature of the paramagnetic cation (section 4.2.2). When the nucleus of the cation is paramagnetic, its spin I can take high values such as $^3/_2$, $^5/_2$, or $^7/_2$ (table 2.2) and its interaction with unpaired electrons creates a spectacular hyperfine pattern (figure 2.7). Interaction between unpaired electrons and nuclei of some ligands can also create a "superhyperfine" structure on the spectrum.

In general, the values of I and of the g_{iso} and A_{iso} parameters deduced from analysis of the spectrum in the isotropic regime are enough to identify the cation, and sometimes even the ligands, but they provide no information on the *structure* of the complex. To obtain this information, the spectrum must be recorded in conditions where motion is slowed to allow all the effects of anisotropy to be observed on the spectrum (chapter 4).

Complement 1 – The paramagnetic radical probes technique

The EPR spectrum shown in figure 2.6a is that of a liquid solution of nitroxide radicals at room temperature. When the temperature is lowered, the Brownian motion causing the molecule to rotate diminishes and anisotropic effects are observed on the spectrum as it progressively changes shape. At very low temperatures, the movements are very slow and the anisotropic effects completely alter the spectrum (figure 2.6c). The interpretation of these features will be given in chapter 4 (figure 4.13).

We can exploit the relationship between the *mobility* of a radical and the shape of its spectrum to derive information on its environment. For example, if a radical is inserted into a biological membrane, the shape of the spectrum provides information on the membrane viscosity. Similarly, if a radical is attached onto a macromolecule, the speed and anisotropy of the radical's rotational movement, which determine the shape of the EPR spectrum, provide information on the geometry of the linkage site (figure 2.6b). The *spin label* technique has been used for many years to study biological systems. Since the emergence of molecular biology techniques that allow a radical to be attached to specifically determined sites, interest in spin labelling has been renewed. Using this technique, it was possible to define veritable "mobility maps" by successively attaching a probe to different positions in a protein or along a pore traversing a membrane.

Complement 2 – "Pascal's triangles"

When unpaired electrons interact with two equivalent nuclei of spin $I_1 = I_2 = ½$, the number of ways the three possible values of ΔE can be obtained is (1:2:1) (figure 2.8a). These numbers determine the relative intensities of the three hyperfine lines in the spectrum (figure 2.9). For three equivalent nuclei of spin $I = ½$, ΔE can take four different values and the number of ways of obtaining them is (1:3:3:1). In these numbers we recognise the coefficients:

$$C_n^p = \frac{n!}{p!\,(n-p)!}$$

which appear in the development of the binomial $(1 + x)^n$ for $n = 2$ and $n = 3$. They verify the recurrence relation

$$C_n^p = C_{n-1}^{p-1} + C_{n-1}^p$$

which makes it possible to calculate them step by step. To do so, we start by plotting a vertical line of 1. In each line, each number is added to the one located to its right, and the result is placed under the arrow. The new line is completed by including 1:

$$
\begin{array}{lll}
 & p =\ 0\ \ 1\ \ 2\ \ 3\ \ 4\ \ 5\ \ 6 & \\
 & 1 & 1 \\
n = 1 & 1 + 1 & 1\quad 1 \\
 & \quad\ \ \downarrow & \\
2 & 1 + 2 + 1 & 1\quad 2\quad 1 \\
 & \quad\ \ \downarrow\quad \downarrow & \\
3 & 1 + 3 + 3 + 1 & 1\quad 3\quad 3\quad 1 \\
 & \quad\ \ \downarrow\quad \downarrow\quad \downarrow & \\
4 & 1 + 4 + 6 + 4 + 1 & 1\quad 4\quad 6\quad 4\quad 1 \\
 & \quad\ \ \downarrow\quad \downarrow\quad \downarrow\quad \downarrow & \\
5 & 1 + 5 + 10 + 10 + 5 + 1 & 1\quad 5\quad 10\quad 10\quad 5\quad 1 \\
 & \quad\ \ \downarrow\quad \downarrow\quad \downarrow\quad \downarrow\quad \downarrow & \\
6 & 1 + 6 + 15 + 20 + 15 + 6 + 1 & 1\quad 6\quad 15\quad 20\quad 15\quad 6\quad 1
\end{array}
$$

This construction is known as "Pascal's triangle" [*Treatise on Arithmetic Triangle*, 1654]. To show its equivalence with the diagram in figure 2.8a, we have reproduced it on the right in the form of a pyramid.

When the spin of the equivalent nuclei is greater than ½, we can determine the relative intensities of the hyperfine lines by constructing "generalised Pascal's triangles". For example, for $I = 1$, the diagram in figure 2.8b shows that we must add the 3 numbers in rows $(p - 1, p, p + 1)$ of line $(n - 1)$ to obtain the number in row p of line n. We thus get the following triangle:

							1						
n = 1						1	1	1					
2					1	2	3	2	1				
3				1	3	6	7	6	3	1			
4			1	4	10	16	19	16	10	4	1		
5		1	5	15	30	45	51	45	30	15	5	1	
6	1	6	21	50	90	126	141	126	90	50	21	6	1

For high values of n, these triangles are more aesthetic (and fun) than practical. Indeed, when a molecule is large enough to contain a large number of equivalent nuclei, it often also contains other non-equivalent nuclei and we must resort to numerical simulation to interpret the spectrum.

References

ATHERTON N.M. (1993) *Principles of Electron Spin Resonance*, Ellis Horwood PTR Prentice Hall, New York.

AYANT Y. & BELORIZKY E. (2000) *Cours de Mécanique Quantique*, Dunod, Paris.

COHEN-TANNOUDJI C., DIU B. & LALOË F. (2015) *Quantum Mechanics*, Wiley-VcH, New York.

EL HASSAN I. *et al.* (2006) *Mendeleev Communications*: 149-151.

FESSENDEN R.W. & SCHULER R.H. (1963) *Journal of Chemical Physics* **39**: 2147-2195.

FREEMANN A.J. & WATSON R.E. (1965) "Hyperfine Interactions in Magnetic Materials" in *Magnetism* vol II A, RADO G.T. & SUHL H. eds, Academic Press, New York.

JEUNET A., NICKEL B. & RASSAT A. (1986) *New Journal of Chemistry* **10**: 123-132.

LIDE D.R. (2010) *Handbook of Chemistry and Physics*, section 1, CRC Press, Boca Raton.

MCCONNELL H.M.& CHESNUT D.B. (1958) *Journal of Chemical Physics* **28**: 107-117.

WERTZ J.E. & BOLTON J.R (1971) *Electron Spin Resonance. Elementary Theory and Practical Applications*, McGraw-Hill, New York.

Exercises

2.1. The normalised form of the $1s$ orbital for the hydrogen atom is $\Phi(r) = (1/\sqrt{\pi})a_0^{-3/2}\exp(-r/a_0)$ where $a_0 = 4\pi\varepsilon_0\hbar^2/m_e e^2$ is the "Bohr radius", which is numerically equal to 5.292×10^{-11} m [Ayant and Belorizky, 2000; Cohen-Tannoudji, 2015].What is the expression for the electron probability density at the nucleus? Calculate its numerical value. Deduce the hyperfine constant A given by equation [2.5] (The experimental value is $A = 1420$ MHz). Because of this hyperfine interaction, the first excited level for the hydrogen atom has an energy of A. Determine the wavelength of the radiation emitted by the atom when it passes from this excited state to its ground state. On earth we can detect the radiation emitted by the hydrogen atoms in interstellar space, and study of this radiation (radioastronomy) has been used to demonstrate, for example, that our galaxy is spiral.

2.2. The allyl radical has the formula $H_2C^{\bullet} - CH = CH_2$. The unpaired electron is delocalised on the three carbon atoms and it interacts:

▹ with the CH proton: hyperfine constant A_2.

▹ with the 4 CH_2 protons, which we assume to be equivalent: hyperfine constant A_1, with $|A_1| > |A_2|$.

a) Draw the general course of the EPR spectrum and indicate how we can measure $|A_1|$ and $|A_2|$.

b) On the experimental spectrum we measure $|A_1| = 40.3$ MHz, $|A_2| = 11.4$ MHz [Fessenden and Schuler, 1963]. Using the McConnell relation and the value $Q = -64$ MHz determined for the methyl radical, assess the spin populations on the carbon atoms. Compare them to those predicted by the simplified model presented in appendix 5.

c) What would the spectrum look like if $|A_2| \gg |A_1|$?

2.3. Determine the parameters g_{iso} and $A_{iso}(^{15}N)$ [in MHz] on the spectrum in figure 2.3. What would be the value of the hyperfine constant for ^{14}N nuclei? We note that the two lines of the doublet have different amplitudes even though their intensities are equal (section 5.4.2). How can this be explained?

2.4. Theoretical calculations show that the interaction between the electron in a $2p$ orbital in the nitrogen atom and the ^{14}N nucleus produces a hyperfine constant $A_s \approx 80$ MHz [Wertz and Bolton, 1971]. From the values of g_{iso} and A_{iso} measured on the spectrum in figure 2.6a, assess the spin population ρ_N for the $2p_z$ orbital in the nitrogen atom from the nitroxide radical.

2.5. Determine the parameters g_{iso} and A_{iso} on the spectrum in figure 2.7. What would be the position of the lines at Q-band (ν = 35 GHz)?

2.6. Mesure the parameter A_{iso} on the spectrum in figure 2.12a and compare it to that of a nitroxide radical (exercises 2.3 and 2.4).

2.7. Demonstrate relations [2.12] and [2.13].

Answers to exercises

2.1. $\rho(0) = [\Phi(0)]^2 = (1/\pi)\, a_0^{-3} = 2.148 \times 10^{30}\ \mathrm{m}^{-3}$. A = 1423 MHz. λ = 0.21 m.

2.2. a) The four equivalent protons produce a pattern of five lines split by $|A_1|/g_{iso}\beta$, with relative intensities (1:4:6:4:1) (complement 2). Interaction with the fifth proton causes each line to split to form two lines of the same intensity separated by $|A_2|/g_{iso}\beta$.

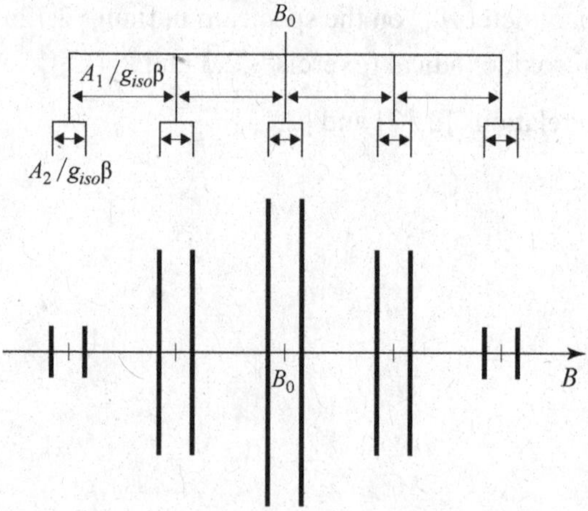

b) We obtain $|\rho(C_1)| = |\rho(C_3)| = 0.63$ and $|\rho(C_2)| = 0.18$. As the sum of the populations must be close to 1, we must have $\rho(C_1) = \rho(C_3) = 0.63$ and $\rho(C_2) = -0.18$. The signs were confirmed by analysis of the spectrum produced by a single crystal. A model described in appendix 5 predicts $\rho(C_1) = \rho(C_3) = 0.66$ and $\rho(C_2) = -0.33$. Even though the agreement is not very good, this simple model does correctly predict the *signs* observed.

c) Two identical isolated patterns should be observed with intensities (1:4:6:4:1).

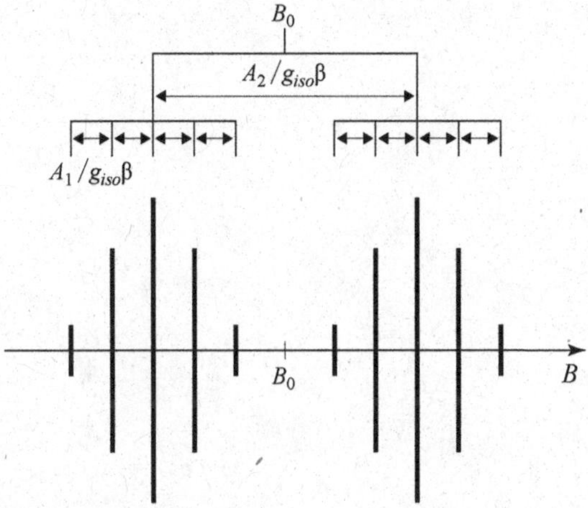

2.3. $g_{iso} = 2.008$ and $A_{iso}(^{15}N)/g_{iso}\beta = 2.135$ mT are measured, and as a result $A_{iso}(^{15}N) = 60$ MHz. Since $g_N(^{14}N)/g_N(^{15}N) = 0.7143$ (table 2.2), this value corresponds to $A_{iso}(^{14}N) = 43$ MHz. Closer examination of the spectrum shows that the line centred at around 345.5 mT is slightly broader than the other line. The source of this difference is discussed in section 5.4.3.

2.4. $g_{iso} = 2.006$ and $A_{iso} = 44$ MHz can be measured, from which we deduce $\rho_N \approx 0.55$.

2.5. Measurement gives $g_{iso} = 2.008$ and $A_{iso}/g_{iso}\beta = 9.56$ mT, thus $A_{iso} = 269$ MHz. The change in frequency shifts the position of the centre for the hyperfine pattern from 334 mT to 1245 mT but the splitting between the lines remains unchanged.

2.6. $A_{iso} = 22$ MHz is found, which corresponds to half the value measured for a nitroxide radical. The spin density $\rho_N \approx 0.5$ for the nitroxide radical (exercise 2.4) is equally distributed over the two equivalent nitrogen atoms in the nitronyl-nitroxide radical. Therefore, $\rho_N \approx 0.25$ for this radical.

2.7. Each group of n_i equivalent nuclei of spin I_i produces $(2n_iI_i + 1)$ hyperfine lines (section 2.4.1), and the total number of lines is the product of all these numbers. According to [2.11], the resonance field takes the form

$$B = B_0 - \sum_{k=1}^{n} M_k (A_{iso})_k / g_{iso}\beta.$$

As its extreme values are $B_0 \pm \sum_{k=1}^{n} I_k (A_{iso})_k / g_{iso}\beta$, the field range covered is $\Delta B = \sum_{k=1}^{n} 2I_k (A_{iso})_k / g_{iso}\beta$. Equation [2.13] is obtained by collecting the contributions of equivalent nuclei.

Chapter 3

Introduction to the formalism of the space of spin states. The Hamiltonian operator

3.1 – Introduction

In the previous chapter we dealt with a paramagnetic centre of spin S, in which the unpaired electrons undergo *isotropic magnetic interactions* due to the field $\mathbf{B}$ and to nuclei of spin I_k, with the following form:

$$H = g\beta \mathbf{S}\cdot\mathbf{B} + \Sigma_k A_k \mathbf{S}\cdot\mathbf{I}_k \qquad [3.1]$$

From H, we have determined the position of the resonance lines by applying an *ad hoc* procedure to deduce the energy levels for the paramagnetic centre and by recognising that only some transitions are allowed according to "selection rules". This procedure and these rules are the result of the formalism of quantum mechanics, which alone can correctly describe a spectroscopic experiment. To treat situations that are more general than those considered up to now, the part of this formalism that is useful in EPR must be used. We will first clarify the significance of the quantity H given by equation [3.1]: the lowest energy level of the paramagnetic centre is degenerate, and H can be used to determine the splitting induced by the magnetic interactions that the unpaired electrons engage in (figure 3.1). The value of the spin S and the g and A_k parameters in equation [3.1] are therefore characteristic of the *ground state* of the paramagnetic centre. This can be illustrated by two examples:

- ▹ In a paramagnetic atom or free ion, the electrons are attracted by the nucleus and repulsed by the other electrons due to electrostatic interactions; they are also subjected to a "*spin-orbit*" interaction. All of these interac-

P. Bertrand, *Electron Paramagnetic Resonance Spectroscopy*,
https://doi.org/10.1007/978-3-030-39663-3_3

tions produce a ground state characterised by the total angular momentum **J** and a $g_{Landé}$ number that can be calculated (appendix 1). If the nucleus of the atom or ion is paramagnetic and characterised by a spin I, an isotropic hyperfine interaction exists that can be described by a term $A\mathbf{J}\cdot\mathbf{I}$, where A is determined by the distribution of the unpaired electrons in this ground state. Equation [3.1] can then be written:

$$H = g_{Landé}\beta\mathbf{J}\cdot\mathbf{B} + A\,\mathbf{J}\cdot\mathbf{I}$$

▹ In a transition ion complex, all of the electrostatic interactions involving the cation's electrons produce a ground state characterised by a spin $S = n/2$, where n is the number of unpaired electrons. Interactions between these electrons and the magnetic field and with the paramagnetic nuclei are *anisotropic*, and the g and A_k numbers in equation [3.1] are replaced by *matrices* for which the elements are determined by the electronic structure of the ground state.

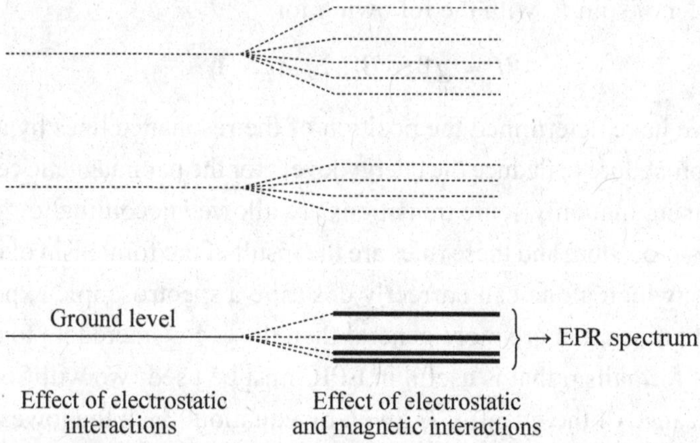

Figure 3.1 – Splitting of the energy levels for a paramagnetic centre caused by the magnetic interactions to which its unpaired electrons are exposed. The EPR spectrum is produced by the lowest energy levels.

From H, the energy levels and the probabilities of the transitions between these levels can be calculated. These are all the elements that determine the *shape* and *intensity* of the EPR spectrum. In this chapter, we will learn to perform these calculations in simple cases, which will help to elucidate the origins of the rules used in the previous chapters, and to deal with the case of anisotropic magnetic moments. The consequences of this anisotropy on the shape of the spectrum will be examined in detail in chapter 4.

3.2 – Space of spin states associated with an angular momentum

This section presents how the angular momenta that will be used repeatedly hereafter should be used. A justification and many developments can be found in works on quantum mechanics [Ayant and Belorizky, 2000; Cohen-Tannoudji *et al.*, 2015].

Consider a paramagnetic entity (nucleus, atom, molecule) for which the ground state is characterised by the angular momentum **J** and the number J. *Spin states* that can be "occupied" by the paramagnetic entity are associated with this angular momentum. These spin states have similar properties to those of the *vectors* in the Euclidean space and their ensemble $\mathcal{E}_J$ is said to constitute a "vector space". A fundamental property of a vector space is that any vector can be expressed as a linear combination of *basis* vectors, the number of basis vectors being the *dimension* of the space. Whereas the Euclidean space is of dimension 3, the space of spin states $\mathcal{E}_J$ is of dimension $N = (2J + 1)$. To distinguish the elements of $\mathcal{E}_J$ from the vectors of the Euclidean space, they are represented by the symbol $|u\rangle$ which recalls the arrow of vectors; these elements are known as "kets" (the etymology is given in section 3.2.2). Hereafter, we briefly describe the operations that will be performed on these kets.

3.2.1 – Construction of linear operators from J – Specific bases of $\mathcal{E}_J$

In the 3-dimensional Euclidean space we select an orthonormal reference frame with axes $\{X, Y, Z\}$ and denote (J_X, J_Y, J_Z) the components of **J** in this frame. In the $\mathcal{E}_J$ space, these components are *linear spin operators* which transform the kets into other kets. To indicate, for example, that the action of $\hat{J}_X$ on the $|u\rangle$ ket gives the $|v\rangle$ ket, we write

$$\hat{J}_X |u\rangle = |v\rangle$$

Here, the symbol above the letter indicates that $\hat{J}_X$ is an *operator*. Due to linearity

$$\hat{J}_X \big(\lambda |u_1\rangle + \mu |u_2\rangle\big) = \lambda \hat{J}_X |u_1\rangle + \mu \hat{J}_X |u_2\rangle$$

where λ and μ are any two *complex numbers*. From $(\hat{J}_X, \hat{J}_Y, \hat{J}_Z)$, other linear operators can be constructed. For example, the action of $\hat{J}_X^2$ consists in applying $\hat{J}_X$ twice in succession:

$$\hat{J}_X^2 |u\rangle = \hat{J}_X \big(\hat{J}_X |u\rangle\big) \qquad [3.2]$$

$\hat{J}_Y^2$ and $\hat{J}_Z^2$ are defined in the same way and the operator $\hat{\mathbf{J}}^2$ is defined by:

$$\hat{\mathbf{J}}^2 = \hat{J}_X^2 + \hat{J}_Y^2 + \hat{J}_Z^2$$

We will see that the magnetic interactions in which the paramagnetic entities engage can be expressed using the components of **J** or their squares. In the space $\mathcal{E}_J$, these interactions become *spin operators* which are expressed as a function of $(\hat{J}_X, \hat{J}_Y, \hat{J}_Z, \hat{J}_X^2, \hat{J}_Y^2, \hat{J}_Z^2)$.

It can be shown that there exists a set of $(2J + 1)$ kets of $\mathcal{E}_J$ denoted $\{|J, M_J\rangle\}$ which verify the two equations:

$$\begin{aligned} \hat{\mathbf{J}}^2 \,|J, M_J\rangle &= J(J+1)\,|J, M_J\rangle \\ \hat{J}_Z \,|J, M_J\rangle &= M_J\,|J, M_J\rangle \qquad M_J = -J, -J+1, \dots J-1, J \end{aligned} \tag{3.3}$$

The action of the $\hat{\mathbf{J}}^2$ and $\hat{J}_Z$ operators on these kets therefore produces the same kets, if we ignore a multiplicative factor. The equations [3.3] have a precise physical signification: when the paramagnetic entity occupies the spin state described by the ket $|J, M_J\rangle$, the value of $\mathbf{J}^2$ is equal to $J(J + 1)$ and that of J_Z is equal to M_J. This property was presented in chapters 1 and 2 in the simplified form "$\mathbf{J}^2$ and J_Z can take the values $J(J + 1)$ and M_J, respectively".

The kets $\{|J, M_J\rangle\}$ make up *a basis* of $\mathcal{E}_J$: any ket $|u\rangle$ can be written in the form:

$$|u\rangle = \sum_{M_J=-J}^{J} c(M_J)|J, M_J\rangle$$

$c(M_J)$ are complex numbers known as the *components* of $|u\rangle$ in the $\{|J, M_J\rangle\}$ basis. This expression can be used to determine the action of $\hat{\mathbf{J}}^2$ and $\hat{J}_Z$ on $|u\rangle$:

$$\begin{aligned} \hat{\mathbf{J}}^2|u\rangle &= \sum_{M_J=-J}^{J} c(M_J)\hat{\mathbf{J}}^2|J, M_J\rangle \\ &= J(J+1)\sum_{M_J=-J}^{J} c(M_J)|J, M_J\rangle \\ &= J(J+1)|u\rangle \end{aligned}$$

$$\begin{aligned} \hat{J}_z|u\rangle &= \sum_{M_J=-J}^{J} c(M_J)\,\hat{J}_z|J, M_J\rangle \\ &= \sum_{M_J=-J}^{J} c(M_J)\,M_J|J, M_J\rangle \end{aligned}$$

The action of $\hat{\mathbf{J}}^2$ on any ket $|u\rangle$ amounts to multiplying it by $J(J + 1)$, but that of $\hat{J}_Z$ is more complicated. The action of the operators $\hat{J}_X$ and $\hat{J}_Y$ on the kets $\{|J, M_J\rangle\}$ will be detailed below (equations [3.9]).

Selection of a direction Z in the Euclidean space determines the component J_Z of **J**, and consequently a particular basis $\{|J, M_J\rangle\}$ of $\mathcal{E}_J$. This is expressed by stating that Z is a "quantization axis". In practice, we will see that this basis, and thus this direction, are chosen to simplify the calculations as far as possible.

3.2.2 – The scalar product

In the Euclidean space, *orthonormal* bases are generally used. As a result, if $\mathbf{v}_1$ and $\mathbf{v}_2$ are two basis vectors, the scalar product $\mathbf{v}_1 \cdot \mathbf{v}_2$ is equal to 1 if these vectors are identical, and to zero if they are different. In the first case, $\mathbf{v}_1$ is said to be *normalised*, in the second $\mathbf{v}_1$ and $\mathbf{v}_2$ are said to be *orthogonal*. In the space $\mathcal{E}_J$, a "scalar product" of two kets, $|v_1\rangle$ and $|v_2\rangle$, which we note $\langle v_1|v_2\rangle$, is also defined. In general, this product is a complex number, such that:

$$\langle v_2|v_1\rangle = (\langle v_1|v_2\rangle)^*$$

The sign * indicates the *complex conjugate*. In general, the scalar product is therefore *not commutative*. The basis $\{|J, M_J\rangle\}$ defined above is *orthonormal* in the sense that the scalar product $\langle J, M_J|J, M_J'\rangle$ is equal to 1 if $M_J = M_J'$ and to 0 if $M_J \neq M_J'$. In the first case, the ket $|J, M_J\rangle$ is said to be *normalised*, in the second the kets $|J, M_J\rangle$ and $|J, M_J'\rangle$ are said to be *orthogonal*.

In practice, the scalar product of any two kets $|v_1\rangle$ and $|v_2\rangle$ is calculated by expressing them in an *orthonormal basis* $\{|u_1\rangle, |u_2\rangle, \ldots |u_N\rangle\}$:

$$\begin{aligned} |v_1\rangle &= c_{11}\,|u_1\rangle + c_{12}\,|u_2\rangle + \ldots + c_{1N}\,|u_N\rangle \\ |v_2\rangle &= c_{21}\,|u_1\rangle + c_{22}\,|u_2\rangle + \ldots + c_{2N}\,|u_N\rangle \end{aligned}$$

The scalar product $\langle v_2|v_1\rangle$ is equal to:

$$\langle v_2|v_1\rangle = c_{11}(c_{21})^* + c_{12}(c_{22})^* + \ldots + c_{1N}\,(c_{2N})^*$$

Note – In the scalar product $\langle v_2|v_1\rangle$, $\langle v_2|$ is said to be the conjugated "bra" of the ket $|v_2\rangle$. The notation and terminology ("bra", "ket") were created by the physicist Paul Dirac, and are derived from the word "bracket".

3.2.3 – Representation of an operator by a matrix

In the Euclidean space, *matrices* are used to easily perform changes of basis. The matrix is a table which reproduces, for example, the components of the old basis vectors in the new basis. In the $\mathcal{E}_J$ space, matrices are used to perform changes of basis, and to describe *the action of an operator* on the kets of a basis. Thus, for example let $\{|u_1\rangle, \ldots |u_N\rangle\}$, where $N = (2J + 1)$, be an orthonormal basis of $\mathcal{E}_J$. Assume that the action of the operator $\hat{A}$ on the ket $|u_j\rangle$ in this basis produces a ket denoted $|v_j\rangle$:

$$\hat{A}\,|u_j\rangle = |v_j\rangle$$

Since $|v_j\rangle$ belongs to $\mathcal{E}_J$, it can be written as a linear combination of the kets from the basis $\{|u_1\rangle, \ldots, |u_N\rangle\}$:

$$|v_j\rangle = A_{1j}\,|u_1\rangle + \ldots + A_{ij}\,|u_i\rangle + \ldots + A_{Nj}\,|u_N\rangle \qquad [3.4]$$

The A_{ij} "components" are generally complex numbers, that can be placed in column j of a table containing N rows and N columns. By performing this operation for all the kets $\{|u_j\rangle\}$ in the basis, a square *matrix* $\tilde{\mathbf{A}}$ of dimension N is obtained. To help remember that the element A_{ij} located in row i and column j represents the *component* of ket $(\hat{A}\,|u_j\rangle)$ on ket $|u_i\rangle$, we can include the kets *above* and to the *right* of the matrix:

	$|u_1\rangle$	…	$|u_j\rangle$	…	$|u_N\rangle$	
$\langle u_1|$	A_{11}	…	A_{1j}	…	A_{1N}	$|u_1\rangle$
…	…	…	…	…	…	…
$\langle u_i|$	A_{i1}	…	A_{ij}	…	A_{iN}	$|u_i\rangle$
…	…	…	…	…	…	…
$\langle u_N|$	A_{N1}	…	A_{Nj}	…	A_{NN}	$|u_N\rangle$

Thanks to this matrix, the action of the operator $\hat{A}$ on any ket $|w\rangle$ of $\mathcal{E}_J$ can be determined. Indeed, if we write $|w\rangle$ in the basis $\{|u_1\rangle, \ldots, |u_N\rangle\}$

$$|w\rangle = c_1|u_1\rangle + \ldots + c_i\,|u_i\rangle + \ldots + c_N\,|u_N\rangle$$

the matrix $\tilde{\mathbf{A}}$ can be used to express the ket $\hat{A}\,|w\rangle$ in this basis. We therefore say that $\tilde{\mathbf{A}}$ "represents" the operator $\hat{A}$ in the basis $\{|u_1\rangle, \ldots, |u_N\rangle\}$.

The elements of the $\tilde{\mathbf{A}}$ matrix can be interpreted as *scalar products*. Indeed, if we perform the scalar product of both sides of equation [3.4] by the ket $|u_i\rangle$, we obtain:

$$\langle u_i|v_j\rangle = A_{1j}\,\langle u_i|u_1\rangle + \ldots + A_{ij}\,\langle u_i|u_i\rangle + \ldots + A_{Nj}\,\langle u_i|u_N\rangle$$

As the basis is orthonormal, all the scalar products on the second side of the equation are null except $\langle u_i|u_i\rangle$ which is equal to 1. We therefore obtain:

$$\langle u_i|v_j\rangle = A_{ij}$$

Thus, when $|v_j\rangle$ is replaced by its expression:

$$\langle u_i|(\hat{A}\,|u_j\rangle) = A_{ij}$$

As a reminder of this property, the elements of the matrix are conventionally written in the form

$$A_{ij} = \langle u_i\,|\hat{A}\,|u_j\rangle$$

and a column of bras $\{\langle u_i|\}$ is included to the *left* of the matrix, as indicated above.

The operators that we will use hereafter are such that their matrix elements satisfy:

$$A_{ij} = (A_{ji})^*$$

The operator $\hat{A}$ and the matrix $\tilde{\mathbf{A}}$ representing it in a given basis are said to be *Hermitian*.

Finally, it can be shown that the matrix $\tilde{\mathbf{B}}$ which represents the operator $\hat{A}^2$ in a basis is equal to the square of the matrix $\tilde{\mathbf{A}}$ representing the operator $\hat{A}$ in the same basis (exercise 3.1):

$$\tilde{\mathbf{B}} = (\tilde{\mathbf{A}})^2 = \tilde{\mathbf{A}}\cdot\tilde{\mathbf{A}}$$

3.2.4 – Eigenvectors and eigenvalues of an operator

Any operator $\hat{A}$ has a set of *eigenvectors* $\{|v_k\rangle\}$ associated with *eigenvalues* $\{a_k\}$ which verify:

$$\hat{A}\,|v_k\rangle = a_k\,|v_k\rangle \qquad [3.5]$$

The eigenvectors are only defined within a multiplicative factor, for which the modulus is equal to 1 if $|v_k\rangle$ is *normalised*. In this case, the previous equation gives rise to:

$$\langle v_k\,|\hat{A}\,|v_k\rangle = a_k$$

It can be shown that the eigenvalues of a Hermitian operator are *real* (exercise 2) and that its eigenvectors can always be chosen to create an *orthonormal basis*.

In line with this statement, the kets $\{|J, M_J\rangle\}$ satisfying the equations [3.3] are *eigenvectors shared* by the operators $\hat{\mathbf{J}}^2$ and $\hat{J}_Z$. The physical signification of equation [3.5] is therefore the same as that of equations [3.3]. It can be stated as follows:

The possible values of the physical quantity A which corresponds to the operator $\hat{A}$ are the eigenvalues of $\hat{A}$. A takes the value a_k when the paramagnetic entity occupies the state described by the eigenvector $|v_k\rangle$.

This basic property gives the notions of eigenvectors and eigenvalues all their significance. When seeking to determine the eigenvectors and eigenvalues of an operator $\hat{A}$, we first represent it by the matrix $\tilde{\mathbf{A}}$ in any basis $\{|u_1\rangle, \ldots |u_N\rangle\}$ and seek the eigenvectors in the form:

$$|v\rangle = c_1 |u_1\rangle + \ldots + c_i |u_i\rangle + \ldots + c_N |u_N\rangle$$

Equation [3.5] is equivalent to the matrix equation:

$$(\tilde{\mathbf{A}} - a\,\tilde{\mathbf{1}})\,\tilde{\mathbf{C}} = 0 \qquad [3.6]$$

where $\tilde{\mathbf{1}}$ is the "unit matrix" which has ones on the main diagonal and zeros everywhere else, and $\tilde{\mathbf{C}}$ is the column matrix of the components $(c_1, c_2, \ldots c_N)$:

$$\tilde{\mathbf{C}} = \begin{bmatrix} c_1 \\ \ldots \\ c_N \end{bmatrix}$$

The matrix equation [3.6] constitutes a system of N simultaneous linear equations for which the unknowns are $(c_1, \ldots c_N)$. As the second side is null, this system only allows a solution different to the trivial solution $(0, \ldots, 0)$ if *its determinant is null*:

$$\begin{vmatrix} A_{11} - a & \ldots & A_{1N} \\ \ldots & \ldots & \ldots \\ A_{N1} & \ldots & A_{NN} - a \end{vmatrix} = 0$$

The eigenvalues $\{a_k\}$ for the operator $\hat{A}$ (and for the $\tilde{\mathbf{A}}$ matrix) are the N roots of this equation. By using the properties of determinants, the sum of the eigenvalues can be shown to be equal to the sum T of the diagonal elements (exercise 3.2):

$$a_1 + a_2 + \ldots + a_N = A_{11} + A_{22} + \ldots + A_{NN} = T$$

This sum is independent of the representation selected; it is known as the *trace* of the operator $\hat{A}$ and of the $\tilde{\mathbf{A}}$ matrix. To identify the eigenvector corresponding to an eigenvalue a_k, a is replaced by a_k in the system of equations [3.6]. As the determinant of this system is null, the system is reduced to $(N - 1)$ independent equations which can be used to determine the ket $|v_k\rangle$ to within a multiplicative factor.

In the specific case where the $\tilde{\mathbf{A}}$ matrix is "diagonal" in the basis $\{|u_1\rangle, \ldots |u_N\rangle\}$, i.e., when all its off-diagonal elements are null, the eigenvalues are the diagonal elements and the eigenvectors are the kets $\{|u_1\rangle, \ldots |u_N\rangle\}$. This explains why the search for the eigenvalues and eigenvectors of a matrix is sometimes called its "diagonalisation".

3.2.5 – Application to a centre characterised by $J = ½$

To familiarise ourselves with the previous definitions, we will examine the specific case where $J = ½$. The space of spin states $\mathcal{E}_{½}$ is associated with the angular momentum **J**. We first select any direction Z in the Euclidean space, and add two directions (X, Y) to obtain an orthonormal reference frame $\{X, Y, Z\}$. We know that in the $\mathcal{E}_{½}$ space there exists a basis $\{|½, ½\rangle, |½, -½\rangle\}$ for which the kets satisfy the equations [3.3], which can be written as follows for the case we are studying:

$$\hat{\mathbf{J}}^2 \,|½, ½\rangle = ¾\,|½, ½\rangle \qquad \hat{\mathbf{J}}^2 \,|½, -½\rangle = ¾\,|½, -½\rangle \tag{3.7}$$

$$\hat{J}_Z \,|½, ½\rangle = ½\,|½, ½\rangle \qquad \hat{J}_Z \,|½, -½\rangle = -½\,|½, -½\rangle \tag{3.8}$$

◇ Construction of the matrices representing the different operators

We will first construct the matrices representing $(\hat{J}_X, \hat{J}_Y, \hat{J}_Z)$ in the basis $\{|½, ½\rangle, |½, -½\rangle\}$. Equations [3.8] directly give the matrix $\tilde{\mathbf{J}}_Z$:

	$\|½, ½\rangle$	$\|½, -½\rangle$
$\langle ½, ½\|$	½	0
$\langle ½, -½\|$	0	−½

The *trace* of the $\hat{J}_Z$ operator is observed to be *null*. This result holds whatever the value of J. Indeed, since $\hat{J}_Z$ is represented in the basis $\{|J, M_J\rangle\}$ by a diagonal matrix for which the elements are $\langle J, M_J|\, \hat{J}_Z \,|J, M_J\rangle = M_J$, where $M_J = -J, -J+1, \ldots J$, its trace T is:

$$T = \sum_{M_J=-J}^{J} M_J = 0$$

To obtain the matrices representing $\hat{J}_X$ and $\hat{J}_Y$, their effect on the kets of the basis must be known. By applying the general properties of angular momentum operators, it can be shown that $\hat{J}_X$ and $\hat{J}_Y$ act on the kets $\{|J, M_J\rangle\}$, defined by the equations [3.3], as follows [Ayant and Belorizky, 2000; Cohen-Tannoudji, 2015]:

$$\hat{J}_X|J,M_J\rangle = \tfrac{1}{2}[J(J+1)-M_J(M_J+1)]^{1/2}|J,M_J+1\rangle + \tfrac{1}{2}[J(J+1)-M_J(M_J-1)]^{1/2}|J,M_J-1\rangle$$
$$\hat{J}_Y|J,M_J\rangle = -\mathrm{i}/2[J(J+1)-M_J(M_J+1)]^{1/2}|J,M_J+1\rangle + \mathrm{i}/2[J(J+1)-M_J(M_J-1)]^{1/2}|J,M_J-1\rangle \quad [3.9]$$

For $J = ½$, these equations become:

$$\hat{J}_X|½, ½\rangle = ½\,|½, -½\rangle\,; \qquad \hat{J}_X|½, -½\rangle = ½\,|½, ½\rangle$$
$$\hat{J}_Y|½, ½\rangle = \mathrm{i}/2\,|½, -½\rangle\,; \qquad \hat{J}_Y|½, -½\rangle = -\mathrm{i}/2\,|½, ½\rangle$$

From this set of equations, the matrices $\tilde{\mathbf{J}}_\mathbf{X}$ and $\tilde{\mathbf{J}}_\mathbf{Y}$ can be deduced; they also have a null trace:

	$|½, ½\rangle$	$|½, -½\rangle$
$\langle ½, ½|$	0	½
$\langle ½, -½|$	½	0

	$|½, ½\rangle$	$|½, -½\rangle$
$\langle ½, ½|$	0	$-\mathrm{i}/2$
$\langle ½, -½|$	$\mathrm{i}/2$	0

We will now examine the operators $(\hat{J}_X^2, \hat{J}_Y^2, \hat{J}_Z^2)$ which act on the kets of $\mathcal{E}_{½}$, as defined by equation [3.2]. For $\hat{J}_X^2$, we can thus write:

$$\hat{J}_X^2\,|½, ½\rangle = \hat{J}_X\left(\hat{J}_X|½, ½\rangle\right) = \hat{J}_X\left(½\,|½, -½\rangle\right) = ¼\,|½, ½\rangle$$

By performing the same operations on the ket $|½, -½\rangle$, the $\tilde{\mathbf{J}}_\mathbf{X}^2$ matrix is obtained:

	$|½, ½\rangle$	$|½, -½\rangle$
$\langle ½, ½|$	¼	0
$\langle ½, -½|$	0	¼

Alternatively, $\tilde{\mathbf{J}}_\mathbf{X}^2$ can be directly calculated by multiplying the $\tilde{\mathbf{J}}_\mathbf{X}$ matrix by itself. To determine the product $\tilde{\mathbf{C}} = \tilde{\mathbf{A}} \cdot \tilde{\mathbf{B}}$ it should be remembered that the matrices must be arranged as follows:

$$\begin{matrix} & \begin{bmatrix} b_{11} & b_{12} \\ b_{21} & b_{22} \end{bmatrix} \\ \begin{bmatrix} a_{11} & a_{12} \\ a_{21} & a_{22} \end{bmatrix} & \begin{bmatrix} c_{11} & c_{12} \\ c_{21} & c_{22} \end{bmatrix} \end{matrix}$$

The matrix element located in the first row of the first column of $\tilde{\mathbf{C}}$ is obtained by "multiplying" the first row of $\tilde{\mathbf{A}}$ by the first column of $\tilde{\mathbf{B}}$, *as follows*:

$$c_{11} = a_{11}\, b_{11} + a_{12}\, b_{21}$$

By applying this procedure, the above matrix is produced. We can readily verify that the matrices $\tilde{\mathbf{J}}_{\mathbf{Y}}{}^2$ and $\tilde{\mathbf{J}}_{\mathbf{Z}}{}^2$ are identical to $\tilde{\mathbf{J}}_{\mathbf{X}}{}^2$. We deduce that the matrix representing $\hat{\mathbf{J}}^2 = \hat{J}_X^2 + \hat{J}_Y^2 + \hat{J}_Z^2$, which is equal to $\tilde{\mathbf{J}}_{\mathbf{X}}{}^2 + \tilde{\mathbf{J}}_{\mathbf{Y}}{}^2 + \tilde{\mathbf{J}}_{\mathbf{Z}}{}^2$, can be written:

	$\lvert ½, ½\rangle$	$\lvert ½, -½\rangle$
$\langle ½, ½\rvert$	¾	0
$\langle ½, -½\rvert$	0	¾

This matrix could also have been directly deduced from equations [3.7]. All of the matrices developed in this example are Hermitian.

Appendix 7 includes a table listing the expressions of the matrix elements defined from the components of any angular momentum **S**; these expressions will be used repeatedly hereafter.

◇ *Determining the eigenvalues and eigenvectors*

Since the operators $\hat{J}_Z$, $\hat{J}_X^2$, $\hat{J}_Y^2$, $\hat{J}_Z^2$ and $\hat{\mathbf{J}}^2$ are represented by *diagonal matrices* in the basis $\{\lvert ½, ½\rangle, \lvert ½, -½\rangle\}$, these kets are eigenvectors shared by all these operators, and the associated eigenvalues are the respective diagonal elements of these matrices. As might be expected, the eigenvalues of $\hat{J}_X^2$, $\hat{J}_Y^2$, $\hat{J}_Z^2$ and $\hat{\mathbf{J}}^2$ are *positive*. For the operators $\hat{J}_X$ and $\hat{J}_Y$, for which the matrices are not diagonal in this basis, the general method described in section 3.2.4 must be applied. As an example we will describe the calculation for $\hat{J}_X$. We seek its eigenvectors $\{\lvert u\rangle\}$ in the form:

$$\lvert u\rangle = c_1\lvert ½, ½\rangle + c_2\lvert ½, -½\rangle$$

The eigenvalues λ satisfy the matrix equation:

$$(\tilde{\mathbf{J}}_{\mathbf{X}} - \lambda\tilde{\mathbf{1}})\begin{bmatrix} c_1 \\ c_2 \end{bmatrix} = 0 \qquad [3.10]$$

where $(\tilde{\mathbf{J}}_X - \lambda\tilde{\mathbf{1}})$ is the matrix:

	$\vert ½, ½\rangle$	$\vert ½, -½\rangle$
$\langle ½, ½\vert$	$-\lambda$	½
$\langle ½, -½\vert$	½	$-\lambda$

The two-equation system with two unknowns which corresponds to equation [3.10] only allows a solution (c_1, c_2) different from zero if its determinant is null:

$$\lambda^2 - ¼ = 0$$

The two eigenvalues are therefore $\lambda_1 = ½$ and $\lambda_2 = -½$. To identify the eigenvector associated with λ_1, we use one of the two equations from system [3.10] in which we have replaced λ by λ_1:

$$-½\, c_1 + ½\, c_2 = 0$$

We therefore have $c_1 = c_2$. An example of a *normalised* ket which satisfies this relation is:

$$1/\sqrt{2}\ \left(|½, ½\rangle + |½, -½\rangle\right)$$

For the eigenvalue $\lambda_2 = -½$, the same method produces the ket:

$$1/\sqrt{2}\ \left(|½, ½\rangle - |½, -½\rangle\right)$$

These two kets can readily be shown to be *orthogonal*. Similarly, we obtain the eigenvalues and eigenvectors of $\hat{J}_Y$:

$$\text{eigenvalue } ½\text{:}\ \ 1/\sqrt{2}\ \left(|½, ½\rangle + \mathrm{i}|½, -½\rangle\right)$$

$$\text{eigenvalue } -½\text{:}\ \ 1/\sqrt{2}\ \left(|½, ½\rangle - \mathrm{i}|½, -½\rangle\right)$$

Note – As the three directions (X, Y, Z) of the Euclidean space are *equivalent*, the operators $(\hat{J}_X, \hat{J}_Y, \hat{J}_Z)$ have the same eigenvalues. The same is true for $(\hat{J}_X^2, \hat{J}_Y^2, \hat{J}_Z^2)$. This property is linked to the symmetry of the space, it is therefore valid for any value of J. Moreover, we can use it to demonstrate a result which will be useful hereafter (exercise 3.4):

$$\sum_{M_J=-J}^{J} (M_J)^2 = \frac{J(J+1)(2J+1)}{3} \qquad [3.11]$$

3.2.6 – How can we use the formalism of the space of spin states associated with an angular momentum?

When the ground state of a paramagnetic centre is characterised by an angular momentum **J**, the sum H of the magnetic interactions to which the unpaired electrons are exposed can be expressed as a function of the components of **J** in any Cartesian reference frame $\{X, Y, Z\}$. In the space of spin states $\mathcal{E}_J$, H is an *operator* $\hat{H}$ which is expressed as a function of $(\hat{J}_X, \hat{J}_Y, \hat{J}_Z)$. This operator is known as the *spin Hamiltonian* of the paramagnetic centre. The corresponding physical quantity is the **energy** E of the centre, which explains the fundamental role that this operator plays in the interpretation of the spectrum. Indeed:

- ▹ the *eigenvalues* of $\hat{H}$ are the possible values of E: they correspond to the *energy levels* of the paramagnetic centre,
- ▹ from the *eigenvectors* associated with these energy levels, we can calculate the *transition probabilities* between levels and consequently the allowed transitions and the intensity of the resonance lines.

To determine the EPR spectrum from H, we must therefore go back and forth between the physical quantities associated with the paramagnetic centre and the abstract space of spin states which is used to perform the calculations. This procedure will be used repeatedly, and can be illustrated by the following diagram:

Paramagnetic centre		Space of spin states
ground state characterised by J	$\Rightarrow$	space of states $\mathcal{E}_J$, dimension $(2J + 1)$
selection of a reference frame $\{X, Y, Z\}$: components (J_X, J_Y, J_Z) of $\mathbf{J}$	$\Rightarrow$	operators $(\hat{J}_X, \hat{J}_Y, \hat{J}_Z)$
magnetic interactions to which the centre is exposed: $H(J_X, J_Y, J_Z)$	$\Rightarrow$	hamiltonian operator $\hat{H}(\hat{J}_X, \hat{J}_Y, \hat{J}_Z)$
		$\downarrow$ diagonalisation
allowed energy levels	$\Leftarrow$	eigenvalues
transition probabilities	$\Leftarrow$	eigenvectors
$\downarrow$		
EPR spectrum		

3.3 – Spin states and allowed energy levels for a paramagnetic centre placed in a magnetic field

3.3.1 – Interaction between a centre with an isotropic magnetic moment and a field B

To become more familiar with the above procedure diagram, we will first apply it to a paramagnetic centre of spin S for which the *isotropic* magnetic moment takes the form:

$$\boldsymbol{\mu} = -g\beta\mathbf{S}$$

When the centre is placed in a field $\mathbf{B}$, it is subjected to the interaction $-\boldsymbol{\mu}\cdot\mathbf{B}$ which can be written:

$$H_{Zeeman} = g\beta\mathbf{B}\cdot\mathbf{S} \qquad [3.12]$$

The notation H_{Zeeman} reminds us that the effect of a magnetic field on the energy levels of a paramagnetic entity, which was observed on atomic gases as early as the end of the 19th century, is called the " Zeeman effet". By expressing the scalar product in *any* orthonormal reference frame $\{X, Y, Z\}$, we obtain:

$$H_{Zeeman} = g\beta(B_X S_X + B_Y S_Y + B_Z S_Z)$$

The operators $(\hat{S}_X, \hat{S}_Y, \hat{S}_Z)$ act on the kets of the space $\mathcal{E}_S$ associated with the spin S. Since their action on the kets $\{|S, M_S\rangle\}$ in the basis made up of the eigenvectors shared by $(\hat{\mathbf{S}}^2, \hat{S}_Z)$ is known (equations [3.3] and [3.9]), the matrix representing the Hamiltonian $\hat{H}_{Zeeman}$ in this basis can be constructed. This matrix, of dimension $(2S + 1)$, includes diagonal elements due to $\hat{S}_Z$ and off-diagonal elements due to $\hat{S}_X$ and $\hat{S}_Y$. The eigenvalues and eigenvectors are difficult to calculate if the $\{X, Y, Z\}$ axes are taken at random (complement 1), but the calculation is considerably simplified if Z is taken *parallel* to $\mathbf{B}$. Indeed, in this condition, $B_X = B_Y = 0$, and equation [3.12] is simply written:

$$H_{Zeeman} = g\beta B S_Z$$

where B is the modulus of $\mathbf{B}$. The Hamiltonian becomes:

$$\hat{H}_{Zeeman} = g\beta B\hat{S}_Z$$

The matrix representing this Hamiltonian in the basis $\{|S, M_S\rangle\}$ of eigenvectors shared by $(\hat{\mathbf{S}}^2, \hat{S}_Z)$ is *diagonal*. Its eigenvectors are therefore the kets $\{|S, M_S\rangle\}$, and its eigenvalues $E(M_S)$ are the diagonal elements $g\beta BM_S$. We once again find expression [1.11] from section 1.4.2, which produces $(2S + 1)$ equidistant energy levels separated by $\Delta E = g\beta B$.

This example clearly shows the advantage of choosing a basis such that the matrix representing the Hamiltonian is as simple as possible. The eigenvectors that we have found will be used in section 3.4 to determine the allowed transitions between energy levels.

3.3.2 – When the magnetic moment is anisotropic

We will now examine the case where the interaction between the unpaired electrons of the centre and the magnetic field is *anisotropic*. The magnetic moment then takes the form:

$$\boldsymbol{\mu} = -\beta\,\tilde{\mathbf{g}}\,\mathbf{S} \qquad [3.13]$$

where $\tilde{\mathbf{g}}$ is a matrix which takes the diagonal form

$$\begin{bmatrix} g_x & 0 & 0 \\ 0 & g_y & 0 \\ 0 & 0 & g_z \end{bmatrix}$$

in an $\{x, y, z\}$ Cartesian system of axes known as the *principal axes* of the matrix. These axes are *linked to the centre*, and are sometimes called its *magnetic axes*. The three numbers (g_x, g_y, g_z) are the *principal values* of the $\tilde{\mathbf{g}}$ matrix. In chapter 4, we will see that this triplet constitutes a veritable *signature* of the centre that can often be determined from the EPR spectrum.

To simplify the writing of the magnetic moment, we will use the principal axes $\{x, y, z\}$ as the reference frame. Equation [3.13] then becomes equivalent to the matrix equation:

$$\begin{bmatrix} \mu_x \\ \mu_y \\ \mu_z \end{bmatrix} = -\beta \begin{bmatrix} g_x & 0 & 0 \\ 0 & g_y & 0 \\ 0 & 0 & g_z \end{bmatrix} \begin{bmatrix} S_x \\ S_y \\ S_z \end{bmatrix} \quad [3.14]$$

The interaction $-\boldsymbol{\mu}\cdot\mathbf{B}$ is written:

$$H_{Zeeman} = -(\mu_x B_x + \mu_y B_y + \mu_z B_z)$$

Using equation [3.14], we obtain the Hamiltonian:

$$\hat{H}_{Zeeman} = \beta(g_x B_x \hat{S}_x + g_y B_y \hat{S}_y + g_z B_z \hat{S}_z) \quad [3.15]$$

As the trace of the ($\hat{S}_x$, $\hat{S}_y$, $\hat{S}_z$) operators is null (section 3.2.5), the trace of $\hat{H}_{Zeeman}$ is also null, and thus the *mean* of its eigenvalues is null. Determining the eigenvalues and eigenvectors of the Hamiltonian is more difficult than when the magnetic moment is isotropic. Indeed, the matrix represented by $\hat{H}_{Zeeman}$ in the basis of the eigenvectors of ($\hat{\mathbf{S}}^2$, $\hat{S}_z$) is not diagonal because of the matrix elements of $\hat{S}_x$ and $\hat{S}_y$. This also occurs if the direction Z of $\mathbf{B}$ is chosen as the quantization axis. Nevertheless, a simple procedure can be used to resolve the problem. We will start by writing $\mathbf{B}$ in the form

$$\mathbf{B} = B\,\mathbf{u}$$

where $\mathbf{u}(u_x, u_y, u_z)$ is the *unit* vector in the direction of $\mathbf{B}$ (figure 3.2). H_{Zeeman} then becomes:

$$H_{Zeeman} = \beta B\,(g_x u_x S_x + g_y u_y S_y + g_z u_z S_z) \quad [3.16]$$

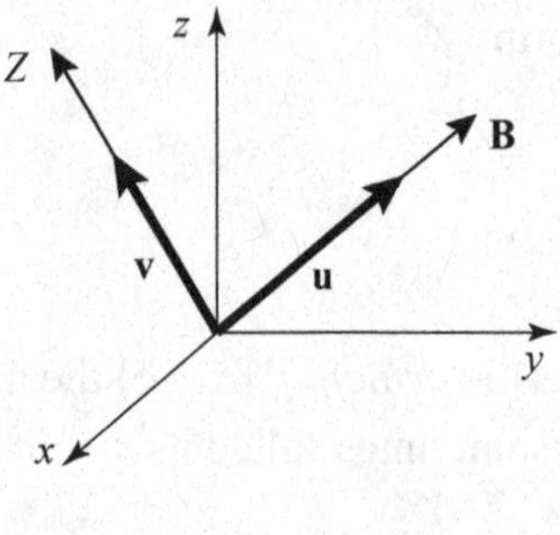

Figure 3.2 – Important directions for "diagonalisation" of the Hamiltonian [3.15]. $\{x, y, z\}$ are the principal axes of the $\tilde{\mathbf{g}}$ matrix, $\mathbf{B}$ is the magnetic field. The unit vectors $\mathbf{u}$ and $\mathbf{v}$ define the direction of $\mathbf{B}$ and the quantization axis Z, respectively.

We now show that it can be written in the form

$$H_{Zeeman} = g'\beta B\,(v_x S_x + v_y S_y + v_z S_z) \quad [3.17]$$

where g' is a *positive* number and (v_x, v_y, v_z) are the components of a unit vector **v**, to be determined. The equivalence of equations [3.16] and [3.17] implies that:

$$g'v_x = g_x u_x\,,\; g'v_y = g_y u_y\,,\; g'v_z = g_z u_z \quad [3.18]$$

g' and (v_x, v_y, v_z) are determined by writing that the sums of the squares of the two sides of these equations are equal:

$$g'^2\,[v_x^2 + v_y^2 + v_z^2] = g_x^2 u_x^2 + g_y^2 u_y^2 + g_z^2 u_z^2$$

As **v** is a unit vector, we obtain

$$g' = [g_x^2 u_x^2 + g_y^2 u_y^2 + g_z^2 u_z^2]^{1/2} \quad [3.19]$$

and the components (v_x, v_y, v_z) of **v** are deduced from equations [3.18].

In equation [3.17], the quantity $(v_xS_x + v_yS_y + v_zS_z)$ is the *scalar product* $\mathbf{v}\cdot\mathbf{S}$, which is written S_Z if Z is *the direction of* **v** (figure 3.2). H_{Zeeman} therefore becomes:

$$H_{Zeeman} = g'\beta B\, S_Z \quad [3.20]$$

and the Hamiltonian [3.15] takes the very simple form:

$$\hat{H}_{Zeeman} = g'\beta B\hat{S}_Z$$

In the basis of $\mathcal{E}_S$ composed of the eigenvectors $\{|S, M_S\rangle\}$ shared by $(\hat{\mathbf{S}}^2, \hat{S}_Z)$ (the quantization axis Z is the direction of **v**), the Hamiltonian is represented by a *diagonal* matrix where the eigenvalues are the diagonal elements $E(M_S) = g'\beta BM_S$ and the eigenvectors are the kets $\{|S, M_S\rangle\}$. As predicted, the mean of the eigenvalues is null.

A pattern of $(2S + 1)$ *equidistant* energy levels is obtained, as in the case of an isotropic magnetic moment. However, the splitting $\Delta E = g'\beta B$ between these levels now depends on the *direction of* **B** *relative to the axes* $\{x, y, z\}$ and therefore relative to the centre through the intermediary of components (u_x, u_y, u_z) in equation [3.19]. The possible spin states of the paramagnetic centre, which are the eigenvectors of $\hat{H}_{Zeeman}$, also depend on this direction (exercise 3.6). The consequences of these *anisotropies* on the shape and intensity of the EPR spectrum will be examined in detail in chapters 4 and 5, but we can already note the following points:

▹ When **B** is oriented along the magnetic axis x (figure 3.2), $u_x = 1$, $u_y = 0$, $u_z = 0$. For this orientation, equations [3.18] and [3.19] show that $g' = g_x$ ($\Delta E = g_x\beta B$) and that Z is parallel to x. Naturally, equivalent results are obtained when **B** is oriented in the y or z direction.

▹ If the symmetry properties of the molecule are such that $g_x = g_y = g_z = g$, the relations [3.18] and [3.19] indicate that $g' = g$ and $\mathbf{v} = \mathbf{u}$: the quantization axis Z is the direction of **B**, and we return to the case of the *isotropic* magnetic moment. This result was predictable, since equation [3.15] is then identical to equation [3.12].

3.4 – Transition probabilities and allowed transitions

The formalism of the space of spin states makes it possible to justify the *selection rules* given in section 1.4.3 for a paramagnetic centre with an *isotropic* magnetic moment. In section 3.3.1, we showed that this centre can occupy one of the $(2S + 1)$ states represented by the kets $\{|S, M_S\rangle\}$ from the $\mathcal{E}_S$ basis defined by the direction Z of **B**. If we expose the centre to electromagnetic radiation with a frequency ν, the interaction between its magnetic moment $\boldsymbol{\mu}$ and the magnetic component $\mathbf{b}_1(t) = \mathbf{B}_1\cos 2\pi\nu t$ of the radiation can induce transitions between some of these states. We will see in section 5.2.1 that the *probability per unit time* of transitions between two states represented by $|S, M_S\rangle$ and $|S, M_S'\rangle$ (with $M_S \neq M_S'$) is proportional to $|V|^2$, where V is the matrix element

$$V = \langle S, M_S' | \hat{H}_1 | S, M_S\rangle$$

Where:

$$\hat{H}_1 = -\boldsymbol{\mu}\cdot\mathbf{B}_1 = g\beta\mathbf{B}_1\cdot\mathbf{S}$$

To express this scalar product, we add two axes (X, Y) to Z to obtain an orthonormal reference frame $\{X, Y, Z\}$:

$$\hat{H}_1 = g\beta(B_{1X}\hat{S}_X + B_{1Y}\hat{S}_Y + B_{1Z}\hat{S}_Z)$$

The matrix element V is written:

$$V = g\beta\,\langle S, M_S'|B_{1X}\hat{S}_X + B_{1Y}\hat{S}_Y + B_{1Z}\hat{S}_Z|S, M_S\rangle$$

We will now examine the matrix elements obtained by developing this expression.

▹ The matrix element of $\hat{S}_Z$ is *null*. Indeed, we have:

$$\hat{S}_Z\,|S, M_S\rangle = M_S\,|S, M_S\rangle$$

The orthogonality of the kets $|S, M_S\rangle$ and $|S, M_S'\rangle$ leads to $\langle S, M_S'|\hat{S}_Z|S, M_S\rangle = 0$. As a result, the component B_{1Z} along the field **B** *is not involved* in the transition probability. For a given value of B_1, this probability is therefore maximum when $\mathbf{B}_1$ is *perpendicular* to **B**. We mentioned in section 1.5.2 that standard resonance cavities are designed to satisfy this condition.

▹ Let us now consider the matrix elements of $\hat{S}_X$ and $\hat{S}_Y$. According to equations [3.9], the action of these operators on the ket $|S, M_S\rangle$ gives a linear combination of the kets $|S, M_S + 1\rangle$ and $|S, M_S - 1\rangle$. The matrix elements $\langle S, M_S'|\hat{S}_X|S, M_S\rangle$ and $\langle S, M_S'|\hat{S}_Y|S, M_S\rangle$ are therefore different from zero only if $M_S' = M_S + 1$ or $M_S' = M_S - 1$: on the energy level diagram, only transitions towards *adjacent* levels are possible. The *selection rule* is said to be:

$$\Delta M_S = \pm 1$$

For transitions from $|S, M_S\rangle$ towards $|S, M_S + 1\rangle$, from equations [3.9] we deduce:

$$V = \tfrac{1}{2} g\beta \,[S(S+1) - M_S(M_S+1)]^{1/2}\,(B_{1X} - \mathrm{i}\,B_{1Y})$$

As component $\mathbf{B}_1$ is perpendicular to the direction Z of **B**, we have $(B_{1X})^2 + (B_{1Y})^2 = (B_1)^2$ which produces:

$$|V|^2 = g^2\beta^2 (B_1)^2\,[S(S+1) - M_S(M_S+1)]/4 \qquad [3.21]$$

The value of $|V|^2$ for the transitions from $|S, M_S\rangle$ towards $|S, M_S - 1\rangle$ is obtained by changing $(M_S + 1)$ to $(M_S - 1)$. In both cases, the transition probability is proportional to $(B_1)^2$.

The rule presented in section 1.4 is found.

This calculation was performed for an *isotropic* magnetic moment. When the moment is *anisotropic*, the transition probability is also maximal when $\mathbf{B}_1$ is perpendicular to **B** (exercise 3.7) and it is also proportional to $(B_1)^2$ (equation [5.12]).

3.5 – Possible spin states and allowed transitions in the presence of hyperfine interaction

Up to now, we considered only the interaction between the moment of the paramagnetic centre and the field **B**. As this interaction is expressed as a function of the components of the angular momentum **S** of the centre, we used the space $\mathcal{E}_S$ of spin states associated with **S**. When the magnetic interactions involve the components of other angular momenta, the relevant spaces must be used. For example, consider a paramagnetic centre of spin S the unpaired electrons of which are exposed to *isotropic* magnetic interactions with the field **B** and with a nucleus of spin I. The sum of these interactions is written:

$$H = H_{Zeeman} + H_{hyperfine} \qquad [3.22]$$

with

$$H_{Zeeman} = g\beta\mathbf{B}\cdot\mathbf{S}\ ;\ H_{hyperfine} = A\mathbf{S}\cdot\mathbf{I}$$

To describe the states of the system {paramagnetic centre + nucleus}, we use the kets $\{|u\rangle\}$ of the space $\mathcal{E}_S$ of dimension $(2S + 1)$ associated with **S** and the kets $\{|v\rangle\}$ of the space $\mathcal{E}_I$ of dimension $(2I + 1)$ associated with **I**. The states of this system are described by kets noted $|u\rangle\,|v\rangle$.

3.5.1 – Determining the energy levels

In general, it is not possible to obtain the exact expressions for the eigenvalues and eigenvectors of the Hamiltonian $\hat{H}$ defined by equation [3.22]. However, approximate expressions can be obtained in the frequent situation where the H_{Zeeman} term is much larger than $H_{hyperfine}$, by applying "perturbation theory". The principle of this theory is as follows: when the Hamiltonian $\hat{H}$ takes the form:

$$\hat{H} = \hat{H}_0 + \hat{H}_P$$

where $\hat{H}_0$ is an operator for which the eigenvalues and eigenvectors are known, and $\hat{H}_P$ is small relative to $\hat{H}_0$, we obtain approximations of the eigenvalues and eigenvectors of $\hat{H}$ by adding corrective terms to the eigenvalues and eigenvectors of $\hat{H}_0$. These terms are smaller when the $\hat{H}_P$ "perturbation" is weak. The method used to calculate these corrective terms is presented in complement 2, and we apply it here to the Hamiltonian [3.22]. It involves two steps:

▹ First, the eigenvalues and eigenvectors of the principal term $\hat{H}_{Zeeman}$ must be determined. This problem was already resolved in section 3.3.1: if we choose the direction Z of **B** as quantization axis, $\hat{H}_{Zeeman}$ is written:

$$\hat{H}_{Zeeman} = g\beta B\hat{S}_Z$$

Its eigenvalues are $E(M_S) = g\beta BM_S$ and the associated eigenvectors are the kets $\{|S, M_S\rangle\}$, eigenvectors shared by $(\hat{\mathbf{S}}^2, \hat{S}_Z)$. To describe the spin states of the nucleus, it is convenient to choose the eigenvectors $\{|I, M_I\rangle\}$ shared by $(\hat{\mathbf{I}}^2, \hat{I}_Z)$ as the basis for the vector space $\mathcal{E}_I$. As the $\hat{H}_{Zeeman}$ term does not involve the angular momentum **I**, it imposes no constraints on the possible states of the nucleus. Consequently, the ket $\{|S, M_S\rangle\}$ and any one of the $(2I + 1)$ kets $\{|I, M_I\rangle\}$ are associated with the eigenvalue $E(M_S)$. This produces $(2I + 1)$ kets described by:

$$\{|S, M_S\rangle|I, M_I\rangle\} \qquad M_I = -I, -I+1, \dots I$$

Each eigenvalue $E(M_S)$ is therefore $(2I + 1)$-fold *degenerate*. To simplify the expression, these kets are written $\{|M_S\rangle|M_I\rangle\}$.

▹ As the eigenvalues of $\hat{H}_{Zeeman}$ are degenerate, perturbation theory prescribes constructing, for each value of M_S, the $(2I + 1)$-dimensional matrix representing the $\hat{H}_{hyperfine}$ perturbation in the sub-space defined by the kets $\{|M_S\rangle|M_I\rangle\}$. The elements of this matrix take the form:

$$\langle M_S|\langle M_I\,|\hat{H}_{hyperfine}\,|M_S\rangle|M_I'\rangle$$

To express the $\hat{H}_{hyperfine}$ operator, 2 axes (X, Y) are added to Z to create an orthonormal $\{X, Y, Z\}$ reference frame:

$$\hat{H}_{hyperfine} = A\,(\hat{S}_X\hat{I}_X + \hat{S}_Y\hat{I}_Y + \hat{S}_Z\hat{I}_Z) \qquad [3.23]$$

For example, we will examine the case $S = ½$, $I = ½$. The possible states of the system are written $\{|M_S\rangle|M_I\rangle\}$, where $M_S = \pm ½$ and $M_I = \pm ½$, and the matrices representing the $\hat{H}_{hyperfine}$ operator must be constructed for the following two sub-spaces:

$$M_S = -½\text{:} \quad \{|-½\rangle|-½\rangle, |-½\rangle|½\rangle\}$$

$$M_S = ½\text{:} \quad \{|½\rangle|-½\rangle, |½\rangle|½\rangle\}$$

To cause the $\hat{S}_X\hat{I}_X$, $\hat{S}_Y\hat{I}_Y$ and $\hat{S}_Z\hat{I}_Z$ operators to act on a ket $|M_S\rangle|M_I\rangle$, we apply $(\hat{S}_X, \hat{S}_Y, \hat{S}_Z)$ to $|M_S\rangle$ and $(\hat{I}_X, \hat{I}_Y, \hat{I}_Z)$ to $|M_I\rangle$. Using the results from section 3.2.5, this produces:

- $M_S = -½$:

$$A\hat{S}_X\hat{I}_X|-½\rangle|-½\rangle = (A/4)\,|½\rangle|½\rangle\,; \quad A\hat{S}_X\hat{I}_X|-½\rangle|½\rangle = (A/4)|½\rangle|-½\rangle$$

$$A\hat{S}_Y\hat{I}_Y|-½\rangle|-½\rangle = -\,(A/4)\,|½\rangle|½\rangle\,; \quad A\hat{S}_Y\hat{I}_Y|-½\rangle|½\rangle = (A/4)|½\rangle|-½\rangle$$

$$A\hat{S}_Z\hat{I}_Z|-½\rangle|-½\rangle = (A/4)|-½\rangle|-½\rangle\,; \quad A\hat{S}_Z\hat{I}_Z|-½\rangle|½\rangle = (-A/4)|-½\rangle|½\rangle$$

We can readily verify that in the $\{|-½\rangle|M_I\rangle\}$ sub-space, only the term $A\hat{S}_Z\hat{I}_Z$ of $\hat{H}_{hyperfine}$ gives a non-null contribution to the matrix which is written:

	$|-½\rangle|-½\rangle$	$|-½\rangle|½\rangle$
$\langle -½|\langle -½|$	$A/4$	0
$\langle -½|\langle ½|$	0	$-A/4$

As this matrix is diagonal, its eigenvalues and eigenvectors are given by:

$$A/4\ ;\ |-½\rangle|-½\rangle$$

$$-A/4\ ;\ |-½\rangle|½\rangle$$

- $M_S = ½$:

$$A\hat{S}_X\hat{I}_X|½\rangle|-½\rangle = (A/4)|-½\rangle|½\rangle\ ;\quad A\hat{S}_X\hat{I}_X|½\rangle|½\rangle = (A/4)|-½\rangle|-½\rangle$$

$$A\hat{S}_Y\hat{I}_Y|½\rangle|-½\rangle = (A/4)|-½\rangle|½\rangle\ ;\quad A\hat{S}_Y\hat{I}_Y|½\rangle|½\rangle = (-A/4)|-½\rangle|-½\rangle$$

$$A\hat{S}_Z\hat{I}_Z|½\rangle|-½\rangle = (-A/4)|½\rangle|-½\rangle\ ;\quad A\hat{S}_Z\hat{I}_Z|½\rangle|½\rangle = (A/4)|½\rangle|½\rangle$$

The matrix of $\hat{H}_{hyperfine}$ is also diagonal:

	$|½\rangle|-½\rangle$	$|½\rangle|½\rangle$
$\langle ½|\langle -½|$	$-A/4$	0
$\langle ½|\langle ½|$	0	$A/4$

Its eigenvalues and eigenvectors are therefore:

$$-A/4\ ;\ |½\rangle|-½\rangle$$

$$A/4\ ;\ |½\rangle|½\rangle$$

To the *first order* of perturbation theory, the 4 eigenvalues and the 4 eigenvectors of the Hamiltonian $\hat{H}$ (equation [3.22]) can be written:

$$-½g\beta B + A/4\ ;\ |-½\rangle|-½\rangle$$

$$-½g\beta B - A/4\ ;\ |-½\rangle|½\rangle$$

$$½g\beta B - A/4\ ;\ |½\rangle|-½\rangle$$

$$½g\beta B + A/4\ ;\ |½\rangle|½\rangle$$

These results can readily be generalised to the case where S and I have any values. Indeed, for fixed M_S, the matrix elements of the $\hat{H}_{hyperfine}$ operator in the sub-space $\{|M_S\rangle|M_I\rangle\}$ take the form:

$$\langle M_S|\langle M_I|\, A\,(\hat{S}_X\hat{I}_X + \hat{S}_Y\hat{I}_Y + \hat{S}_Z\hat{I}_Z)|M_S\rangle|M_I'\rangle$$

Because the ket $|M_S\rangle$ is present on either side of this matrix element, and due to the properties of the $\hat{S}_X$ and $\hat{S}_Y$ operators (equations [3.9]), only the term $A\hat{S}_Z\hat{I}_Z$ gives a non-null contribution, equal to AM_SM_I. The $\hat{H}_{hyperfine}$ matrix is therefore *diagonal* in the $\{|M_S\rangle|M_I\rangle\}$ sub-space, and its eigenvalues are the diagonal elements. To the first order of perturbation theory, the eigenvalues and eigenvectors of $\hat{H}$ (equation [3.22]) are therefore written:

$$g\beta BM_S + AM_SM_I\,;\ |M_S\rangle|M_I\rangle \qquad [3.24]$$

We find the energy levels given by equation [2.7]. In section 4.4.2 we deal with the case where the magnetic moment and the hyperfine interactions are *anisotropic*.

3.5.2 – Allowed transitions

To determine the allowed transitions, we proceed as in section 3.4. The matrix element which determines the transition probability between the kets $|M_S\rangle|M_I\rangle$ and $|M_S'\rangle|M_I'\rangle$ is written:

$$V = \langle M_S'|\langle M_I'\,|\hat{H}_1\,|M_S\rangle|M_I\rangle$$

where

$$\hat{H}_1 = g\beta(B_{1X}\hat{S}_X + B_{1Y}\hat{S}_Y + B_{1Z}\hat{S}_Z)$$

Since $\hat{H}_1$ *only* acts *on the* $\{|M_S\rangle\}$ kets, V can be factorised:

$$V = \langle M_I'|M_I\rangle\langle M_S'|\hat{H}_1|M_S\rangle$$

As the kets of the $\{|M_I\rangle\}$ basis are orthogonal, the first factor is different from zero only if $M_I' = M_I$. As for the second, it is identical to the matrix element obtained in the absence of hyperfine interaction. The selection rule is therefore written:

$$\Delta M_I = 0 \quad \text{and} \quad \Delta M_S = \pm 1$$

We find the rule announced in section 2.3.2.

3.6 – Points to consider in applications

3.6.1 – *Why is a spin Hamiltonian used in EPR and magnetic spectroscopies?*

The EPR spectrum produced by a paramagnetic molecule is determined by its lowest energy levels and their corresponding states. These levels and states, determined by all of the interactions involving the unpaired electrons, are very difficult to calculate. However, strong *electrostatic interactions* (with the nuclei and other electrons) cause more extensive splitting of energy levels than weak *magnetic interactions*. According to perturbation theory (complement 2), the lowest energy levels for the molecule can be considered to result from the effects of magnetic interactions on the ground level (see figure 3.1, which is not to scale). Thanks to the properties of angular momenta, this effect can be reproduced by substituting an *operator*, such as $\hat{H}_{Zeeman}$ or $\hat{H}_{hyperfine}$, for each magnetic interaction. These operators include the following:

- *Spin operators* defined from the angular momenta characterising the molecule's ground state and the paramagnetic entities interacting with the unpaired electrons, such as **S** for $\hat{H}_{Zeeman}$ or (**S**, **I**) for $\hat{H}_{hyperfine}$.
- *Parameters*, such as the elements of the $\tilde{\mathbf{g}}$ matrix for $\hat{H}_{Zeeman}$ or those of the $\tilde{\mathbf{A}}$ matrix for $\hat{H}_{hyperfine}$.

3.6.2 – *What does the spin Hamiltonian do?*

The shape and intensity of the experimental EPR spectrum for a paramagnetic molecule can be interpreted by constructing a "spin Hamiltonian"; this is done by adding the operators describing the magnetic interactions involving its unpaired electrons. When the spectrum calculated from this Hamiltonian is compared to the experimental spectrum, it is possible to determine the values of the spins and of the parameters included in the operators. These values provide information on the *nature* and *structure* of the molecules studied. More detailed information can be obtained if the values of the parameters can be reproduced by a *molecular model* of the ground state and of the magnetic interactions.

3.6.3 – *Looking back to the procedure described in chapters 1 and 2*

In chapters 1 and 2, we considered the case of *isotropic* interactions between electrons and the magnetic field or the paramagnetic nuclei, and we used an *ad hoc* procedure to determine the shape of the EPR spectrum from the Zeeman and hyperfine terms. In this chapter, we have demonstrated that this procedure is a consequence of the formalism of the space of spin states, where these terms are taken as *operators*. This formalism also allows us to treat *anisotropic* interactions. In this case, the possible states of the paramagnetic centre and the splitting of the energy levels depend on the direction of **B** relative to the molecule. These anisotropic effects have very significant consequences on the *shape* and *intensity* of the EPR spectrum, which will be examined in chapters 4 and 5.

Complement 1 – Diagonalisation of $\tilde{\mathbf{H}}_{Zeeman}$ in any basis

To demonstrate the importance of the basis chosen to calculate the eigenvalues and eigenvectors for the Hamiltonian, we will consider the case of a paramagnetic centre of spin $S = ½$, with an isotropic magnetic moment. In *any* reference frame $\{x, y, z\}$, the $\hat{H}_{Zeeman}$ Hamiltonian is written:

$$\hat{H}_{Zeeman} = g\beta\,(B_x\hat{S}_x + B_y\hat{S}_y + B_z\hat{S}_z)$$

Using the results from section 3.2.5, we can readily construct the $\tilde{\mathbf{H}}_{Zeeman}$ matrix representing it in the $\{|½, ½\rangle_z, |½, -½\rangle_z\}$ basis of the $\mathcal{E}_{½}$ space made up of the eigenvectors shared by $(\hat{\mathbf{S}}^2, \hat{S}_z)$ (the subscript z recalls the quantization axis):

	$\lvert ½, ½\rangle_z$	$\lvert ½, -½\rangle_z$
$\langle ½, ½\rvert_z$	$g\beta B_z/2$	$g\beta(B_x - \mathrm{i}B_y)/2$
$\langle ½, -½\rvert_z$	$g\beta(B_x + \mathrm{i}B_y)/2$	$-g\beta B_z/2$

We are looking for eigenvectors $|u\rangle$ with the form:

$$|u\rangle = c_+\,|½, ½\rangle_z + c_-\,|½, -½\rangle_z$$

The equation

$$\hat{H}_{Zeeman}\,|u\rangle = E\,|u\rangle$$

is written in the following matrix form:

$$\tilde{\mathbf{H}}_{Zeeman}\begin{bmatrix} c_+ \\ c_- \end{bmatrix} = E\begin{bmatrix} c_+ \\ c_- \end{bmatrix} \qquad [1]$$

By developing equation [1], we obtain a two-equation system from which to determine (c_+, c_-). This system allows a solution different to the trivial solution (0, 0) if its determinant is null:

$$\begin{vmatrix} (g\beta B_z/2) - E & g\beta(B_x - \mathrm{i}B_y)/2 \\ g\beta(B_x + \mathrm{i}B_y)/2 & (-g\beta B_z/2) - E \end{vmatrix} = 0$$

This eigenvalue equation is therefore written:

$$(E - g\beta B_z/2)\,(E + g\beta B_z/2) - (g\beta)^2\,(B_x^{\,2} + B_y^{\,2})/4 = 0$$

i.e.,

$$E^2 = (g\beta B)^2/4\,;\; B^2 = B_x^{\,2} + B_y^{\,2} + B_z^{\,2}$$

We deduce the two eigenvalues:

$$E_1 = g\beta B/2\,;\; E_2 = -g\beta B/2$$

To obtain the eigenvector $|u_1\rangle$ associated with E_1, we replace E with E_1 in equation [1]. Since the determinant is null for this value of E, the system is reduced to

$$(B_z - B)\, c_{1+} + (B_x - \mathrm{i}B_y)\, c_{1-} = 0 \qquad [2]$$

which only gives *the ratio* c_{1+}/c_{1-}. If we require $|u\rangle$ to be normalised, the components also verify:

$$|c_{1+}|^2 + |c_{1-}|^2 = 1$$

The ket $|u_1\rangle$ is therefore defined up to a multiplicative factor of modulus one. For the ket $|u_2\rangle$, which corresponds to the eigenvalue E_2, equation [2] is replaced by:

$$(B_x + \mathrm{i}B_y)\, c_{2+} - (B_z - B)\, c_{2-} = 0$$

It can be easily verified that the scalar product

$$\langle u_1|u_2\rangle = (c_{1+})^*\, c_{2+} + (c_{1-})^*\, c_{2-}$$

is null, which demonstrates that the two eigenvectors are *orthogonal*. But since the kets $|u_1\rangle$ and $|u_2\rangle$ are eigenvectors of $\hat{H}_{Zeeman}$, they verify:

$$(g\beta\mathbf{B}\cdot\mathbf{S})\,|u\rangle = E\,|u\rangle$$

which can also be written

$$g\beta B\,\hat{S}_Z|u\rangle = E\,|u\rangle$$

where Z is the direction of **B**. These kets are therefore eigenvectors of $\hat{S}_Z$ which we can note $\{|½, ½\rangle_Z, |½, -½\rangle_Z\}$ (section 3.3.1).

Complement 2 – The principle of perturbation theory

We very often need to resort to approximation methods to obtain closed-form expressions for the eigenvalues and eigenvectors of the Hamiltonian operator. In this volume, we use "perturbation theory", which applies when the Hamiltonian takes the form:

$$\hat{H} = \hat{H}_0 + \hat{H}_P \qquad [1]$$

- ▹ $\hat{H}_0$ is an operator with known eigenvalues $\{E_{0k}\}$ and eigenvectors $\{|u_k\rangle\}$.
- ▹ The operator $\hat{H}_P$ is such that its matrix elements in the $\{|u_k\rangle\}$ basis are *small* relative to the *differences* between the E_{0k} eigenvalues. $\hat{H}_P$ is said to constitute a *perturbation* of $\hat{H}_0$.

We present the principle of perturbation theory in a simplified manner, for 2- and 3-dimensional matrices. A more formal and more complete presentation of this method can be found in works on quantum mechanics [Ayant and Belorizky, 2000; Cohen-Tannoudji, 2015].

1 – Approximation of the eigenvalues and eigenvectors of a symmetric matrix for which the off-diagonal elements are "small"

We will first show how we can approximate the eigenvalues and eigenvectors of a matrix when the off-diagonal elements are small relative to the differences between the diagonal elements, starting with a matrix of dimension 2. An operator $\hat{H}$ is represented in an orthonormal basis $\{|u_1\rangle, |u_2\rangle\}$ by the matrix

$$\tilde{\mathbf{H}} = \begin{bmatrix} H_1 & h_{12} \\ h_{21} & H_2 \end{bmatrix}$$

We seek its eigenvectors in the form:

$$|v\rangle = c_1\,|u_1\rangle + c_2\,|u_2\rangle$$

The equation

$$\hat{H}\,|v\rangle = E\,|v\rangle$$

is equivalent to the matrix equation:

$$\begin{bmatrix} H_1 & h_{12} \\ h_{21} & H_2 \end{bmatrix}\begin{bmatrix} c_1 \\ c_2 \end{bmatrix} = E\begin{bmatrix} c_1 \\ c_2 \end{bmatrix} \qquad [2]$$

If this equation is developed, we obtain a system of two simultaneous equations which provides a solution different to the trivial (0,0) solution if its determinant is null:

$$\begin{vmatrix} H_1 - E & h_{12} \\ h_{21} & H_2 - E \end{vmatrix} = 0 \qquad [3]$$

The roots of this second-degree equation are the eigenvalues. For these values of E, equation [2] is reduced to a single equation from which the ratio c_1/c_2 can be determined.

When the operator $\hat{H}$ is *Hermitian*, the diagonal elements H_1 and H_2 are real and the off-diagonal elements are complex conjugates: $h_{21} = h_{12}{}^*$. To simplify the writing, we will assume hereafter that the matrix is *symmetric*. The eigenvalue equation [3] is thus written:

$$(E - H_1)(E - H_2) - (h_{12})^2 = 0 \qquad [4]$$

When $h_{12} = 0$, the eigenvalues are $E_1 = H_1$ and $E_2 = H_2$. When h_{12} is *small* relative to the difference between the diagonal elements:

$$|h_{12}| \ll |H_2 - H_1| \qquad [5]$$

we can obtain *approximations* of E_1 and E_2 without explicitly solving equation [4]. Indeed, by stating $E_1 = H_1 + e_1$ and $\Delta H_{12} = H_2 - H_1$, this equation becomes:

$$e_1 (e_1 - \Delta H_{12}) - (h_{12})^2 = 0 \qquad [6]$$

If the condition

$$|e_1| \ll |\Delta H_{12}| \qquad [7]$$

is satisfied, we can neglect e_1 relative to ΔH_{12} in the first term, which produces:

$$e_1 \approx -(h_{12})^2/\Delta H_{12}$$

$$E_1 \approx H_1 - (h_{12})^2/\Delta H_{12} \qquad [8]$$

In the same way, we obtain:

$$E_2 \approx H_2 + (h_{12})^2/\Delta H_{12} \qquad [9]$$

Condition [7] is satisfied if

$$(h_{12})^2 \ll (\Delta H_{12})^2 \qquad [10]$$

i.e., the off-diagonal elements are much smaller than the difference between the diagonal elements (inequality [5]). The conditions imposed by inequalities [5] and [10] imply that the second term on the second side of equation [8] is much smaller than the first. H_1 and $(H_1 - (h_{12})^2/\Delta H_{12})$ are said to represent the first and *second order* approximations of E_1, *respectively*. The same definition naturally applies to the different terms in E_2 (equation [9]).

We will now move on to the eigenvectors:

▷ **In the first order approximation**, we have $E_1 \approx H_1$ and $E_2 \approx H_2$. In this approximation, everything happens as if the matrix $\tilde{\mathbf{H}}$ were *diagonal*, and the eigenvectors are the kets $\{|u_1\rangle, |u_2\rangle\}$.

▷ **In the second order approximation**, the eigenvalues are given by equations [8] and [9]. To obtain the corresponding eigenvectors, we replace E successively by E_1, then by E_2 in equation [2]. This operation produces:

$$\begin{aligned} |v_1\rangle &\approx |u_1\rangle - (h_{12}/\Delta H_{12})\,|u_2\rangle \\ |v_2\rangle &\approx |u_2\rangle + (h_{12}/\Delta H_{12})\,|u_1\rangle \end{aligned} \qquad [11]$$

The precision can be improved by continuing the calculation at higher orders. For example, the third-order approximation of E_1 can be obtained by replacing E by $(H_1 - (h_{12})^2/\Delta H_{12} + e'_1)$ in equation [4], where $e'_1 \ll (h_{12})^2/\Delta H_{12}$.

When condition [10] is not satisfied, expressions [8] and [9] are invalid and equation [4] must be explicitly solved to determine the eigenvalues and eigenvectors.

The method that we have just described can be applied to a matrix of *any dimension*. We will now examine the case of a matrix of dimension 3. Consider an operator $\hat{H}$ represented in the basis $\{|u_1\rangle, |u_2\rangle, |u_3\rangle\}$ by the symmetric matrix:

$$\begin{bmatrix} H_1 & h_{12} & h_{13} \\ h_{12} & H_2 & h_{23} \\ h_{13} & h_{23} & H_3 \end{bmatrix}$$

The eigenvalue equation is written:

$$(E-H_1)(E-H_2)(E-H_3)-(h_{23})^2(E-H_1)-(h_{13})^2(E-H_2)-(h_{12})^2(E-H_3)-2h_{23}h_{13}h_{12} = 0$$

By setting $E_1 = H_1 + e_1$, $\Delta H_{12} = H_2 - H_1$, $\Delta H_{13} = H_3 - H_1$, this equation becomes:

$$e_1(e_1-\Delta H_{12})(e_1-\Delta H_{13})-(h_{23})^2e_1-(h_{13})^2(e_1-\Delta H_{12})-(h_{12})^2(e_1-\Delta H_{13})-2h_{23}h_{13}h_{12}=0 \qquad [12]$$

The previous method allows us to approximate the roots of this third-degree equation in two cases:

- When all the off-diagonal elements are small relative to the differences $|\Delta H_{12}|$ and $|\Delta H_{13}|$ between the diagonal elements, we can neglect e_1 relative to ΔH_{12} and ΔH_{13}; this produces:

$$E_1 \approx H_1 - (h_{12})^2 / \Delta H_{12} - (h_{13})^2 / \Delta H_{13} \qquad [13]$$

and similar expressions for E_2 and E_3.

In the first order approximation, we have $E_1 \approx H_1$, $E_2 \approx H_2$, $E_3 \approx H_3$. Everything happens as if the matrix were diagonal, and the eigenvectors are the kets $\{|u_1\rangle, |u_2\rangle, |u_3\rangle\}$.

- When all the off-diagonal elements are small relative to ΔH_{13}, but not relative to ΔH_{12}, equation [12] produces the following second-degree equation:

$$e\,(e - \Delta H_{12}) - (h_{12})^2 = 0 \qquad [14]$$

which is in fact equation [6]. This equation must be solved to determine (E_1, E_2). In the first order approximation, the third eigenvalue is given by:

$$E_3 \approx H_3$$

2 – Application to perturbation theory

We will now apply these results to perturbation theory in the case where the Hamiltonian operator acts in a 3-dimensional vector space. In the $\{|u_1\rangle, |u_2\rangle, |u_3\rangle\}$ basis of eigenvectors of $\hat{H}_0$, the matrix representing the Hamiltonian $\hat{H}$ (equation [1]) is the sum of the diagonal matrix representing $\hat{H}_0$ and the matrix representing $\hat{H}_P$. It therefore takes the form:

$$\begin{bmatrix} E_{01} + h_{11} & h_{12} & h_{13} \\ h_{12} & E_{02} + h_{22} & h_{23} \\ h_{13} & h_{23} & E_{03} + h_{33} \end{bmatrix}$$

where $\{E_{01}, E_{02}, E_{03}\}$ are the eigenvalues of $\hat{H}_0$ and $h_{ij} = \langle u_i | \hat{H}_P | u_j \rangle$. Based on the statements above, we can approximate the eigenvalues and eigenvectors for this matrix in the following two cases:

1. When all the eigenvalues $\{E_{01}, E_{02}, E_{03}\}$ of $\hat{H}_0$ are distinct (we say that they are non-degenerate) and all the off-diagonal elements are small relative to the differences:

$$\Delta H_{12} = (E_{02} - E_{01}) + (h_{22} - h_{11})$$

$$\Delta H_{13} = (E_{03} - E_{01}) + (h_{33} - h_{11})$$

the eigenvalues of $\hat{H}$ are given to the first order by:

$$E_1 \approx E_{01} + h_{11} = E_{01} + \langle u_1 | \hat{H}_P | u_1 \rangle$$

$$E_2 \approx E_{02} + h_{22} = E_{02} + \langle u_2 | \hat{H}_P | u_2 \rangle$$

$$E_3 \approx E_{03} + h_{33} = E_{03} + \langle u_3 | \hat{H}_P | u_3 \rangle$$

In this approximation, the eigenvectors are the kets $\{|u_1\rangle, |u_2\rangle, |u_3\rangle\}$ (figure 3.3a).

2. When $E_{01} = E_{02} = E_0$, the eigenvalue E_0 is said to be degenerate. In this case, the off-diagonal elements are not small relative to $\Delta H_{12} = (h_{22} - h_{11})$. But if they are small compared to the $(E_{03} - E_0)$ difference, in the first order approximation we can write:

$$E_3 \approx E_{03} + \langle u_3 | \hat{H}_P | u_3 \rangle$$

The two other eigenvalues are the solutions to equation [14] which can be written as follows:

$$[E - (E_0 + h_{11})]\,[E - (E_0 + h_{22})] - (h_{12})^2 = 0 \qquad [15]$$

As equation [14] is identical to equation [6], the solutions to equation [15] can be written in the form $E = E_0 + e$, where e is an eigenvalue of the matrix:

$$\begin{bmatrix} h_{11} & h_{12} \\ h_{12} & h_{22} \end{bmatrix}$$

This matrix is none other than the representation of the perturbation $\hat{H}_P$ *in the* $\{|u_1\rangle, |u_2\rangle\}$ *sub-space* associated with the degenerate eigenvalue E_0. The associated eigenvectors are linear combinations of these two kets (figure 3.3b).

These results can readily be generalised to the case of a *Hermitian* matrix of any dimension. When the elements of the matrix of the perturbation $\hat{H}_P$ are much smaller that the differences between the energy levels given by the principal term $\hat{H}_0$, which will be the case in most situations presented in this volume, the first order approximation will be sufficient. Otherwise, the calculation must be extended to the second order. In these conditions, correction terms similar to those included in equations [8], [9], and [13] for the energy levels, and in equations [11] for the eigenvectors, emerge.

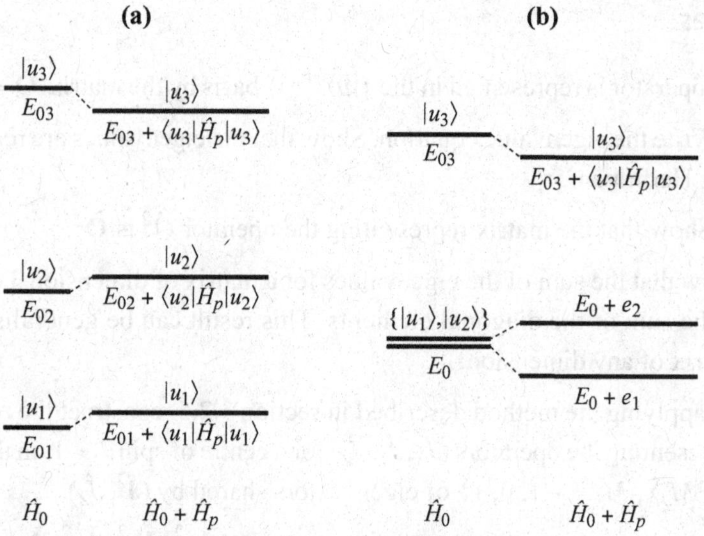

Figure 3.3 – Results from first order perturbation theory in a vector space of dimension 3. **(a)** The three eigenvalues are non-degenerate. **(b)** One eigenvalue is degenerate.

References

AYANT Y. & BELORIZKY E. (2000) *Cours de Mécanique Quantique*, Dunod, Paris.

COHEN-TANNOUDJI C., DIU B. & LALOË F. (2015) *Quantum Mechanics*, Wiley-VcH, New York.

Exercises

3.1. An operator is represented in the $\{|u\rangle, |v\rangle\}$ basis by the matrix $\tilde{\mathbf{O}} = \begin{bmatrix} a & b \\ c & d \end{bmatrix}$

a) Write the eigenvalue equation. Show that the eigenvalues are real when $\hat{O}$ is Hermitian.

b) Show that the matrix representing the operator $\hat{O}^2$ is $\tilde{\mathbf{O}}^2$.

3.2. Show that the sum of the eigenvalues for a matrix of dimension 3 is equal to the sum of the diagonal elements. This result can be generalised to a matrix of any dimension.

3.3. By applying the method described in section 3.2.5, construct the matrices representing the operators $(\hat{J}_X, \hat{J}_Y, \hat{J}_Z)$ for a centre of spin $J = 1$, in the basis $\{|1, M_J\rangle, M_J = -1, 0, 1\}$ of eigenvectors shared by $(\hat{\mathbf{J}}^2, \hat{J}_Z)$.

3.4. Use the expression of the trace of the matrices $(\tilde{\mathbf{J}}_X{}^2, \tilde{\mathbf{J}}_Y{}^2, \tilde{\mathbf{J}}_Z{}^2, \tilde{\mathbf{J}}^2)$ to show that the trace T of $\tilde{\mathbf{J}}_Z{}^2$ is equal to:

$$T = \sum_{M_J=-J}^{J} (M_J)^2 = \frac{J(J+1)(2J+1)}{3}$$

3.5. We say that the $\tilde{\mathbf{g}}$ matrix for a paramagnetic centre is *axial, with axis z*, when its principal values verify $g_x = g_y \neq g_z$, and we then write $g_x = g_y = g_\perp$, $g_z = g_{//}$. This type of centre is placed in a magnetic field **B**, and θ is the angle between the axis z and **B**, φ the angle between the axis x and the projection of **B** in the (x, y) plane (figure 3.4).

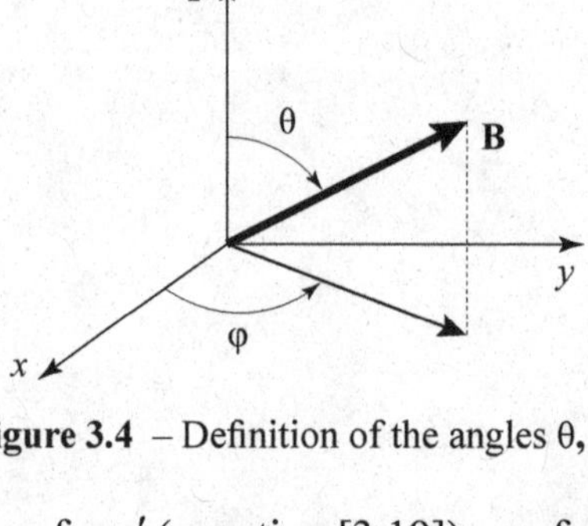

Figure 3.4 – Definition of the angles θ, φ.

Give the expression for g' (equation [3.19]) as a function of $g_{//}$, $g_\perp$ and θ. g' is observed to be independent of φ. What can be deduced about the principal axes x and y?

3.6. A paramagnetic centre of spin $S = ½$ characterised by an axial $\tilde{\mathbf{g}}$ matrix with axis z is placed in a magnetic field $\mathbf{B}$ forming the angle θ with z (see exercise 3.5). x is taken in the $(z, \mathbf{B})$ plane (figure 3.5). In section 3.3.2, we saw that the Zeeman Hamiltonian can be written

$$\hat{H}_{Zeeman} = g'\beta B\hat{S}_Z$$

where g' is given by equation [3.19] and Z is the direction of the vector $\mathbf{v}$ defined by equations [3.18]. We wish to express the eigenvectors of $\hat{H}_{Zeeman}$ in the $\{|½, ½\rangle_z, |½, -½\rangle_z\}$ basis of eigenvectors shared by $(\hat{\mathbf{S}}^2, \hat{S}_z)$.

a) Show that $\mathbf{v}$ is in the $(z, \mathbf{B})$ plane. We set $\psi = (z, Z)$ (figure 3.5).

b) What are the eigenvalues E_1 and E_2 for $\hat{H}_{Zeeman}$? Express $\hat{H}_{Zeeman}$ as a function of $\hat{S}_x$, $\hat{S}_z$, ψ.

c) Let $|w_1\rangle$ and $|w_2\rangle$ be the eigenvectors associated with E_1 and E_2. We write them in the following form:

$$|w\rangle = \cos\gamma\ |½, ½\rangle_z + \sin\gamma\ |½, -½\rangle_z$$

where γ is an angle to be determined. By applying $\hat{H}_{Zeeman}$ to the ket $|w\rangle$ and by using the results from section 3.2.5, show that $\gamma = \psi/2$ for $E_1 = g'\beta B/2$ and $\gamma = \psi/2 + \pi/2$ for $E_2 = -g'\beta B/2$.

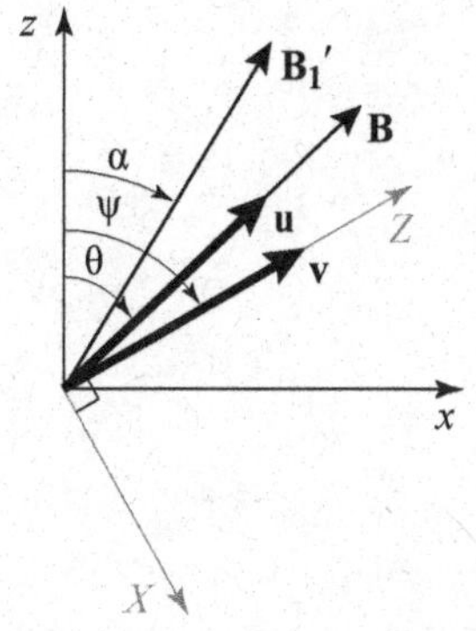

Figure 3.5 – The y and Y axes are perpendicular to the plane of the figure.

3.7. The aim of this exercise is to show that the transition probability is maximal when the fields $\mathbf{B}$ and $\mathbf{B}_1$ are perpendicular, even when the magnetic moment is anisotropic.

A paramagnetic centre of spin S characterised by an axial $\tilde{\mathbf{g}}$ matrix with axis z (see exercise 3.5) is placed in a field $\mathbf{B}$ which forms the angle θ

with z. x is chosen in the ($\mathbf{B}$, z) plane and the reference frame is completed with y perpendicular to the (x, z) plane (figure 3.5).

a) Write the components of the unit vector $\mathbf{u}$ in the direction of $\mathbf{B}$, and of the unit vector $\mathbf{v}$ which defines the quantization axis Z of $\hat{H}_{Zeeman}$ (section 3.3.2). The reference frame (X, Y, Z) is completed by placing X in the (x, z) plane, and defining $\psi = (z, Z)$ (figure 3.5). What is the expression for $\tan\psi$?

b) The transition probability between two states $|S, M_S\rangle$ and $|S, M_S'\rangle$ is proportional to the square of the modulus of $V = \langle S, M_S' | \hat{H}_1 | S, M_S\rangle$, with $\hat{H}_1 = \beta(g_\perp B_{1x}\hat{S}_x + g_\perp B_{1y}\hat{S}_y + g_{//}B_{1z}\hat{S}_z)$. By expressing ($\hat{S}_x$, $\hat{S}_y$, $\hat{S}_z$) as a function of ($\hat{S}_X$, $\hat{S}_Y$, $\hat{S}_Z$), write $\hat{H}_1$ in the form:

$$\hat{H}_1 = \beta(b_X\hat{S}_X + b_Y\hat{S}_Y + b_Z\hat{S}_Z)$$

Deduce that only transitions between $|S, M_S\rangle$ and $|S, M_S \pm 1\rangle$ are possible and that their probability is proportional to

$$|V|^2 = (\beta^2/4)\,[S(S+1) - M_S(M_S \pm 1)](b_X^2 + b_Y^2)$$

c) $\mathbf{B}_1'$ is the projection of $\mathbf{B}_1$ in the plane (x, z) and $\alpha = (z, \mathbf{B}_1')$ (figure 3.5). Show that $(b_X^2 + b_Y^2)$ is maximal when $\mathbf{B}_1$ is perpendicular to $\mathbf{B}$.

Answers to exercises

3.1. a) $\lambda^2 - (a+d)\lambda + ad - bc = 0$.

$\Delta = (a+d)^2 - 4(ad-bc) = (a-d)^2 + 4bc$. When the matrix is Hermitian, $bc = |b|^2$ and as a result $\Delta > 0$: the eigenvalues are real.

b) $\hat{O}^2|u\rangle = \hat{O}(a\,|u\rangle + c\,|v\rangle) = (a^2+bc)\,|u\rangle + (ac+cd)\,|v\rangle$

$\hat{O}^2|v\rangle = \hat{O}(b\,|u\rangle + d\,|v\rangle) = (ab+bd)\,|u\rangle + (d^2+bc)\,|v\rangle$

$$\tilde{\mathbf{O}}^2 = \begin{bmatrix} a^2+bc & ab+bd \\ ac+cd & d^2+bc \end{bmatrix}$$

It is easy to verify that this matrix is equal to $\tilde{\mathbf{O}}\cdot\tilde{\mathbf{O}}$.

3.2. $\tilde{\mathbf{A}}$ is the matrix and $\{a_{ij}\}$ are its elements. The eigenvalue equation is written:

$$\begin{vmatrix} a_{11}-\lambda & a_{12} & a_{13} \\ a_{21} & a_{22}-\lambda & a_{23} \\ a_{31} & a_{32} & a_{33}-\lambda \end{vmatrix} = 0$$

▹ When this determinant is developed, the terms in λ^3 and λ^2 which necessarily derive from the product of the diagonal elements $(a_{11}-\lambda)(a_{22}-\lambda)(a_{33}-\lambda)$, are written $-\lambda^3 + \lambda^2(a_{11}+a_{22}+a_{33}) + \ldots = 0$.

▹ In addition, if $(\lambda_1, \lambda_2, \lambda_3)$ are the eigenvalues, this equation is written $(\lambda_1-\lambda)(\lambda_2-\lambda)(\lambda_3-\lambda) = 0$. The terms in λ^3 and λ^2 are therefore

$$-\lambda^3 + \lambda^2(\lambda_1+\lambda_2+\lambda_3) + \ldots = 0.$$

By identification, we deduce:

$$\lambda_1+\lambda_2+\lambda_3 = a_{11}+a_{22}+a_{33}$$

3.3. Matrices of $\hat{J}_X$ and $\hat{J}_Y$:

	$\|1,-1\rangle$	$\|1,0\rangle$	$\|1,1\rangle$
$\langle 1,-1\|$	0	$\frac{1}{\sqrt{2}}$	0
$\langle 1,0\|$	$\frac{1}{\sqrt{2}}$	0	$\frac{1}{\sqrt{2}}$
$\langle 1,1\|$	0	$\frac{1}{\sqrt{2}}$	0

	$\|1,-1\rangle$	$\|1,0\rangle$	$\|1,1\rangle$
$\langle 1,-1\|$	0	$\frac{i}{\sqrt{2}}$	0
$\langle 1,0\|$	$-\frac{i}{\sqrt{2}}$	0	$\frac{i}{\sqrt{2}}$
$\langle 1,1\|$	0	$-\frac{i}{\sqrt{2}}$	0

3.4. In the $\{|J, M_J\rangle\}$ basis made of the eigenvectors shared by $(\tilde{\mathbf{J}}^2, \hat{J}_Z)$, the matrix $\tilde{\mathbf{J}}_Z{}^2$ is diagonal and its trace T is equal to:

$$T = \sum_{M_J=-J}^{J} (M_J)^2$$

As the matrices $\tilde{\mathbf{J}}_X{}^2$, $\tilde{\mathbf{J}}_Y{}^2$, $\tilde{\mathbf{J}}_Z{}^2$ have the same eigenvalues (section 3.2.5), their traces are equal. The equation $\tilde{\mathbf{J}}_X{}^2 + \tilde{\mathbf{J}}_Y{}^2 + \tilde{\mathbf{J}}_Z{}^2 = \tilde{\mathbf{J}}^2$ leads to $T = ⅓\,[\text{trace}\,(\tilde{\mathbf{J}}^2)]$, i.e.:

$$T = \frac{1}{3} \sum_{M_J=-J}^{J} J(J+1) = \frac{J(J+1)(2J+1)}{3}$$

This identity can also be demonstrated by mathematical induction.

3.5. $g' = (g_\perp{}^2 \sin^2\theta + g_{//}{}^2 \cos^2\theta)^{½}$. Since g' does not depend on φ, we can arbitrarily choose x and y in the plane perpendicular to z.

3.6. a) In the $\{x, y, z\}$ reference frame, we have $\mathbf{u}\,(\sin\theta, 0, \cos\theta)$ and $\mathbf{v}\,(\sin\theta\, g_\perp/g', 0, \cos\theta\, g_{//}/g')$.

b) $E_1 = g'\beta B/2$, $E_2 = -g'\beta B/2$. $\hat{H}_{Zeeman} = g'\beta B\,(\hat{S}_x \sin\psi + \hat{S}_z \cos\psi)$.

c) γ is determined by writing $g'\beta B(\hat{S}_x \sin\psi + \hat{S}_z \cos\psi)|w\rangle = E\,|w\rangle$ where $E = E_1$ or E_2. We therefore obtain:

$$|w_1\rangle = \cos(\psi/2)\,|½, ½\rangle_z + \sin(\psi/2)\,|½, -½\rangle_z$$

$$|w_2\rangle = -\sin(\psi/2)\,|½, ½\rangle_z + \cos(\psi/2)\,|½, -½\rangle_z$$

3.7. a) $\mathbf{u}(\sin\theta, 0, \cos\theta)$; $\mathbf{v}(\sin\theta\, g_\perp/g', 0, \cos\theta\, g_{//}/g')$; $\tan\psi = v_x/v_z = (g_\perp/g_{//})\tan\theta$.

b) $S_x = S_X \cos\psi + S_Z \sin\psi$; $S_y = S_Y$; $S_z = -S_X \sin\psi + S_Z \cos\psi$.

$$b_X = g_\perp B_{1x} \cos\psi - g_{//} B_{1z} \sin\psi;$$

$$b_Y = g_\perp B_{1y};$$

$$b_Z = g_\perp B_{1x} \sin\psi + g_{//} B_{1z} \cos\psi.$$

Only $\Delta M_S = \pm 1$ transitions are allowed.

c) Using the relation between $\tan\psi$ and $\tan\theta$ obtained in question a), we deduce

$$(b_X{}^2 + b_Y{}^2) = (g_{//}\sin\psi/\sin\theta)^2\,(B_{1x}\cos\theta - B_{1z}\sin\theta)^2 + (g_\perp B_{1y})^2$$

Since $B_{1x} = B_1'\sin\alpha$, $B_{1z} = B_1'\cos\alpha$ (figure 3.5), we can write

$$(B_{1x}\cos\theta - B_{1z}\sin\theta)^2 = (B_1')^2[\sin(\alpha - \theta)]^2.$$

This quantity is maximal for $\alpha = \theta \pm \pi/2$, i.e., when **B** is perpendicular to $\mathbf{B}_1'$ and therefore to $\mathbf{B}_1$.

Chapter 4

How anisotropy of the $\tilde{g}$ and $\tilde{A}$ matrices affects spectrum shape for radicals and transition ion complexes

4.1 – Introduction

In the previous chapter, we saw that it is possible to calculate the lowest energy levels for a paramagnetic centre using a spin Hamiltonian constructed from the components of one or more angular momenta. Among these angular momenta, the one that is included in the expression for the *magnetic moment* of the centre plays the principal role. For an atom or a free ion, this is the sum **J** of the total angular momenta **L** and **S** of the electrons in the incomplete subshell, and the value of J is determined by the spin-orbit coupling (appendix 1). In complexes of rare earth and actinide ions, the incomplete subshell of the cation – $4f$ for rare earths and $5f$ for actinides – is *internal*. Its electrons interact weakly with the ligands and, as a first approximation, we can use a spin Hamiltonian expressed as a function of the components of the angular momentum **J** of *the free ion* (chapter 8).

In contrast, in free radicals and transition ion complexes – where the orbitals occupied by the unpaired electrons and their energies are considerably different from those of an ion or free atom – the magnetic moment of the ground state is written

$$\boldsymbol{\mu} = -\beta\,\tilde{g}\cdot\mathbf{S} \qquad [4.1]$$

and the spin S is equal to $n/2$, where n is the number of unpaired electrons in the molecule. We saw in section 3.3.2 that the interaction of this moment with the field **B** causes splitting of the energy levels which depends on:

P. Bertrand, *Electron Paramagnetic Resonance Spectroscopy*,
https://doi.org/10.1007/978-3-030-39663-3_4

▹ the principal values of the $\tilde{\mathbf{g}}$ matrix,

▹ the direction of **B** relative to the principal axes of this matrix.

The principal axes and the principal values of the $\tilde{\mathbf{g}}$ matrix are determined by the electronic structure of the molecule's ground state, and a number of *molecular models* have been proposed to calculate the values of these parameters. For example, in the case of complexes of transition ions, the simplest models are based on ligand field-type descriptions. More elaborate models are based on methods such as the Hartree-Fock method or Density Functional Theory (DFT). In section 4.2, we describe the main factors which determine the principal values (g_x, g_y, g_z) for free radicals and transition ion complexes. This triplet of numbers is a true *signature* of the molecule's ground state, and we will see in section 4.3 how it can be determined from the EPR spectrum by considering various types of sample: single crystal, polycrystalline powder or frozen solution. Similarly, in section 4.4 we will see that the features produced on the spectrum by anisotropic hyperfine interactions can be used in some cases to measure the principal values of the $\tilde{\mathbf{A}}$ matrix. We will conclude this chapter in section 4.5 by describing how *movement* of the molecules affects the shape of the spectrum.

4.2 – The $\tilde{\mathbf{g}}$ matrix

Expression [4.1] of the magnetic moment appears to suggest that this quantity is only related to the *spin angular momentum* of the electrons. However, if this were the case the principal values (g_x, g_y, g_z) of the $\tilde{\mathbf{g}}$ matrix would be equal to $g_e = 2.0023$, and EPR spectroscopy would lose much of its interest. In fact, the principal values differ from g_e because of the *orbital angular momentum* of the electrons and the *spin-orbit* coupling. This effect is described for a transition ion complex in appendix 2. In this section, we identify the characteristics of the $\tilde{\mathbf{g}}$ matrix which are useful when analysing spectra.

4.2.1 – How the molecule's symmetry properties affect the $\tilde{\mathbf{g}}$ matrix

Since the principal axes and the principal values of the $\tilde{\mathbf{g}}$ matrix are determined by the electronic structure of the molecule's ground state, they depend on its symmetry properties, in particular the existence of *axes of symmetry*. In general, a figure is said to have a *k*-order rotation axis when it is invariant by a rotation of $2\pi/k$ around this axis. Several situations can be encountered:

- ▹ The molecule has no axis of symmetry. In this case, the three principal values are *on principle* different, and the $\tilde{g}$ matrix is said to be *rhombic*. However, the principal values may be equal for reasons which have nothing to do with the symmetry properties.
- ▹ When the molecule has an axis of symmetry, this axis is a principal axis of the $\tilde{g}$ matrix. If the order of this axis z is higher than or equal to 3, the principal values g_x and g_y are equal. In this case, the $\tilde{g}$ matrix is said to be *axial*, and x and y (orthogonal) can be arbitrarily chosen in the plane perpendicular to z (exercise 3.5).
- ▹ When the molecule has 3 equivalent directions (x, y, z), it is said to have *cubic* symmetry. This is the case of transition ion complexes with octahedral (O_h) or tetrahedral (T_d) symmetry. The three principal values are then equal ($g_x = g_y = g_z = g$) and we can write

$$\tilde{g} = g\,\tilde{1}$$

where $\tilde{1}$ is the *unit matrix* defined by:

$$\tilde{1} = \begin{bmatrix} 1 & 0 & 0 \\ 0 & 1 & 0 \\ 0 & 0 & 1 \end{bmatrix}$$

The $\tilde{g}$ matrix is said to be *isotropic*.

4.2.2 – The principal values of the $\tilde{g}$ matrix

In general, the principal values are more different from g_e when

- ▹ The spin-orbit coupling constant ζ for the atoms onto which the unpaired electrons are delocalised is large. ζ is a positive quantity that is generally expressed in cm^{-1}; it increases rapidly with the atomic number Z. Its value has been determined for numerous atoms from their atomic absorption spectra.
- ▹ The molecule's excited states are close to the ground state. The energy of these levels depends on the *nature of the atoms* but also on their *interactions* in the molecule.

These properties explain why the triplet (g_x, g_y, g_z) is very sensitive to the electronic structure of the molecules and why it is a veritable *spectral signature*. We will now further examine the factors determining this signature in free radicals and transition ion complexes.

◇ *Free radicals*

In *organic* free radicals, the unpaired electron is generally delocalised onto light atoms such as C, N, or O which have a relatively small spin-orbit coupling constant (table 4.1).

Table 4.1 – Spin-orbit coupling constants for atoms in organic molecules.

atom	Z	ζ [cm^{-1}]
C	6	29
N	7	76
O	8	151
F	9	272
S	16	382
Cl	17	587

In addition, the energy of the excited states, which corresponds to transitions in the ultraviolet or visible radiation range, is generally high. As a result, the difference between $g_e = 2.0023$ and the three principal values, that are generally ranked in the order $g_x > g_y > g_z$, is *small* in this type of molecule. However, this difference increases when the unpaired electron is delocalised onto heavier atoms, as shown by the data in table 4.2 which relate to the radicals of biological origin shown in figure 4.1. In this table we have also indicated the *mean* of the principal values which, as we will see in section 4.5, is none other than the number g_{iso}; this quantity is involved in the isotropic regime (equation [4.22]).

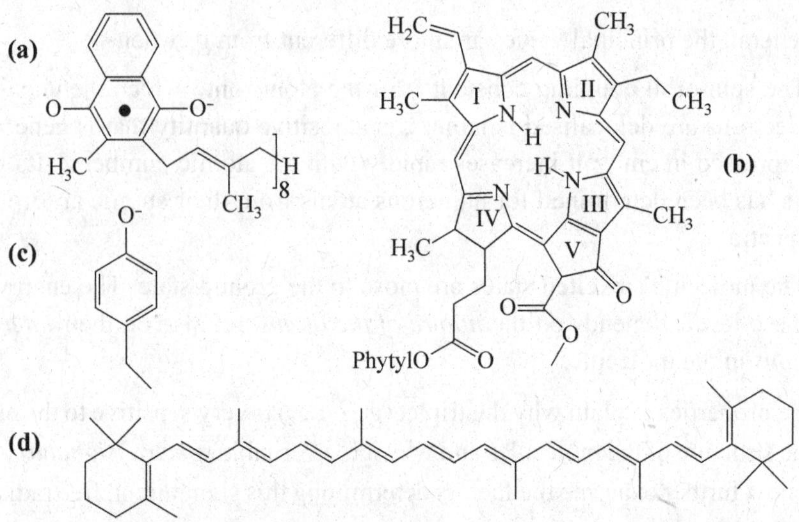

Figure 4.1 – The radicals listed in table 4.2: **(a)** semiquinone Q_A, **(b)** pheophytin, **(c)** tyrosine, **(d)** carotenoid.

Table 4.2 – Principal values of the $\tilde{\mathbf{g}}$ matrix for radicals in photosystem 2 determined by high-field EPR [Un *et al.*, 2001].

radical	g_x	g_y	g_z	mean
carotenoid	2.0032	2.0025	2.0021	2.0026
pheophytin	2.0042	2.0032	2.0024	2.0033
semiquinone Q_A	2.0061	2.0051	2.0021	2.0044
tyrosine D	2.0076	2.0043	2.0021	2.0047

While the three principal values are very close to g_e for the carotenoid radical, which is a purely hydrocarbon molecule, g_x and g_y are significantly different from it in pheophytin, where the spin density is partially delocalised onto nitrogen atoms. The difference increases further in semiquinone and tyrosine where an oxygen atom has a significant spin population, of around 0.28 in the case of tyrosine [Dole *et al.*, 1997]. In all cases, the smallest principal value g_z remains practically equal to g_e, as predicted by the theoretical models [Stone, 1963; Stone, 1964]. Even greater differences are observed in radicals where *sulfur atoms* have a notable spin population. For example, this is the case with the cysteinyl radical, for which the high value of the spin-orbit coupling constant (table 4.1) and the proximity of the excited states produce $g_x \approx 2.3$, $g_y = g_z = 2.008$ [Van Gastel *et al.*, 2004].

For *inorganic radicals*, the difference between the principal values of the $\tilde{\mathbf{g}}$ matrix and g_e also increases with the atomic number of the atoms, as shown by the data in table 4.3 for radicals trapped in crystals. These principal values were reproduced using simplified molecular orbitals models [Ovenall and Whiffen, 1961; Atherton, 1993] and by DFT-type methods [Schreckenbach and Ziegler, 1996].

Table 4.3 – Principal values of the $\tilde{\mathbf{g}}$ matrix for inorganic radicals trapped in crystals [a] [Marshall *et al.*, 1964]; [b] [Zeldes and Livingston, 1961]; [c] [Morton, 1964]; [d] [Reuveni *et al.*, 1970].

	g_x	g_y	g_z	mean
CO_2^-	2.0032	2.0016	1.9972	2.0007 [a]
NO_2	2.0057	2.0015	1.9910	1.9994 [b]
NO_2^{2-}	2.0099	2.0070	2.0038	2.0069 [c]
SO_2^-	2.0120	2.0057	2.0019	2.0065 [d]
ClO_2	2.0183	2.0088	2.0036	2.0102 [c]

◇ *Transition ion complexes*

Only *mononuclear* complexes are considered here, the case of polynuclear complexes is specifically dealt with in chapter 7. In mononuclear complexes, a cation which has an incomplete $3d$, $4d$ or $5d$ subshell is coordinated to a certain number of anionic or polar ligands. Unpaired electrons are mainly localised on cations, which have a larger spin-orbit coupling constant than the light atoms in table 4.1. Its value is indicated in table 4.4 for cations in the first transition series.

Table 4.4 – Spin-orbit coupling constant for ions with a $3d^n$ configuration [Bendix *et al.*, 1993].

$3d^n$ ion	Ti^{3+}	V^{3+}	Cr^{3+}	Mn^{3+}	Mn^{2+}	Fe^{2+}	Co^{2+}	Ni^{2+}	Cu^{2+}
n	1	2	3	4	5	6	7	8	9
ζ [cm^{-1}]	155	210	275	355	300	400	515	630	830

In addition, the excited states of these complexes, which produce "$d-d$" transitions in the infrared or visible region, are often closer than those of the radicals. As a result, the difference between the principal values of the $\tilde{\mathbf{g}}$ matrix and g_e is generally greater than in radicals.

The spin S of a transition ion complex is equal to $n/2$, where n is the number of unpaired electrons in the ground state. Its value depends on the number of electrons in the incomplete subshell, but also on the strength of the electrostatic interactions between these electrons and the ligands. Indeed, whereas the five d orbitals are degenerate in the free ion, their energy in the complex depends on their orientation relative to the ligands (complement 1). The mode of filling of the orbitals which minimises the energy of the complex is the result of competition between two effects:

- ▹ The energy due to electrostatic interactions *between the electrons* is minimal when the electrons occupy different orbitals with a maximum total spin S, as in the free ion (appendix 1).
- ▹ The energy due to the electrostatic interactions *with ligands* is minimal when the electrons are grouped together in lower energy orbitals, which favours minimal S.

A range of situations can therefore be produced depending on the strength of the interaction with the ligands:

▷ *High-spin situation*

When the interaction with ligands is *weak*, the splitting of the *d* orbitals is small and the mode of filling the orbitals which minimises the energy of the complex is identical to that which corresponds to the ground term for the free cation. It is given by Hund's rule, which indicates that *S* takes the *maximum* value allowed by the Pauli principle (appendix 1). This so-called *high-spin* situation only occurs with cations from the *first transition series*. It is observed in all complexes with tetrahedral coordination and with some ligands in octahedral complexes. It is illustrated in figure 4.2a for an octahedral coordination.

In the high-spin situation, ligand field theory produces expressions for the principal values of the $\tilde{\mathbf{g}}$ matrix with the form (appendix 2):

$$g_i = g_e + \Delta g_i \, ; \Delta g_i = -\lambda/\Delta_i \qquad i = x, y, z \qquad [4.2]$$

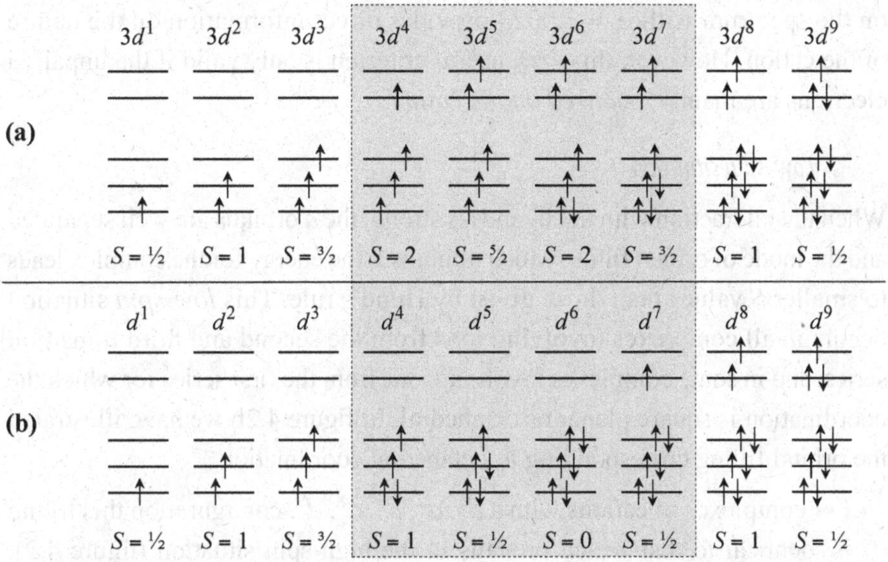

Figure 4.2 – Spin value for cations in the $3d^n$ configuration in octahedral coordination in **(a)** high-spin and **(b)** low-spin situations. The configurations for which *S* differs in the two situations are highlighted in the grey box.

The parameter λ is linked to the spin-orbit coupling constant ζ for the cation by $\lambda = \pm\,\zeta/2S$, and Δ_i is a *positive* quantity which has the dimension of energy; this energy is smaller when the excited states are close. The absolute value of Δg_i is small. It is, for example, generally less than 0.3 in complexes of cations from the first transition series. But the important point is that its *sign* depends only on that of λ, which is itself determined by the number of $3d$ electrons:

▹ If the $3d$ subshell is less than half-full, $\lambda = \zeta/2S$ is positive and all three g_i values are *less than* g_e.

This is the case for Ti^{3+} and V^{4+} ($3d^1$), V^{3+} ($3d^2$), Cr^{3+} and Mn^{4+} ($3d^3$), Cr^{2+} and Mn^{3+} ($3d^4$).

▹ If the $3d$ subshell is more than half-full, $\lambda = -\zeta/2S$ is negative and all three g_i values are *greater than* g_e.

This is the case for Fe^{2+} ($3d^6$), Co^{2+} and Ni^{3+} ($3d^7$), Co^+ and Ni^{2+} ($3d^8$), Ni^+ and Cu^{2+} ($3d^9$).

▹ For cations with a $3d^5$ configuration like Mn^{2+} and Fe^{3+}, the ground term is characterised by $S = 5/2$ and $L = 0$. Spin-orbit coupling has no impact and the (g_x, g_y, g_z) values are equal to g_e (appendix 2).

Thus, for high-spin complexes, comparison of values of (g_x, g_y, g_z) measured on the spectrum with $g_e = 2.0023$ provides direct information on the nature of the cation. However, this very useful criterion is only valid if the unpaired electrons are mainly *localised on the cation*.

▹ *Low-spin situation*

When the interaction with the ligands is strong, the d orbitals are well separated and the mode of orbital filling which minimises the energy for the complex leads to smaller S values than those given by Hund's rule. This *low-spin* situation occurs in all complexes involving ions from the second and third transition series, and in some complexes involving ions from the first series for which the coordination is square planar or octahedral. In figure 4.2b we have illustrated the orbital filling corresponding to octahedral coordination.

▹ For complexes of cations with a d^1, d^2, d^3, d^8, d^9, configuration this filling is identical to that which prevails in the high-spin situation (figure 4.2). The principal values of the $\tilde{\mathbf{g}}$ matrix are therefore described by equation [4.2] and the rule relating to the difference between (g_x, g_y, g_z) and g_e applies. For example, complexes of Mo^{5+} ($4d^1$), Tc^{4+} ($4d^3$), W^{5+} ($5d^1$) and Re^{4+} ($5d^3$) generally have (g_x, g_y, g_z) values that are less than g_e. Values greater than g_e observed for some complexes of Mo^{5+} are the result of significant delocalisation of the spin density onto the ligands [Hanson *et al.* 1987; Cosper *et al.*, 2005]. Conversely, the values of (g_x, g_y, g_z) are greater than g_e in complexes of Pd^+ and Ag^{2+} ($4d^9$).

▷ For transition ion complexes with a d^4 ($S = 1$), d^5 ($S = ½$) and d^7 ($S = ½$) configuration in the low-spin situation, expressions giving the principal values for the $\tilde{g}$ matrix have been established using ligand field theory. In the case of cations of configuration d^5:

- first transition series: Mn^{2+} and Fe^{3+}
- second transition series: Rh^{4+}, Ru^{2+}, Tc^{2+}, Mo^{+}
- third transition series: Ir^{4+}

these expressions involve two dimensionless parameters describing the axial and rhombic distortions. They adequately reproduce the experimental values of (g_x, g_y, g_z) for ferric haeme complexes, which have a large anisotropy due to a significant contribution from the orbital angular momentum of the electrons (table 4.5).

Table 4.5 – Comparison of the experimental values of (g_x, g_y, g_z) for low-spin haeme iron complexes and values calculated by applying ligand field theory [More *et al.*, 1990].

Molecule	g_{exp}	g_{calc}
cytochrome P450	1.91; 2.25; 2.42	1.91; 2.245; 2.42
leghaemoglobin: triazido derivative	1.72; 2.20; 2.79	1.70; 2.20; 2.80
horseradish peroxidase	1.63; 2.09; 2.96	1.62; 2.09; 2.96
flavocytochrome b_2[a]	1.45; 2.22; 3.01	1.44; 2.215; 3.01
cytochrome *c*	1.24; 2.24; 3.06	1.20; 2.22; 3.07
horseradish peroxidase: cyanide derivative	1.2; 2.09; 3.19	1.15; 2.08; 3.19
cytochrome *c*: cyanide derivative	*n.d.*[b]; 1.96; 3.34	0.21; 1.93; 3.33

[a] The spectrum is represented in figure 4.8. [b] not determined.

▷ *Intermediate-spin*

In some complexes, the value of S is *intermediate* between those predicted in high-spin and low-spin situations, for example $S = 3/2$ for ferric complexes and $S = 1$ for ferrous complexes [Kahn, 1993]. Complexes also exist where the ground state is a "quantum mixture" of states characterised by different spin values , like $S = 5/2$ and $S = 3/2$ for Fe^{3+} [Bertrand *et al.*, 1983; Kahn, 1993].

In the next section, we will see how the principal values of the $\tilde{g}$ matrix can be measured from the spectrum produced by a sample containing a large number of paramagnetic centres.

4.3 – Shape of the spectrum produced by an ensemble of paramagnetic centres in the absence of hyperfine interaction

In this section we consider a sample containing identical molecules characterised by their spin S and their $\tilde{\mathbf{g}}$ matrix, without hyperfine interaction, and we wish to determine the shape of the EPR spectrum for different modes of organisation of the molecules in the sample.

4.3.1 – Variation in g' values with the direction of **B**

We saw in section 3.3.2 that the interaction between the magnetic moment given by equation [4.1] and a magnetic field **B** produces a pattern of $(2S + 1)$ equidistant energy levels, with splitting:

$$\Delta E = g'\beta B$$

where:

$$g' = [g_x^2 u_x^2 + g_y^2 u_y^2 + g_z^2 u_z^2]^{1/2} \qquad [4.3]$$

The numbers (g_x, g_y, g_z) are the principal values of the $\tilde{\mathbf{g}}$ matrix and (u_x, u_y, u_z) are the components of the unit vector **u** in the direction of **B**, in the system of magnetic axes $\{x, y, z\}$ of the molecule. The "prime" suffix of g' indicates that this number depends on the *direction* of **B** relative to the molecule. Resonance occurs when $\Delta E = h\nu$, where ν is the frequency of the spectrometer. The resonance field is therefore given by:

$$B = h\nu / g'\beta \qquad [4.4]$$

The sample is fixed in the cavity of the spectrometer, but the orientation of the molecules relative to **B** can vary in the sample. Since the EPR spectrum results from superposition of the resonance lines produced by all the molecules, the following questions arise:

- ▹ If the sample is such that all the molecules are oriented in the same way, how does the spectrum vary when the sample is rotated in the field?
- ▹ If the molecules in the sample are oriented differently relative to the magnetic field, what is the shape of the spectrum?

To answer these questions, we must examine how the resonance field B (equation [4.4]), and consequently the g' value (equation [4.3]), vary as a function of the direction of **B**.

▷ We will start with a centre with *cubic symmetry* for which $g_x = g_y = g_z = g$. Equation [4.3] shows that $g' = g$ whatever the direction of **B**. The resonance lines of all the molecules add together to produce a single line at $B = h\nu/g\beta$. The spectrum is therefore reduced to this unique line and it is independent of how the molecules are organised in the sample, and of the orientation of the sample relative to **B**.

▷ We then consider a centre with *axial symmetry*. If z is the axis of symmetry, the two principal values g_x and g_y are equal, and are generally written:

$$g_x = g_y = g_\perp, \; g_z = g_{//}$$

Equation [4.3] can then be written:

$$g' = [g_\perp{}^2 (u_x{}^2 + u_y{}^2) + g_{//}{}^2 u_z{}^2]^{1/2}$$

If θ is the angle between the axis z and the vector **u** (figure 4.3a), we have $u_z{}^2 = \cos^2\theta$ and $u_x{}^2 + u_y{}^2 = \sin^2\theta$, and g' can be written:

$$g' = [g_\perp{}^2 \sin^2\theta + g_{//}{}^2 \cos^2\theta]^{1/2} \qquad [4.5]$$

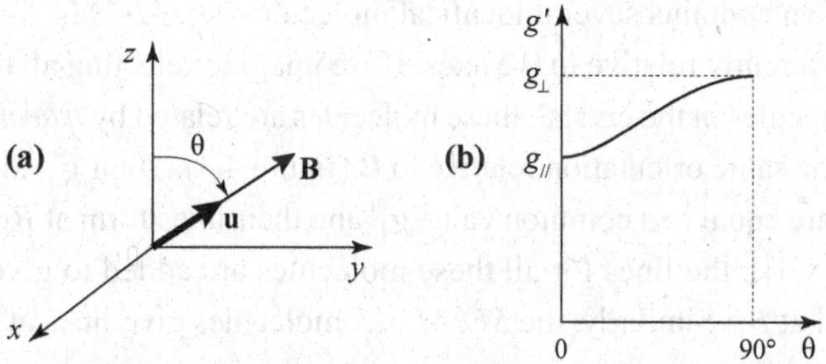

Figure 4.3 – Centre with axial symmetry: **(a)** definition of the angle θ, **(b)** variation of g' as a function of θ for $g_\perp > g_{//}$.

In this case, g' only depends on the angle θ between the molecular axis z and the field **B**. The number g' varies monotonically from $g_{//}$ for θ = 0 (**u** parallel to z) to $g_\perp$ for θ = 90° (**u** perpendicular to z), which explains the notations $g_{//}$ and $g_\perp$ (figure 4.3b).

▷ In the general case where the three numbers (g_x, g_y, g_z) are distinct, g' depends on two parameters, and its variations as a function of the direction of **B** are more difficult to visualise. An idea can be obtained by seeking the directions of **u** which satisfy equation [4.3] for a given value of g'. This problem is dealt with in complement 2 with the help of a simple geometric representation. The principal axes can always be labelled such that:

$$g_x \leq g_y \leq g_z$$

The results can then be stated as follows:

1. When the direction of **u** varies, g' remains within the interval $[g_x, g_z]$.
2. The bounds g_x and g_z are only reached when **u** is parallel to x and z, respectively.
3. In contrast, the intermediate value g_y is obtained when **u** is parallel to y and for a whole ensemble of directions that can be very different from y.

These properties have a very significant influence on the shape of the spectrum produced by a sample containing a large number of molecules randomly oriented relative to the field **B**.

4.3.2 – Shape of the EPR spectrum depending on the nature of the sample

◇ Single crystals

A single crystal is the result of stacking of parallelepipedic cells known as *unit cells*, the edges of which define the fundamental axes **(a, b, c)** of the crystal. The unit cell often contains several identical molecules M_1, M_2, M_3, … which are oriented differently relative to the axes. If we imagine selecting all the equivalent M_1 molecules in the crystal, these molecules are related by *translations* and thus have the same orientation relative to **B** (figure 4.4). Their g' values (equation [4.3]) are equal to a common value g_1' and their lines form at $B_1 = h\nu/g_1'\beta$ (equation [4.4]): the lines for all these molecules are added to give a unique line centred at B_1. Similarly, the M_2, M_3, … molecules give lines at B_2, B_3, …

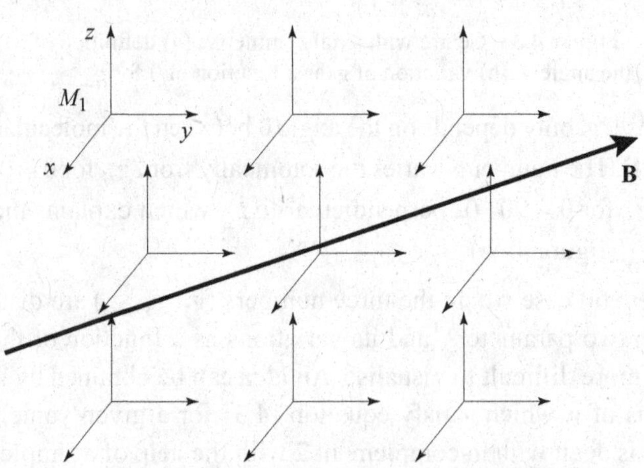

Figure 4.4 – Spatial organisation of paramagnetic molecules in a single crystal. $\{x,y,z\}$ are the principal axes of the $\tilde{\mathbf{g}}$ matrix.

If the crystal is rotated in the field **B**, the values of $g_1', g_2', g_3', \ldots$ and therefore of $B_1, B_2, B_3, \ldots$ will be altered. By studying these variations in detail it is possible to determine the principal values of the $\tilde{\mathbf{g}}$ matrix and the direction of its principal axes relative to the axes of the crystal. If the arrangement of the molecules in the crystal is known, the orientation of the principal axes relative to the molecules can be deduced.

These experiments require crystals containing sufficient paramagnetic centres to produce a detectable EPR signal. In addition, it is not always easy to precisely monitor the position of the resonance lines as a function of the orientation of the crystal when the unit cell contains several molecules. But when possible, these experiments are the only ones that give the direction of the principal axes relative to the molecule. These directions are very useful in some EPR applications (see section 4.6.1).

An example of application to Ti^{3+} centres substituted into a crystal of $LaMgAl_{11}O_{19}$ is presented in complement 5.

◇ *Polycrystalline powders or frozen solutions*

In this type of sample, the molecules are *frozen* in random orientations relative to **B** and each one produces a resonance line for the field value corresponding to its orientation. We will assume that all the orientations are possible and equally probable. Contrary to what might be expected, the spectrum which results from the superposition of all these lines has remarkable features. These features can be used to measure the principal values of the $\tilde{\mathbf{g}}$ matrix, they are determined by the *density of the resonance lines* $D(B)$, defined as follows: $D(B)\mathrm{d}B$ represents the fraction of molecules for which lines form in the $[B, B + \mathrm{d}B]$ range.

- ▹ We will first examine the case where the $\tilde{\mathbf{g}}$ matrix is *axial*. Since g' varies monotonically between $g_{//}$ and $g_{\perp}$ depending on the orientation of the molecules (figure 4.3b), the density $D(B)$ is different from zero if B is in the range $B_{//} = h\nu/g_{//}\beta$ to $B_{\perp} = h\nu/g_{\perp}\beta$. For a molecule to produce a line at $B_{//}$, its symmetry axis *z must* be *parallel* to **B**, whereas for it to produce a line at $B_{\perp}$, it is *sufficient* for its *z* axis to be *perpendicular* to **B**. If all the orientations are equally probable in the sample, the second situation is much more frequent than the first, and as a result $D(B_{\perp})$ is larger than $D(B_{//})$. These considerations can be made quantitative by calculating the density $D(B)$ (complement 3). For $g_{//} > g_{\perp}$, the following expression is obtained:

$$D(B) = \frac{B_{//} B_{\perp}^{2}}{[B_{\perp}^{2} - B_{//}^{2}]^{1/2}} \frac{\cdot 1}{B^{2}[B_{\perp}^{2} - B^{2}]^{1/2}} \quad [4.6]$$

For $g_{//} < g_{\perp}$, $B_{//}$ and $B_{\perp}$ must be switched, as must B and $B_{\perp}$ in the two terms containing square roots.

The density $D(B)$ is represented by the continuous line on figure 4.5a, where we distinguish between two cases depending on whether $g_{//}$ is greater or less than $g_{\perp}$. We will see in chapter 5 that the resonance lines have a certain width, such that the absorption signal is represented by the dashed line in figure 4.5a. The spectrum, which is the derivative ds/dB of the absorption signal, presents characteristic features at the limits $B_{//}$ and $B_{\perp}$ where the density undergoes drastic variations (figure 4.5b). These features can be used to measure $g_{//}$ and $g_{\perp}$.

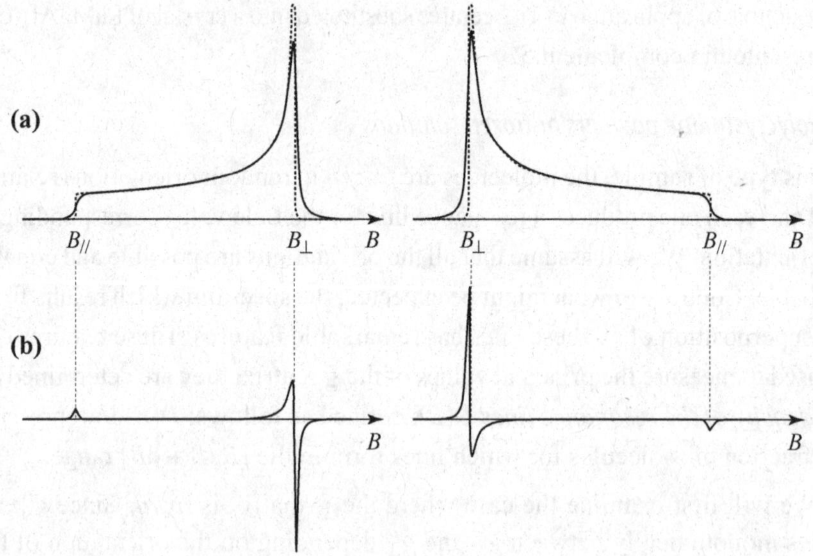

Figure 4.5 – **(a)** Density of the resonance lines (dashed line) and absorption signal (continuous line) for a centre with axial symmetry. $g_{//} > g_{\perp}$ is shown on the left; $g_{//} < g_{\perp}$ on the right. **(b)** The spectrum. The areas under the positive and negative portions of the signal are equal (see text).

Figure 4.6 represents the X-band spectrum for a frozen solution of complexes of Ni^{3+} ($3d^7$, S = ½) with square planar coordination [Ottenwalder *et al.*, 2004]. On the spectrum, $g_{//}$ = 2.018 and $g_{\perp}$ = 2.263 can be measured. These values are higher than g_e, as predicted in section 4.2.2. As the ligands are two atoms of oxygen and two atoms of nitrogen, the symmetry is not strictly axial. Nevertheless, the spectrum shows that the 4 ligands play equivalent roles from the point of view of the magnetic properties.

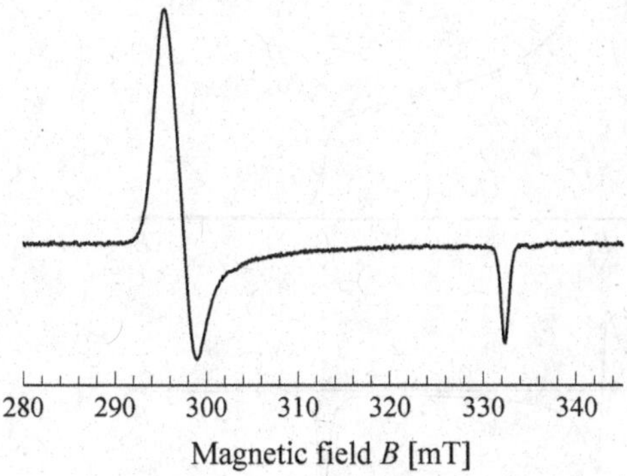

Figure 4.6 – X-band spectrum for a frozen solution of complexes of Ni^{3+} ions with square planar geometry. Experimental conditions: temperature 100 K, microwave frequency 9.3850 GHz, power 1 mW. Modulation: frequency 100 kHz, peak-to-peak amplitude 0.5 mT. [Ottenwaelder *et al.* (2004) *Chemical Communications*, **5**, 504–505. Reproduced with permission from The Royal Society of Chemistry]

▷ When the three principal values (g_x, g_y, g_z) are different, the expression describing the density of the resonance lines is more complicated [Kneubuhl, 1960]. Its shape is represented by the continuous line shown in figure 4.7a. It is important to remember that only molecules for which the x axis is parallel to **B** produce a line at $B_x = h\nu/g_x\beta$ and that only those for which the z axis is parallel to **B** produce a line at $B_z = h\nu/g_z\beta$.

In contrast, a whole ensemble of orientations exist for which g' takes the intermediate value g_y (complement 2). This explains why the density is maximal at $B_y = h\nu/g_y\beta$. We have represented the absorption signal $s(B)$ by the dashed line on figure 4.7a and the $\mathrm{d}s/\mathrm{d}B$ spectrum in figure 4.7b.

In contrast to the "axial" shapes shown in figure 4.5, the shape of the spectrum in figure 4.7 is termed "rhombic". Once again, the features appearing at (B_x, B_y, B_z), where the density is subject to sudden variations, can be used to readily measure the (g_x, g_y, g_z) parameters.

Figure 4.8 shows the spectrum for a frozen solution of flavocytochrome b_2, a biological macromolecule which contains a haeme with an Fe^{3+} ion ($3d^5$) in the weak spin situation $S = ½$. In this type of complex, the orbital angular momentum contributes significantly and the principal values of the $\tilde{\mathbf{g}}$ matrix differ significantly from $g_e = 2.0023$ (table 4.5).

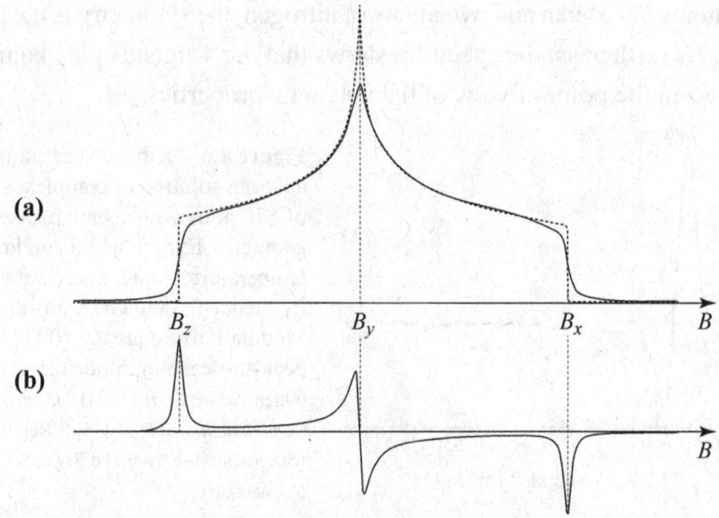

Figure 4.7 – **(a)** Density of the resonance lines (dashed line) and absorption signal (continuous line) for a centre with rhombic symmetry. The notable values of the field are $B_i = h\nu/g_i\beta$, where $i = x, y, z$ and ν is the frequency of the spectrometer. **(b)** The spectrum. The areas under the positive and negative portions of the signal are equal (see text).

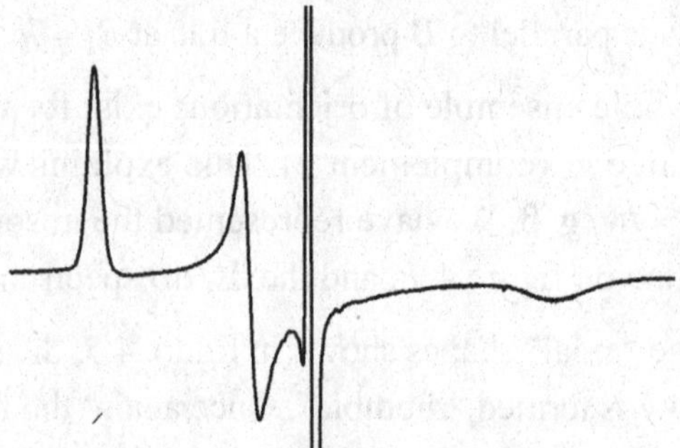

Figure 4.8 – Spectrum for a frozen solution of the enzyme flavocytochrome b_2. The very anisotropic signal due to the haeme is characterised by $g_x = 1.45$, $g_y = 2.22$, $g_z = 3.01$ (table 4.5). The narrow radical signal at $g \approx 2$ is that of flavin in the semi-quinone state. Experimental conditions: temperature 15 K, microwave frequency 9.40917 GHz, power 1 mW. Modulation: frequency 100 kHz, peak-to-peak amplitude 2 mT.

4.3.3 – Notes on the spectra produced by polycrystalline powders or frozen solutions

- ▷ To fully understand the nature of the spectrum appearing on the spectrometer's screen when this type of sample is submitted to a field sweep, it is important to realise that for each value of B, only the molecules *oriented* relative to **B** such that their value of g' verifies equation [4.4] will enter a resonant state. Imagine that the molecules have a pixel emitting a flash at resonance, and that an "EPR demon" sitting in the cavity of the spectrometer observes these flashes. The demon would notice that the field sweep causes the "resonating" molecules to light up successively for each value of B, each molecule lighting up only once during the whole sweep. As we are not as lucky as the demon, we can only detect the absorption signal $s(B)$ which can be considered to represent the sum of all the flashes emitted for each value B of the field. In addition, the intensity of each flash is weighted by its *transition probability*, as we will see in chapter 5.
- ▷ For a polycrystalline sample or a frozen solution, the anisotropy of the magnetic moment causes *the resonance lines to spread over* the field range determined by the smallest and largest of the three numbers (g_x, g_y, g_z). If all orientations of the molecules are possible and equally probable, the spectrum does not depend on the orientation of the sample relative to **B**. This type of spectrum is termed a "powder spectrum".
- ▷ In contrast to what might be expected, it is often easier to determine the numbers (g_x, g_y, g_z) for a powder spectrum than for single crystals where the molecules are perfectly ordered. The field of application for EPR spectrometry is thus considerably extended. However, a sample in which the molecules are *ordered* is required to determine the orientation of the principal axes of the $\tilde{\mathbf{g}}$ matrix for the molecule.
- ▷ The curve representing $\mathrm{d}s/\mathrm{d}B$ is such that the area under its positive part is equal to the area under its negative portion (figures 4.5b and 4.7b). Indeed, the total algebraic area under this curve is equal to $s(\infty) - s(0)$. As these two quantities are null, the area is null. This property can be used to verify that the spectrum was recorded over a sufficiently broad field range and to ensure that it is not deformed by underlying signals produced by other paramagnetic centres.
- ▷ The features of a powder spectrum due to the Zeeman term in the Hamiltonian appear for the values (B_x, B_y, B_z) where the density of the resonance lines changes suddenly (figures 4.5 and 4.7). More generally, we

will see in chapters 6 and 7 that the features produced near to (B_x, B_y, B_z) can often be interpreted by calculating the position of the resonance lines for the three "canonical directions" of the field, i.e., those where one of the principal axes of the $\tilde{\mathbf{g}}$ matrix is parallel to **B**. Although the lines at B_y are produced by molecules with various orientations relative to **B** (complement 2), the features emerging at this field value are mainly determined by molecules for which the *y* axis is close to the direction of **B**.

4.4 – How anisotropic hyperfine interaction affects the shape of the EPR spectrum

In the *isotropic* regime, interaction between the unpaired electrons in a molecule and a nucleus of spin *I* creates a pattern of $(2I + 1)$ equidistant lines on the EPR spectrum. This pattern is centred on the position that the line would have in the absence of interaction (section 2.3.4). We will see that when the hyperfine interaction is anisotropic, the *position* and *splitting* of the hyperfine structure depends on the orientation of the field relative to the molecule.

4.4.1 – The hyperfine matrix $\tilde{\mathbf{A}}$

The interaction between the unpaired electrons from a paramagnetic centre of spin *S* and a nucleus of spin *I* can be described by a spin operator of the form:

$$\hat{H}_{hyperfine} = \mathbf{S} \cdot \tilde{\mathbf{A}} \cdot \mathbf{I}$$

where $\tilde{\mathbf{A}}$ is the "hyperfine matrix". We describe hereafter the characteristics of this matrix that will be used to interpret spectra.

◇ *Symmetry properties of the paramagnetic centre and the $\tilde{\mathbf{A}}$ matrix*

Unlike the $\tilde{\mathbf{g}}$ matrix, which depends on the overall distribution of unpaired electrons, $\tilde{\mathbf{A}}$ is a *local* quantity which is determined by the portion of the distribution which is close enough to the nucleus to interact with it. Its properties are therefore only linked to the symmetry of the paramagnetic centre if the centre is a transition ion complex and the nucleus is that of the cation. In this specific case, we can paraphrase what was stated for the $\tilde{\mathbf{g}}$ matrix (section 4.2.1):

- ▷ When the complex has no axis of symmetry, the three principal values for the $\tilde{\mathbf{A}}$ matrix are generally different.

▹ When the complex has an axis of symmetry, it is one of the principal axes of the $\tilde{\mathbf{A}}$ and $\tilde{\mathbf{g}}$ matrices. If this Z axis is of an order greater than or equal to 3, the principal values A_X and A_Y are equal. We can then arbitrarily choose X and Y (orthogonal) in the plane perpendicular to Z, and we note:

$$A_Z = A_{//} \ , \ A_X = A_Y = A_{\perp}$$

▹ For a complex with octahedral or tetrahedral symmetry, the three principal values are equal: $A_X = A_Y = A_Z = A$, and we can write:

$$\tilde{\mathbf{A}} = A\,\tilde{\mathbf{1}}$$

where $\tilde{\mathbf{1}}$ is the unit matrix. In this case, the $\tilde{\mathbf{A}}$ matrix is said to be *isotropic*.

Beyond this specific case, the symmetry properties of the $\tilde{\mathbf{A}}$ matrix reflect those of the local distribution of electrons "perceived by the nucleus", which generally differ from those of the paramagnetic centre considered as a whole.

◇ *Principal values of the $\tilde{\mathbf{A}}$ matrix*

They take the form:

$$A_i = A_s + (A_{dip})_i \quad i = X, Y, Z$$

▹ The term A_s is due to the interaction between the nucleus and the electrons in the *s* orbitals of its own atom. In section 2.2.2 we described the "core polarisation" and "spin polarisation" mechanisms which create spin density in the *s* orbitals, and produce a term A_s which is proportional to the *spin population* at the nucleus.

▹ The terms $(A_{dip})_i$ are due to the dipolar interactions between the magnetic moment of the nucleus and the magnetic moments of the unpaired electrons. As these interactions are inversely proportional to the cube of the distance $(1/r^3)$, dipolar terms are also determined by the *local* spin density.

As a real example of the foregoing, consider the complex $Cu(NH_3)_4^{2+}$. We note z the 4-fold axis of symmetry and x, y the metal-ligand directions. The Cu^{2+} ion ($3d^9$) is characterised by $S = ½$ and its nucleus by $I_{Cu} = 3/2$ (table 2.2). In the ground state of the complex, the unpaired electron is mainly localised in the $d_{x^2-y^2}$ orbital of the Cu^{2+} ion, but part of the spin density is delocalised on the ligands. The Hamiltonian therefore includes two types of hyperfine operators:

1. Interaction of the Cu nucleus with the unpaired electron is described by an operator $\mathbf{S} \cdot \tilde{\mathbf{A}}^{Cu} \cdot \mathbf{I}_{Cu}$, where $I_{Cu} = 3/2$. The term A_s^{Cu} is due to the core polarisation mechanism, and the terms $(A_{dip}{}^{Cu})_i$ are the result of dipolar

interactions with the electron in the $d_{x^2-y^2}$ orbital. As the x and y directions are equivalent with respect to these interactions, the $\tilde{\mathbf{A}}^{Cu}$ matrix is axial, with z as the symmetry axis.

2. Interaction of the ^{14}N nucleus in a NH_3 ligand with the unpaired electron is described by an operator $\mathbf{S}\cdot\tilde{\mathbf{A}}^{N}\cdot\mathbf{I}_N$ where $I_N = 1$. As the unpaired electron is partially delocalised onto the 2s and $2p_\sigma$ orbitals of the nitrogen, the contact and core polarisation mechanisms produce a term A_s^N and the dipolar interaction with the $2p_\sigma$ orbital gives rise to the axially symmetric terms $(A_{dip}^N)_i$, with the Cu–N direction as the axis of symmetry. There also exists a minor contribution due to the dipolar interaction between the nitrogen nucleus and the $d_{x^2-y^2}$ orbital.

The interactions with the nucleus of the transition ion and the nuclei of the ligands are, respectively, said to create the "hyperfine pattern" and the "superhyperfine pattern" on the spectrum.

◇ *The two contributions to the dipolar terms*

The dipolar terms $(A_{dip})_i$ are produced by interactions between the magnetic moment of the nucleus and the *orbital* and *spin* magnetic moments of the unpaired electrons. Since the principal values of the $\tilde{\mathbf{g}}$ matrix have the following form (section 4.2.2)

$$g_i = g_e + \Delta g_i \quad i = x, y, z$$

where Δg_i and g_e represent the contributions of the orbital and spin magnetic moments of the electrons, the orbital magnetic moments might be expected to give dipolar terms proportional to Δg_i. Detailed calculation performed on free radicals and transition ion complexes show that this is effectively the case. A simple example is dealt with in appendix 2 in the context of ligand field theory. The orbital contribution is negligible in the case of free radicals for which the (g_x, g_y, g_z) components are very close to g_e. In transition ion complexes, this contribution can play an important role in the interaction with the nucleus of the transition ion, but also with the nuclei of the ligands that are often studied by ENDOR spectroscopy [Atherton and Horsewill, 1980].

When the orbital contribution *is negligible*, the principal values of the $\tilde{\mathbf{A}}$ matrix can simply be written:

$$A_i = A_s + T_{ii} \quad i = X, Y, Z \qquad [4.7]$$

where the T_{ii} are the principal values of the dipolar $\tilde{\mathbf{T}}$ matrix characterising the dipolar interactions between the nuclear magnetic moment and the *spin magnetic moments* of the unpaired electrons (appendix 3). The trace of this matrix is null:

$$T_{XX} + T_{YY} + T_{ZZ} = 0 \qquad [4.8]$$

We will see that the A_{iso} quantity, which is involved in the isotropic regime, is the mean of the principal values of the $\tilde{\mathbf{A}}$ matrix (equation [4.22]). Equations [4.7] and [4.8] show that $A_{iso} = A_s$. This relation was mentioned in section 2.2.3 for the case of free radicals.

◇ *The case of complexes of cations in an S state*

For ions with a $3d^5$ configuration such as Mn^{2+} and Fe^{3+}, the dipolar terms are null to the second order in perturbation theory and the $\tilde{\mathbf{A}}$ matrix should be written $A_s\tilde{\mathbf{1}}$ (appendix 2). In practice, mild anisotropy is observed due to higher order terms.

4.4.2 – Expression of the resonance field in the presence of anisotropic hyperfine interaction

We will now examine how anisotropy of the hyperfine interactions affects the shape of the EPR spectrum. Consider a paramagnetic centre characterised by a spin S and a $\tilde{\mathbf{g}}$ matrix; the unpaired electrons from this centre interact with a nucleus of spin I. When the centre is placed in a field $\mathbf{B}$, the Hamiltonian for the {paramagnetic centre-nucleus} system is written:

$$\begin{aligned}\hat{H} &= \hat{H}_{Zeeman} + \hat{H}_{hyperfine}\\ &= \beta\,\mathbf{B}\cdot\tilde{\mathbf{g}}\cdot\mathbf{S} + \mathbf{S}\cdot\tilde{\mathbf{A}}\cdot\mathbf{I}\end{aligned}$$

For simplicity we assume that the principal axes of the $\tilde{\mathbf{g}}$ and $\tilde{\mathbf{A}}$ matrices are identical and name them $\{x, y, z\}$. By expressing the scalar products in this system of axes, we obtain:

$$\begin{aligned}\hat{H}_{Zeeman} &= \beta(g_x B_x \hat{S}_x + g_y B_y \hat{S}_y + g_z B_z \hat{S}_z)\\ \hat{H}_{hyperfine} &= (A_x \hat{S}_x \hat{I}_x + A_y \hat{S}_y \hat{I}_y + A_z \hat{S}_z \hat{I}_z)\end{aligned}$$

The precise expression of the eigenvalues and eigenvectors for the Hamiltonian $\hat{H}$ cannot be determined in the most general case, but approximations can be obtained by applying perturbation theory in the very frequent situation where $\hat{H}_{hyperfine}$ is much smaller than $\hat{H}_{Zeeman}$. The calculation is performed in complement 4 for any direction of **B**. Here, we present the different steps in the specific case where the magnetic field is parallel to one of the principal axes, e.g. the z axis.

1. The $\hat{H}_{Zeeman}$ operator is written $g_z\beta B\hat{S}_z$. As basis, we select the eigenvectors $\{|S, M_S\rangle\}$ shared by $(\hat{\mathbf{S}}^2, \hat{S}_z)$ and the eigenvectors $\{|I, M_I\rangle\}$ shared by $(\hat{\mathbf{I}}^2, \hat{I}_z)$. The $\hat{H}_{Zeeman}$ operator is represented by a diagonal matrix in the basis $\{|S, M_S\rangle|I, M_I\rangle\}$, where M_I can take the values $\{-I, -I+1, \dots I\}$. The eigenvalues $E(M_S) = g_z\beta BM_S$ are therefore $(2I+1)$-fold degenerate.
2. The second step consists in constructing, for each value of M_S, the matrix representing the perturbation operator $\hat{H}_{hyperfine}$ in the sub-space $\{|S, M_S\rangle|I, M_I\rangle\}$, and then seeking its eigenvalues. For a set value of M_S, the elements of this matrix take the form:

$$\langle S, M_S|\langle I, M_I|\, A_x\hat{S}_x\hat{I}_x + A_y\hat{S}_y\hat{I}_y + A_z\hat{S}_z\hat{I}_z\, |S, M_S\rangle|I, M_I'\rangle$$

According to equations [3.9], the operators $\hat{S}_x$ and $\hat{S}_y$ do not contribute to the matrix elements. The remaining term $A_z\hat{S}_z\hat{I}_z$ produces a *diagonal* matrix in which the diagonal elements are the eigenvalues. To first order in perturbation theory, the eigenvalues of $\hat{H}$ are therefore given by:

$$E(M_S, M_I) = g_z\beta BM_S + A_z M_S M_I \qquad [4.9]$$

We naturally obtain symmetric expressions when **B** is parallel to x or y.

For *any direction* of **B** defined by the unit vector **u** (u_x, u_y, u_z), the possible energies of the paramagnetic centre take the form (complement 4):

$$E(M_S, M_I) = g'\beta BM_S + A'M_S M_I \qquad [4.10]$$

where g' is given by equation [4.3] which we recall here:

$$g' = [g_x^2\, u_x^2 + g_y^2\, u_y^2 + g_z^2\, u_z^2]^{1/2}$$

and A' by the following relation:

$$A' = [A_x^2\, (g_x/g')^2\, u_x^2 + A_y^2\, (g_y/g')^2\, u_y^2 + A_z^2\, (g_z/g')^2\, u_z^2]^{1/2} \qquad [4.11]$$

The parameter A' is *positive*, whereas the principal value A_z, which appears in equation [4.9], is an *algebraic* quantity. This is only an apparent contradiction, as we saw in section 2.3.2 that the EPR spectrum can give only the *absolute value* of the hyperfine constants. We can therefore assume, for simplicity, that A' is positive, as is the case in equation [4.11].

The transitions allowed between energy levels are determined by the selection rule

$$\Delta M_S = \pm 1, \Delta M_I = 0$$

The transitions from energy level $E(M_S, M_I)$ therefore can only take place towards the levels $E(M_S + 1, M_I)$ and $E(M_S - 1, M_I)$. Equation [4.10] shows that these two transitions have the same energy:

$$\Delta E(M_I) = g'\beta B + A'M_I$$

For these transitions, identified by (M_I), resonance occurs when the value of B is such that $\Delta E(M_I) = h\nu$, where ν is the spectrometer's frequency. The resonance field is therefore equal to:

$$B(M_I) = B_0' - (A'/g'\beta)\, M_I \qquad M_I = -I, -I+1, \dots I \qquad [4.12]$$

where $B_0' = h\nu/g'\beta$. The "prime" symbol reminds us that this value depends on the *direction* of **B**. Therefore, a pattern is produced with $(2I + 1)$ equidistant lines centred on position B_0', which would be the line's position in the absence of interaction, as in the case where the magnetic moment and the hyperfine interaction are isotropic (equation [2.8]). But because of the anisotropy of the $\tilde{\mathbf{g}}$ and $\tilde{\mathbf{A}}$ matrices, the *position of this pattern* determined by g' and its *splitting* $A'/g'\beta$ depend on the *direction* of **B** relative to the axes $\{x, y, z\}$. The consequences of this anisotropy depend on the nature of the sample:

- ▹ In a *single crystal*, the molecules that are related by translation are oriented in the same direction relative to **B** (figure 4.4). These molecules produce a pattern with $(2I + 1)$ lines for which the position and the splitting vary depending on the orientation of the crystal relative to **B**. The spectrum has as many patterns as there are molecules oriented differently within the unit cell.
- ▹ If the sample is a polycrystalline powder or a frozen solution, the superposition of the hyperfine patterns given by all the molecules randomly oriented relative to **B** results in a "powder spectrum". We will see that its shape is strongly dependent on the *relative anisotropy* of the $\tilde{\mathbf{g}}$ and $\tilde{\mathbf{A}}$ matrices.

4.4.3 – Effect of $\tilde{\mathbf{g}}$ and $\tilde{\mathbf{A}}$ matrix anisotropy on the shape of the powder spectrum

We will use the results from the previous section to examine the limit situations where one of the matrices is isotropic and the other not.

◇ *The $\tilde{\mathbf{g}}$ matrix is anisotropic and the $\tilde{\mathbf{A}}$ matrix is isotropic*

We assume that:

$$g_x \neq g_y \neq g_z$$
$$A_x = A_y = A_z = A$$

Equation [4.11] indicates that $A' = A$ whatever the molecule's orientation relative to **B**. Equation [4.12] is therefore written:

$$B(M_I) = (h\nu - AM_I)/g'\beta$$

This expression is equivalent to equation [4.4] which gives the resonance field in the absence of hyperfine interaction, on the condition that $h\nu$ is replaced by $(h\nu - AM_I)$. The *density $D_{M_I}(B)$ of the resonance lines* for this transition is therefore identical to the density $D(B)$ shown in figure 4.7a, provided $B_x = h\nu/g_x\beta$, $B_y = h\nu/g_y\beta$, $B_z = h\nu/g_z\beta$ are replaced by $B_x(M_I)$, $B_y(M_I)$, $B_z(M_I)$, respectively, which are defined by:

$$B_i(M_I) = B_i - AM_I/g_i\beta \quad i = x, y, z$$

The total density of the resonance lines, which determines the features of the spectrum, is obtained by adding the $(2I + 1)$ densities $D_{M_I}(B)$. This total density is represented by a continuous line in figure 4.9a for $I = 1$, and the corresponding spectrum profile is shown in figure 4.9b. The spectrum contains patterns of three hyperfine lines centred at B_x, B_y and B_z, separated by $A/g_x\beta$, $A/g_y\beta$, $A/g_z\beta$, respectively.

When the principal values (A_x, A_y, A_z) are different, the shape of the spectrum is similar to that shown in figure 4.9 as long as the splittings $A_x/g_x\beta$, $A_y/g_y\beta$, $A_z/g_z\beta$ remain weak relative to the differences between B_x, B_y, B_z. In this case, the parameters (g_x, g_y, g_z) and (A_x, A_y, A_z) can be directly measured on the spectrum. Otherwise, the patterns overlap and numerical simulation must be used to determine these parameters. As the differences between B_x, B_y and B_z are proportional to the frequency of the spectrometer, the hyperfine patterns can be separated by recording the spectrum at a higher frequency (figure 9.6).

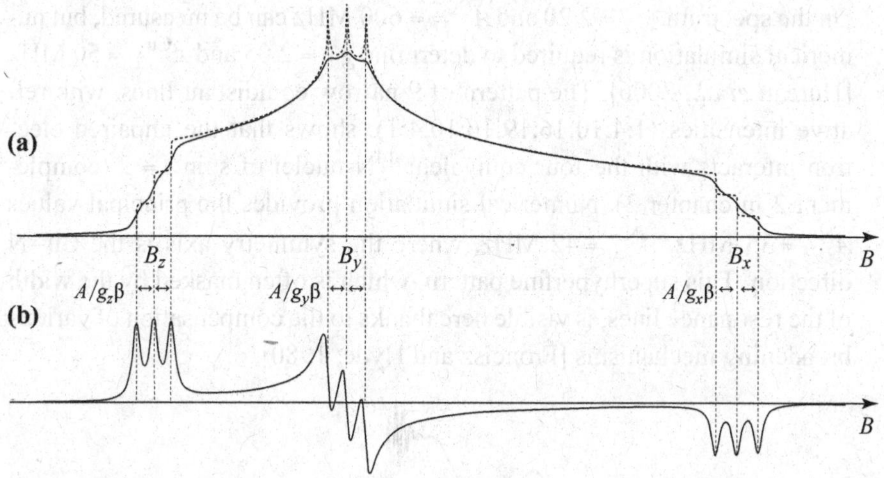

Figure 4.9 – Effect of a hyperfine interaction with a nucleus of spin $I = 1$ on the spectrum. The $\tilde{g}$ matrix is anisotropic, the $\tilde{A}$ matrix is isotropic. **(a)** Total density of the resonance lines (dashed line) and absorption signal (continuous line). **(b)** The spectrum.

To illustrate this situation, we will present two examples:

▷ Figure 4.10 represents the calculated Q-band ($\nu = 35$ GHz) spectrum for a complex where the Cu^{2+} ion is coordinated to 5 ligands in a very distorted geometry, with the parameters $g_x = 2.02$, $g_y = 2.16$, $g_z = 2.25$ and $A_x = 179$ MHz, $A_y = 151$ MHz, $A_z = 315$ MHz [Banci *et al.*, 1981]. The principal values of the $\tilde{g}$ matrix are greater than g_e, as predicted by ligand field theory for a $3d^9$ ion.

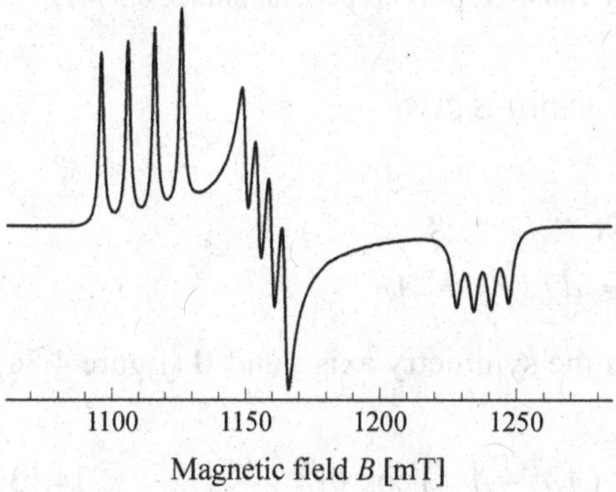

Figure 4.10 – Q-band spectrum for a Cu^{2+} complex with "rhombic" symmetry, calculated for $\nu = 35.0$ GHz with $g_x = 2.02$, $g_y = 2.16$, $g_z = 2.25$, $A_x = 179$ MHz, $A_y = 151$ MHz, $A_z = 315$ MHz. These parameters are those of a complex with distorted trigonal bipyramidal geometry [Banci *et al.*, 1981].

▷ Figure 4.11 represents the X-band spectrum for a complex with square planar geometry in which a Cu^{2+} ion is coordinated with 4 nitrogen atoms.

On the spectrum, $g_{//} = 2.20$ and $A^{Cu}{}_{//} = 600$ MHz can be measured, but numerical simulation is required to determine $g_{\perp} = 2.05$ and $A^{Cu}{}_{\perp} = 50$ MHz [Hureau *et al.*, 2006]. The pattern of 9 narrow equidistant lines, with relative intensities (1:4:10:16:19:16:10:4:1), shows that the unpaired electron interacts with the four equivalent ^{14}N nuclei of spin $I = 1$ (complement 2 in chapter 2). Numerical simulation provides the principal values $A^{N}{}_{//} = 35$ MHz, $A^{N}{}_{\perp} = 42$ MHz, where the symmetry axis is the Cu–N direction. This superhyperfine pattern, which is often masked by the width of the resonance lines, is visible here thanks to the compensation of various broadening mechanisms [Froncisz and Hyde, 1980].

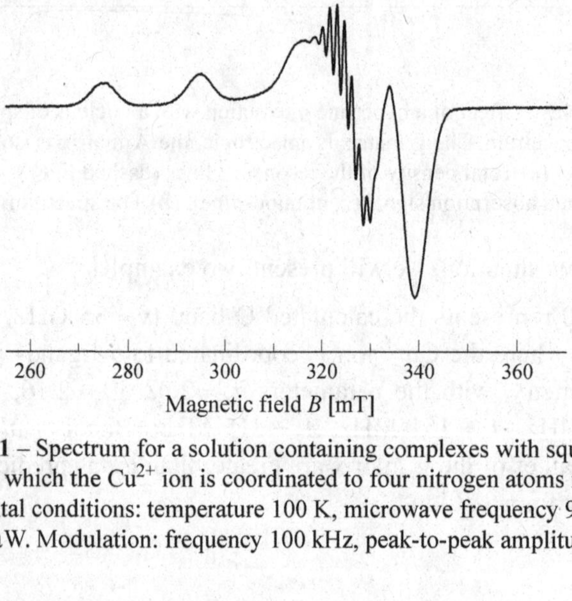

Figure 4.11 – Spectrum for a solution containing complexes with square planar geometry in which the Cu^{2+} ion is coordinated to four nitrogen atoms in a peptide. Experimental conditions: temperature 100 K, microwave frequency 9.380 GHz, power 30 µW. Modulation: frequency 100 kHz, peak-to-peak amplitude 0.5 mT.

◇ *The $\tilde{\mathbf{g}}$ matrix is isotropic and the $\tilde{\mathbf{A}}$ matrix is axial*

In this case, we have:

$$g_x = g_y = g_z = g$$

$$A_x = A_y = A_{\perp},\ A_z = A_{//}$$

If we introduce the angle θ between the symmetry axis *z* and **B** (figure 4.3a) into equation [4.11], it becomes:

$$A'(\theta) = [A_{\perp}^2 + (A_{//}^2 - A_{\perp}^2)\cos^2\theta]^{1/2} \quad [4.13]$$

and equation [4.12] can then be written:

$$B(M_I) = B_0 - M_I A'(\theta)/g\beta \quad [4.14]$$

where $B_0 = h\nu/g\beta$ is the position of the line in the absence of interaction. The *density* $D_{M_I}(B)$ for the resonance lines with positions defined by equations [4.13] and [4.14] can be calculated in a similar manner to that presented in complement 3 (exercise 4.3).

For $A_{//} > A_\perp$, the following normalised density is obtained:

$$D_{M_I}(B) = (2B_0 - B_\perp - B_{//})^{-1/2}(B_\perp - B_{//})^{-1/2} \{(B_0 - B)/[(2B_0 - B_\perp - B)^{1/2}(B_\perp - B)]^{1/2}\} \quad [4.15]$$

where:

$$B_{//} = B_0 - A_{//}M_I/g\beta;\ B_\perp = B_0 - A_\perp M_I/g\beta$$

For $A_\perp > A_{//}$, $B_\perp$ and $B_{//}$ must be switched, as must $B_\perp$ and B in the terms containing square roots. For $I = 1$, the values of $B_{//}$ and $B_\perp$ are given for the 3 transitions by:

$$M_I = -1: B_{//} = B_0 + A_{//}/g\beta;\ B_\perp = B_0 + A_\perp/g\beta$$

$$M_I = 0: B_{//} = B_\perp = B_0$$

$$M_I = 1: B_{//} = B_0 - A_{//}/g\beta;\ B_\perp = B_0 - A_\perp/g\beta$$

The transitions ($M_I = 1$) and ($M_I = -1$) give symmetric densities relative to B_0, and the transition ($M_I = 0$) gives a single line at B_0 (figure 4.12a). As the areas under the 3 densities are equal, the amplitude of the central line is much greater than that of the lateral features which spread over a field range equal to $|A_\perp - A_{//}|/g\beta$. The spectral profile is represented in figure 4.12b.

Although the spectra presented in figures 4.9b and 4.12b are both produced by a paramagnetic centre of spin $S = 1/2$ coupled to a nucleus of spin $I = 1$, their shapes are very different. In particular, the pattern of three hyperfine lines is absent from the spectrum in figure 4.12b. The main cause of anisotropy, which determines the field range over which the spectrum extends, has a different effect on the *shape* depending on whether it is due to the $\tilde{g}$ matrix or the $\tilde{A}$ matrix.

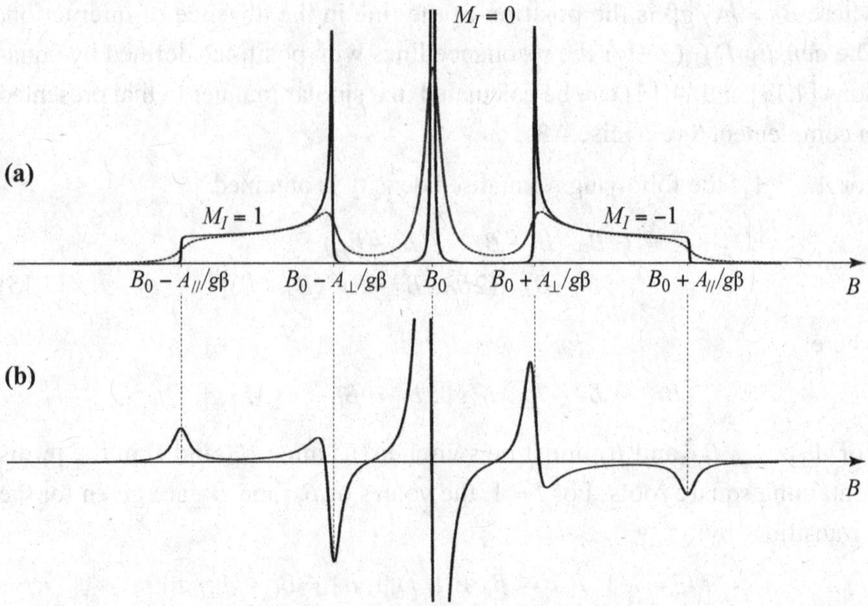

Figure 4.12 – Effect of the hyperfine interaction with a nucleus of spin $I = 1$ on the spectrum. The $\tilde{\mathbf{g}}$ matrix is isotropic and the $\tilde{\mathbf{A}}$ matrix is axial with $A_{//} > A_\perp$. **(a)** Density of the resonance lines (continuous line) and absorption signal (dashed line). **(b)** The spectrum. The lines are attributed for $A_{//}$ and $A_\perp > 0$.

- In the case of free radicals, the anisotropy of the $\tilde{\mathbf{g}}$ matrix is weak. At X-band ($\nu \approx 9$ GHz), the shape of the spectrum at low temperature is therefore mainly determined by the anisotropy of the $\tilde{\mathbf{A}}$ matrix (or matrices). However, a weak anisotropy of the $\tilde{\mathbf{g}}$ matrix is enough to "confuse" this shape, as shown in figure 4.13 where the 3 components from figure 4.12 are represented for a nitroxide radical. Their sum, the shape of which is similar to the spectrum shown in figure 4.7, is a good reproduction of the experimental spectrum in figure 2.6c. The spectrum is much simpler to interpret when it is recorded at high frequency [Möbius *et al.*, 2005]. The spectacular effect of the change in frequency on the spectral resolution for radicals is also clearly apparent in figures 9.5 and 9.6.
- For transition ion complexes a diversity of situations can be encountered at X-band. In the first transition series, the anisotropy of the $\tilde{\mathbf{g}}$ matrix dominates in Cu^{2+} complexes, the two anisotropies are comparable in V^{4+} complexes, and that of the $\tilde{\mathbf{A}}$ matrix is largely dominant in complexes containing manganese ions.

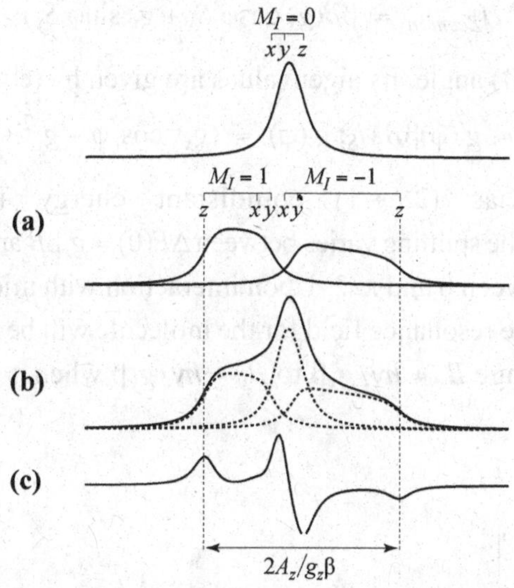

Figure 4.13 – Interpretation of the shape of the X-band spectrum for a frozen solution of a nitroxide radical characterised by $g_x = 2.0089$, $g_y = 2.0064$, $g_z = 2.0027$, $A_x = A_y = 14$ MHz, $A_z = 98$ MHz. **(a)** Shape of the three components from figure 4.12a. For each value of M_I, "stick diagrams" identify the position of the lines for the canonical directions of the field. **(b)** Sum of the three components. **(c)** The spectrum. The lines are assumed to be Lorentzian.

4.5 – How molecular movements affect the spectrum: isotropic and very slow motion regimes

Up to now, we have assumed that the paramagnetic molecules were *frozen* in the sample placed in the field **B**. However, in a crystal, a frozen solution and even more so in a liquid solution, the molecules are in motion. This motion causes their orientation relative to the field to vary, which can alter the shape of the spectrum. Indeed, we have already indicated that the effects of anisotropy *disappear* from the EPR spectrum in the *isotropic regime* when these movements allow the molecules to rapidly explore all the possible orientations relative to the field with an equal probability (section 2.2.3).

4.5.1 – A hypothetical experiment

Consider a paramagnetic molecule of spin S characterised by its $\tilde{g}$ matrix. This molecule is placed in a magnetic field **B** of variable magnitude, but with a fixed direction perpendicular to the principal axis z. Its Hamiltonian can be written:

$$\hat{H}_{Zeeman} = \beta B(g_x \cos\varphi\, \hat{S}_x + g_y \sin\varphi\, \hat{S}_y) \quad [4.16]$$

where φ is the (x, **B**) angle. Its eigenvalues are given by (equation [4.3]):

$$E(M_S) = g'(\varphi)\beta B M_S;\ g'(\varphi) = (g_x^2 \cos^2\varphi + g_y^2 \sin^2\varphi)^{1/2}$$

The molecule has $(2S + 1)$ equidistant energy levels split by $\Delta E(\varphi) = g'(\varphi)\beta B$. The splitting varies between $\Delta E(0) = g_x\beta B$ and $\Delta E(\pi/2) = g_y\beta B$ when φ varies between 0 and $\pi/2$. Upon interaction with microwave radiation at a frequency ν, the resonance field for the molecule will be $B(\varphi) = h\nu/g'(\varphi)\beta$, which is in the range $B_x = h\nu/g_x\beta$ to $B_y = h\nu/g_y\beta$ when φ varies between 0 and $\pi/2$.

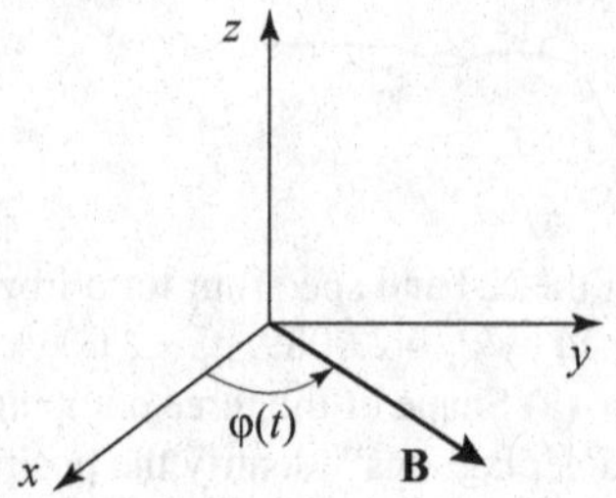

Figure 4.14 – A molecule with magnetic axes $\{x,y,z\}$, placed in a field **B** with a fixed direction, oscillates around the z axis perpendicular to **B**.

Now assume that the molecule oscillates around the z axis such that $\varphi(t)$ varies between 0 and $\pi/2$ with a period T (figure 4.14). As the molecule is subjected to a time-dependent Hamiltonian $\hat{H}_{Zeeman}\,(\varphi(t))$, its spin state changes in a complex manner and the EPR spectrum is difficult to calculate. However its calculation is simple in two limit situations. To define these situations, we must compare two characteristic time constants:

- ▹ $\tau = T/2$, which characterises the rate of reorientation of the molecule.
- ▹ δt, the duration necessary for the separation between energy levels to vary from $\Delta E(0)$ to $\Delta E(\pi/2)$, which is given by the following relation [Ayant and Belorizky, 2000; Cohen-Tannoudji, 2015].

$$|\Delta E(0) - \Delta E(\pi/2)|\delta t = \hbar$$

Thus:

$$\delta t = \hbar/|(g_x - g_y)|\beta B \quad [4.17]$$

This duration depends on the anisotropy of the $\tilde{\mathbf{g}}$ matrix, but also on B, and consequently on the frequency of the spectrometer.

The two limit situations are thus as follows:

- ▷ When τ is much longer than δt, the state of the molecule follows the variations imposed by $\hat{H}_{Zeeman}(\varphi(t))$: at any time t, the splitting of the energy levels is $\Delta E(\varphi(t)) = g'(\varphi(t))\beta B$ and resonance occurs for $B(\varphi(t)) = h\nu/g'(\varphi(t))\beta$. In this "very slow motion" regime, the position of the resonance line is determined by the orientation of the molecule relative to **B**, and the anisotropy of the $\tilde{\mathbf{g}}$ matrix is fully apparent in the spectrum.
- ▷ When τ is much shorter than δt, an averaging effect occurs such that the possible states and the energy levels for the paramagnetic centre are determined by a Hamiltonian similar to that in equation [4.16] but where g_x and g_y are replaced by their average value:

$$g_{iso} = (g_x + g_y)/2 \qquad [4.18]$$

The Hamiltonian [4.16] can therefore be written:

$$\hat{H}_{av} = g_{iso}\beta\,\mathbf{B}\cdot\mathbf{S}$$

Resonance now takes place for $B_{iso} = h\nu/g_{iso}\beta$. In this "very fast motion" regime, the position of the resonance line is independent of the orientation of **B** and the anisotropy effects completely disappear from the spectrum.

These results can be generalised to cover the case where the unpaired electrons also interact with a nucleus of spin I. In this case, the Hamiltonian [4.16] becomes:

$$\hat{H}_{Zeeman} = \beta B\,(g_x \cos\varphi\, \hat{S}_x + g_y \sin\varphi\, \hat{S}_y) + (A_x \hat{S}_x \hat{I}_x + A_y \hat{S}_y \hat{I}_y + A_z \hat{S}_z \hat{I}_z) \qquad [4.19]$$

where we have assumed that the principal axes of the $\tilde{\mathbf{g}}$ and $\tilde{\mathbf{A}}$ matrices are identical. From the results obtained in section 4.4.2, it can readily be shown (exercise 4.4) that, for transition (M_I), the duration δt given by equation [4.17] is replaced by:

$$\delta t(M_I) = \hbar/|(g_x - g_y)\beta B + (A_x - A_y)\,M_I| \qquad [4.20]$$

When the condition $\tau \ll \delta t(M_I)$ holds for all the transitions, the possible states and energy levels for the paramagnetic centre are determined by a Hamiltonian similar to that in equation [4.19], in which g_x and g_y are replaced by the number g_{iso} defined by equation [4.18], and A_x and A_y are replaced by $A_{iso} = (A_x + A_y)/2$. The spectrum is composed of a pattern of $(2I + 1)$ hyperfine lines separated by $A_{iso}/g_{iso}\beta$, and centred on B_{iso}.

4.5.2 – Effects of rotational Brownian motion of paramagnetic molecules

The experiment which we have just described is a simplified view of the situation of molecules in liquid solution. The orientation of these molecules varies continually due to rotational Brownian diffusion. In hydrodynamic models, the rate of reorientation of the molecules is characterised by one or more *correlation times* τ_C. If they are assimilated to spheres of radius r, the single correlation time is given by [Atherton, 1993]:

$$\tau_C = 4\pi\eta r^3/3k_B T \qquad [4.21]$$

where η is the viscosity of the solvent and T is the temperature. The two limit cases defined in the hypothetical experiment can be generalised as follows:

- ▹ When all the molecules in the sample are moving very slowly, their orientation relative to **B** and consequently the value of their resonance field change over time. But, in a sample containing a large number of randomly oriented molecules, all orientations are equally probable at any given time and the density of the resonance lines is therefore not time-dependent. Thus, the shape of the spectrum is identical to that represented in the figures from this chapter. In crystals, the movements of the molecules are correlated but their amplitude is generally too low to cause any changes to appear in the spectrum (in addition, the spectrometer electronics would be unable to keep up).
- ▹ An averaging effect occurs when the motion of the molecules is very fast with respect to all of their interactions, and the Hamiltonian is written

$$\hat{H}_{av} = g_{iso}\,\beta\mathbf{B}\cdot\mathbf{S} + A_{iso}\,\mathbf{S}\cdot\mathbf{I}$$

where g_{iso} and A_{iso} are the means of the principal values of the $\tilde{\mathbf{g}}$ and $\tilde{\mathbf{A}}$ matrices:

$$g_{iso} = (g_x + g_y + g_z)/3;\ A_{iso} = (A_x + A_y + A_z)/3 \qquad [4.22]$$

All the effects of anisotropy disappear from the EPR spectrum: this is the *isotropic regime*. The required rate of reorientation increases with the anisotropy of the principal values of the $\tilde{\mathbf{g}}$ and $\tilde{\mathbf{A}}$ matrices. An example illustrating this point can be given by determining the duration δt calculated using equation [4.20] for three types of paramagnetic centres in which this anisotropy is very different:

- ▹ A *nitroxide radical*. At X-band, the principal contribution to δt is due to the anisotropy of the hyperfine interactions with the ^{14}N nucleus: the $\tilde{\mathbf{A}}$ matrix is practically axial, with principal values close to $A_\perp = 15$ MHz, $A_{//} = 100$ MHz (figure 4.13). Using these numbers, $\delta t = 2 \times 10^{-9}$ s.

▹ An axially symmetric *Cu^{2+} complex* with $g_\perp = 2.0$ and $g_{//} = 2.2$. If the anisotropy of the hyperfine interactions with the Cu nucleus is neglected, $\delta t = 2 \times 10^{-10}$ s for $B = 0.3$ T.

▹ A *Fe^{3+} complex* in the weak spin situation with $g_x = 1.5$, $g_y = 2.2$, $g_z = 3.0$. Using the extreme values g_x and g_z, we obtain $\delta t = 2 \times 10^{-11}$ s for $B = 0.3$ T.

The isotropic regime is achieved when the correlation time τ_C is much shorter than δt. To evaluate τ_C, we use equation [4.21] with $\eta = 10^{-3}$ kg m^{-1} s^{-1}, the viscosity of water at 20 °C.

▹ for a nitroxide radical, $\tau_C = 10^{-10}$ s for $r = 5$ Å.

▹ for a transition ion complex, $\tau_C = 2 \times 10^{-11}$ s for $r = 3$ Å.

These estimates suggest that the isotropic regime is reached at room temperature for the nitroxide radical and the Cu^{2+} complex, but not for the Fe^{3+} complex. To determine whether the spectrum can be *observed* in these conditions, it is also important to consider the *width* of the resonance lines, as we will see in chapter 5.

When the movements are neither *very slow* nor *very rapid*, theoretical models predict intermediate spectral shapes between those for the two limit cases. This is effectively what is observed with solutions of transition ion complexes or free radicals when the temperature is lowered or the viscosity of the solvent is increased. In the case of nitroxide radicals in aqueous solution, the temperature must drop to 100 K for all the effects of anisotropy characterising *very slow* movements to be observable (figure 2.6). In solid medium, paramagnetic centres are generally involved in interactions which extensively limit reorientations relative to **B**, and averaging effects do not occur.

4.6 – Points to consider in applications

4.6.1 – Spectrum for a single crystal

It is often necessary to work on single crystal samples. The need for these types of samples is obvious when studying paramagnetic defects in crystals (impurities, doping agents, vacancies, dangling bonds), but also when investigating radical organic molecules or triplet state molecules generated by irradiation, which have been trapped in molecular crystals at low temperature. Indeed, by studying how the positions of resonance lines vary depending on the orientation of the crystal relative to the magnetic field, the principal axes and principal values (in absolute value) can be determined for the $\tilde{\mathbf{g}}$ and $\tilde{\mathbf{T}}$ matrices. A detailed analysis can be used to deduce the value and sign for the isotropic component A_s and consequently for the spin density [Carrington and McLachlan, 1979; Gordy, 1980].

Numerous studies have been devoted to transition cations or rare earth ions substituted in or inserted into crystals (an example is presented in complement 5). When the crystallographic structure is not overly complex, this method has several advantages over the study of a frozen solution. The structure of the complex is determined by the crystalline environment, even if structural reorganisation can take place. In addition, the spectrum is composed of a limited number of resonance lines, and the broadening caused by unresolved splitting due to intercentre interactions (section 7.3) can be avoided by "magnetic dilution" of the paramagnetic centres. Finally, the detailed study of the spectrum as a function of the crystal's orientation relative to the field can be used to determine not only the principal values of the $\tilde{\mathbf{g}}$ and $\tilde{\mathbf{A}}$ matrices, but also the direction of their principal axes. These directions are required in some applications:

- ▹ By comparing the principal values of the $\tilde{\mathbf{g}}$ and $\tilde{\mathbf{A}}$ matrices for a paramagnetic centre measured on the spectrum to those calculated using a molecular model, the electronic structure of the centre's ground state can be determined. Knowledge of the principal axes provides additional information which can be used to resolve ambiguities. In addition, a detailed interpretation of some pulsed EPR experiments designed to measure weak hyperfine interactions is only possible if the orientation of the magnetic axes relative to the molecule is known [Schweiger and Jeschke, 2001].
- ▹ The EPR spectrum produced by two paramagnetic centres in interaction depends on the relative orientation of the principal axes of their $\tilde{\mathbf{g}}$ matrices (section 7.3). Simulations of the interaction spectrum can help to determine this orientation, and consequently that of the centres, provided the direction of the principal axes relative to the centres is known.
- ▹ Biological membranes contain numerous paramagnetic centres. By studying the variations of the EPR spectrum as a function of the direction of **B** relative to the membrane plane, the orientation of the principal axes of their $\tilde{\mathbf{g}}$ matrix relative to the membrane can be determined. However, the orientation of the centres in the membrane can only be deduced if the direction of the principal axes relative to the centres is known.

4.6.2 – Powder spectrum for centres of spin ½

In this chapter, we discussed the shape of the spectrum produced by a sample containing free radicals or transition ion complexes where the unpaired electrons interact with a nucleus of spin I. In most cases the hyperfine interaction is much weaker than the interaction with the magnetic field, and each molecule produces a pattern of $(2I + 1)$ equidistant lines for which the positions and the splitting are determined by its orientation relative to **B**. Extreme positions are determined by the smallest and largest of the three numbers (g_x, g_y, g_z), which are close to $g_e = 2.0023$, and by (A_x, A_y, A_z) (equation [4.12]). When the sample is a polycrystalline powder or a frozen solution, the powder spectrum resulting from the superposition of multiple patterns is centred near to $B = h\nu/g_e\beta$, i.e., 340 mT at a frequency of 9.40 GHz (figures 4.6 and 4.11). A notable exception is the spectrum produced by complexes of cations with a d^5 configuration in the low-spin situation, in which the anisotropy of the $\tilde{g}$ matrix is particularly strong (table 4.5 and figure 4.8).

When the spin S is greater than ½, we will see that the zero-field splitting terms for the Hamiltonian (appendix 2) either produce new spectral shapes, or cause much more extensive spreading of the resonance lines (chapter 6). This explains why this chapter was illustrated with spectra for centres with spin S = ½. In general, a relatively narrow EPR spectrum centred near to $B = h\nu/g_e\beta$ with a shape similar to those presented in this chapter is typical of a centre of spin ½.

4.6.3 – Spectra for transition ion complexes

The hyperfine structure produced by the nucleus of a transition ion is easy to interpret when the anisotropy of the $\tilde{g}$ matrix considerably exceeds that of the $\tilde{A}$ matrix. It is therefore useful to record the spectrum at a relatively high frequency to avoid overlap between the hyperfine patterns (figure 4.10). The principal values of the $\tilde{g}$ and $\tilde{A}$ matrices deduced by analysing the spectrum directly provide information on the nature of the cation and the structure of the complex (sections 4.2.2 and 4.4.1). The presence of a *superhyperfine* pattern provides information on the identity and possibly the number of ligands (section 4.4.1, figure 4.11). To obtain detailed information on the electronic structure of the complex, these data must be reproduced with a molecular model of the interactions. In the "ligand field" model the unpaired electrons are assumed to be confined to the cation's d orbitals. Based on this hypothesis, it is possible to express the principal values of the $\tilde{\mathbf{g}}$ and $\tilde{\mathbf{A}}$ matrices as a function of a few parameters with simple meanings (appendix 2). However, to achieve satisfactory quantitative agreement, more elaborate models must be used, for example models based on density functional theory (DFT).

4.6.4 – Spectra for free radicals

The spectrum produced by a frozen solution or a polycrystalline sample can be difficult to interpret. When possible, the spectrum should first be recorded in the *isotropic regime* to determine the number and spin of the interacting nuclei and the values of the g_{iso} and A_{iso} parameters (chapter 2). The powder spectrum can be used to study the effects of anisotropy. In these molecules, the anisotropy of the $\tilde{\mathbf{g}}$ matrix is weak (tables 4.2 and 4.3) and the spectrum must therefore be recorded at a very high frequency to allow its principal values to be precisely measured (figures 9.5 and 9.6). In contrast, the principal values of the $\tilde{\mathbf{A}}$ matrix can often be extracted from a spectrum recorded at X-band. By comparing these values to the A_{iso} parameter measured in the isotropic regime, the elements of the dipolar $\tilde{\mathbf{T}}$ matrix can be determined and the spin population deduced (exercise 4.6). When the hyperfine pattern is unresolved on the EPR spectrum, ENDOR (*Electron Nuclear Double Resonance*) spectroscopy or pulsed EPR techniques such as ESEEM (*Electron Spin Echo Envelope Modulation*) and HYSCORE (*Hyperfine Sublevel Correlation Spectroscopy*) must be used.

4.6.5 – The EPR spectrum contains additional information

In this chapter, we deduced the position of the features of the powder spectrum from the *density* of the resonance lines thanks to a purely geometric study based on equations [4.3], [4.4], and [4.12]. However, to analyse the shape of the spectrum in detail (e.g. by simulation) or to exploit the information provided by its intensity, the factors determining the shape, width and intensity *of the resonance lines* must be known. This requires a *physical* study of how the sample's magnetic moments and the microwave radiation interact. This study is presented in chapter 5.

Complement 1 – Splitting of the energy levels for the electrons in an octahedral complex

We will first consider a transition ion surrounded by six negative charges forming a regular octahedron with axes (x, y, z). These negative charges electrostatically repulse the unpaired electrons. The intensity of this repulsion depends on the orbital occupied, but its magnitude is the same, and relatively small, for the (d_{xy}, d_{yz}, d_{xz}) orbitals, the lobes of which are directed *between the charges* (figure 4.15a). It is greater for the d_{z^2} and $d_{x^2-y^2}$ orbitals, the lobes of which are directed *towards the charges* (figure 4.15b), and it can be shown that it has the same magnitude for these two orbitals (figure 4.16a). The energy difference, Δ, between the two groups of orbitals increases when the charges are more negative and closer to the ion. If the two charges located on the z axis are removed to a greater distance, the d_{z^2} and $d_{x^2-y^2}$ orbitals on the one hand, and the d_{xy} and (d_{yz}, d_{xz}) orbitals on the other, become separated (figure 4.16b). When these charges are extensively separated, energy levels equivalent to those of a complex with square planar geometry are obtained (figure 4.16c).

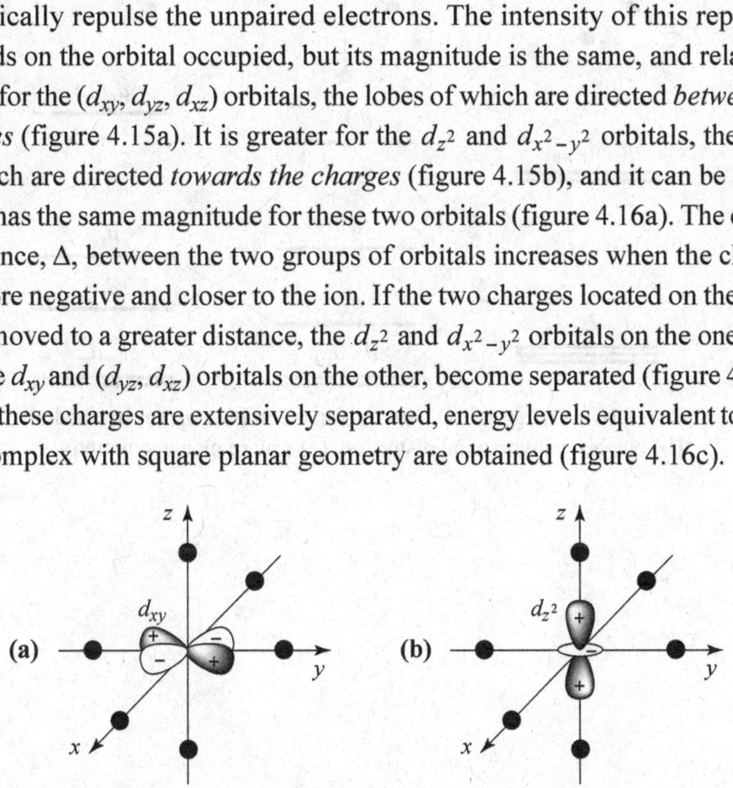

Figure 4.15 – Electrostatic interaction of an electron in a d_{xy} or d_{z^2} orbital with ligands in an octahedral complex.

These considerations, based on symmetry properties, are useful to explain the qualitative effects of electrostatic interactions between the ligands and the cation's d electrons. However, this effect is not reduced to the impact of six negative charges, and it is significantly influenced by the *nature of the ligands*. For example, the value of Δ measured by infrared or UV-visible absorption spectroscopy increases in the "spectrochemical series": Cl^-, F^-, H_2O, NH_3, NO_2^-, CN^-, CO [Gray, 1965]. The CN^- and CO ligands interact very strongly with the cation, producing low-spin complexes.

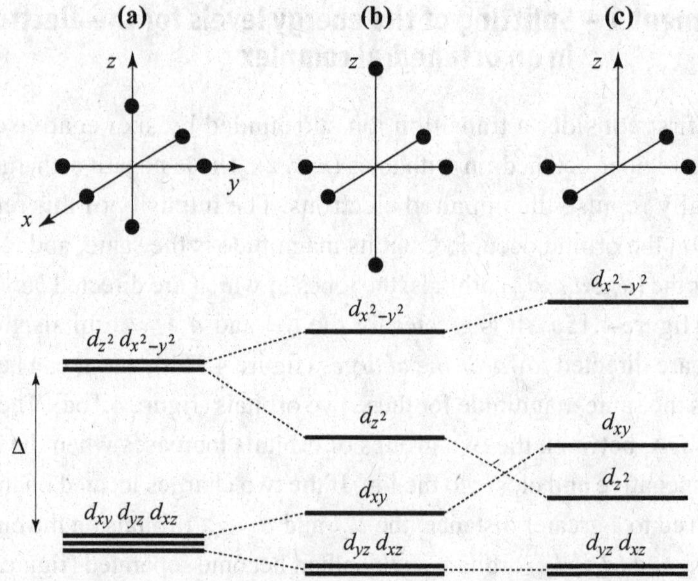

Figure 4.16 – Energy of the *d* orbitals in a complex of: **(a)** octahedral, **(b)** octahedral with axial distortion, **(c)** square planar symmetry.

Complement 2 – Possible values of g' when the $\tilde{g}$ matrix is rhombic

To understand how the parameter

$$g' = [g_x^2\, u_x^2 + g_y^2\, u_y^2 + g_z^2\, u_z^2]^{1/2} \qquad [1]$$

varies depending on the direction of the unit vector $\mathbf{u}(u_x, u_y, u_z)$, *the directions* of $\mathbf{u}$ that make the second side of equation [1] equal to a given value of g' must be identified. If we set $\overrightarrow{OM} = \mathbf{u}$, the problem comes down to seeking the points M for which the (u_x, u_y, u_z) coordinates verify the following two equations:

$$u_x^2 + u_y^2 + u_z^2 = 1 \qquad [2]$$

$$g_x^2\, u_x^2 + g_y^2\, u_y^2 + g_z^2\, u_z^2 = g'^2 \qquad [3]$$

Their geometric interpretation is as follows: M is located at the *intersection* of the sphere of unit radius (equation [2]) and the ellipsoid of revolution, with (x, y, z) axes and principal semi-axes $g'/g_x, g'/g_y, g'/g_z$ (equation [3]). Given the symmetries of the problem, we can focus on the points located in the first octant of the sphere. Hereafter, we assume that the principal axes of the $\tilde{g}$ matrix are labelled such that $g_x < g_y < g_z$.

Figure 4.17 shows how the ellipsoid "dilates" as g' increases from a value less than g_x (figure 4.17a) to a value equal to g_z (figure 4.17f):

- ▹ $g' < g_x$, the ellipsoid *does not intersect* the sphere (figure 4.17a). No direction of $\mathbf{u}$ is appropriate.
- ▹ $g' = g_x$, intersection occurs at point A (figure 4.17b), i.e., for $\mathbf{u}$ parallel to x.

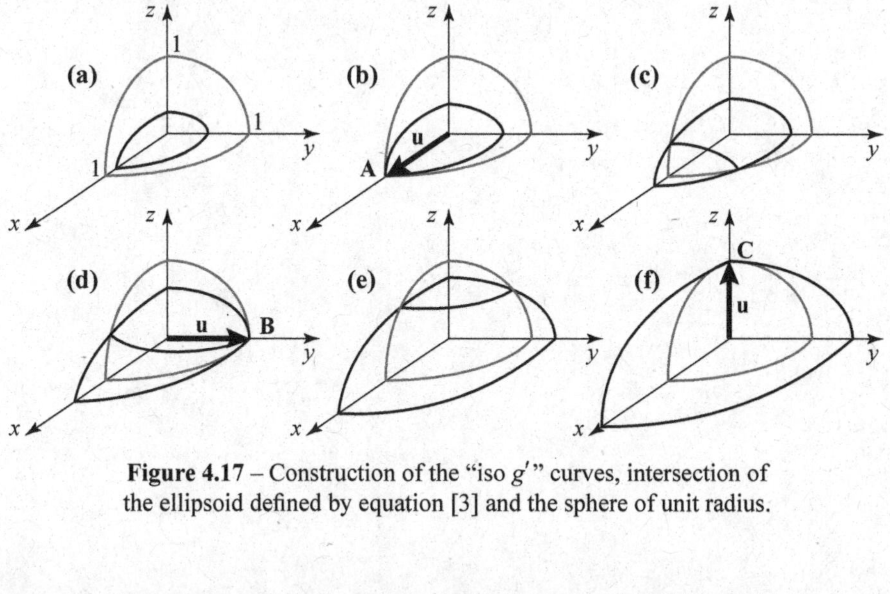

Figure 4.17 – Construction of the "iso g'" curves, intersection of the ellipsoid defined by equation [3] and the sphere of unit radius.

- ▹ $g_x < g' < g_y$, the intersection follows an arc (figure 4.17c): there exists a continuum of directions of **u**, close to the *x* axis, which produce the same value of g'.
- ▹ $g_x < g' = g_y$, point B, which corresponds to **u** parallel to *y*, is obtained along with a whole set of points corresponding to very different directions (figure 4.17d).
- ▹ $g_y < g' < g_z$, the arc of intersection corresponds to directions neighbouring *z* (figure 4.17e).
- ▹ $g' = g_z$, the intersection takes place at point C, i.e., for **u** parallel to *z* (figure 4.17f).
- ▹ $g' > g_z$, the surfaces *do not intersect*.

These results can be summarised as follows:

1. The surfaces only intersect if $g_x \leq g' \leq g_z$.
2. For the lower bound, g_x, and the higher bound, g_z, only one direction is possible (**u** parallel to *x* and **u** parallel to *z*).
3. For values of g' such that $g_x < g' < g_z$, a continuum of directions can satisfy equation [1] and the length of the arc of intersection qualified as "iso g'" gives an idea of their relative number. In particular, there exists a whole ensemble of directions other than **u** parallel to *y* for which $g' = g_y$. In this case, the length of the corresponding "iso g'" arc is maximal (figure 4.17d).

Complement 3 – Expression for the density of resonance lines for a centre with axial symmetry

A sample contains a very large number of molecules randomly oriented relative to **B**, such that all orientations are possible and equally probable. We are interested in the fraction $D(B)\mathrm{d}B$ of molecules for which resonance lines are produced between B and $B + \mathrm{d}B$. It is the same, and conceptually simpler, to consider that a molecule with magnetic axes $\{x, y, z\}$ is placed in a field **B** which can adopt all directions with the same probability, and to consider $D(B)\mathrm{d}B$ as the *probability* that a line will be produced between B and $B + \mathrm{d}B$.

When the centre is axially symmetric, g' only depends on the angle $\theta = (z, \mathbf{u})$, where **u** is the unit vector in the direction of **B** (figure 4.3). Given the symmetries of the problem, we can limit ourselves to making the extremity of **u** sweep out the first octant of the sphere of unit radius, such that θ varies between 0 and $\pi/2$. As all directions of **u** are equally probable, the probability that θ is between θ_0 and $(\theta_0 + \mathrm{d}\theta)$ is equal to the ratio of the area of the elementary spherical segment defined by $(\theta_0, \mathrm{d}\theta)$ (figure 4.18) to the total area of the octant of the sphere, i.e., $\frac{1}{4}(2\pi \sin\theta_0\, \mathrm{d}\theta) / (4\pi/8) = \sin\theta_0\, \mathrm{d}\theta$. The *probability density* of θ is therefore equal to:

$$p(\theta) = \sin\theta \qquad [1]$$

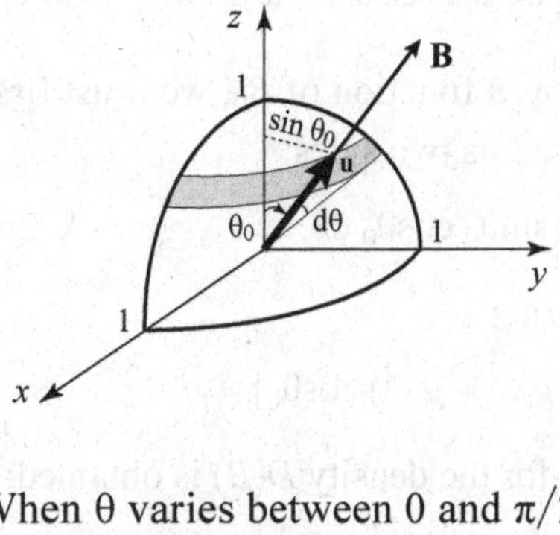

Figure 4.18 – Elementary spherical segment defined by $(\theta_0, \mathrm{d}\theta)$ on an octant of the sphere of unit radius.

When θ varies between 0 and $\pi/2$, the number g' given by equation [4.5], which can be written

$$g' = [g_\perp{}^2 + (g_{//}{}^2 - g_\perp{}^2)\cos^2\theta]^{1/2} \qquad [2]$$

varies from $g_{//}$ to $g_\perp$ and the resonance field

$$B = h\nu/g'\beta \qquad [3]$$

varies from $B_{//} = h\nu/g_{//}\beta$ to $B_\perp = h\nu/g_\perp\beta$ (figure 4.19). Let us assume $g_{//} > g_\perp$. As the curves presented in figure 4.19 are *monotonic*, the probability $D(B_0)\mathrm{d}B$ that a line will be produced between B_0 and $(B_0 + \mathrm{d}B)$ is equal to the probability that θ is between θ_0 and $(\theta_0 + \mathrm{d}\theta)$. We can therefore write:

$$D(B_0)\,\mathrm{d}B = p(\theta_0)\,\mathrm{d}\theta$$

From this, we deduce that:

$$D(B_0) = p(\theta_0)\,(\mathrm{d}\theta/\mathrm{d}B)_0 = p(\theta_0)\,(\mathrm{d}g'/\mathrm{d}B)_0/(\mathrm{d}g'/\mathrm{d}\theta)_0 \quad [4]$$

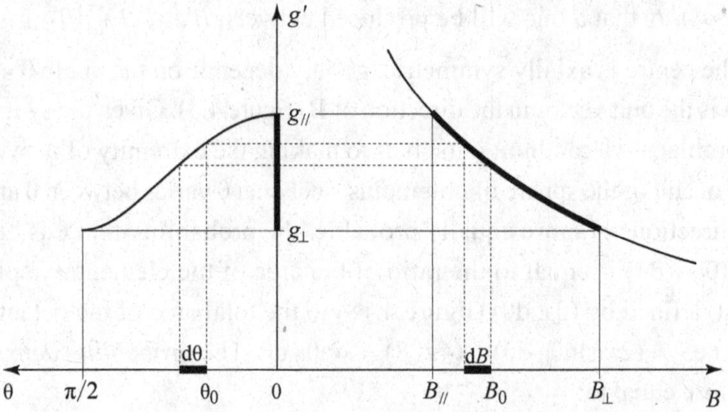

Figure 4.19 – Relations between θ, g' and B (equations [2] and [3]) for $g_{//} > g_\perp$. The fields $B_{//}$ and $B_\perp$ are defined in the text.

To express the second half of equation [4] as a function of B_0, we must first differentiate equation [2] after having squared its two sides:

$$g'_0\,\mathrm{d}g' = -(g_{//}^2 - g_\perp^2)\sin\theta_0\cos\theta_0\,\mathrm{d}\theta$$

Then, using equation [1], it is possible to write:

$$D(B_0) = -(\mathrm{d}g'/\mathrm{d}B)_0\,g'_0/[(g_{//}^2 - g_\perp^2)\cos\theta_0]$$

Using equations [2] and [3], the expression for the density $D(B)$ is obtained:

$$D(B) = B_{//}B_\perp^2(B_\perp^2 - B_{//}^2)^{-1/2}/[B^2(B_\perp^2 - B^2)^{1/2}]$$

Its profile is shown by the continuous line in figure 4.5a. The appropriate expression for $g_{//} < g_\perp$ can be obtained by switching $B_{//}$ and $B_\perp$, and B and $B_\perp$ in the two terms containing square roots. The density $D(B)$ is *normalised*:

$$\int_0^\infty D(B)\,\mathrm{d}B = 1$$

In this calculation, we assumed that the orientation of the molecules relative to **B** could vary continuously. As a result, the value of $D(B)$ for $B = B_\perp$ is infinite, but this singularity disappears as soon as a line width is introduced.

The density of the resonance lines is very useful when seeking to determine the position of the characteristic features of a powder spectrum, but it is not used to *simulate* the experimental spectra. Indeed, we will see in chapter 5 that each line is characterised by a *transition probability*, the value of which depends on the orientation of **B** relative to the molecule. In numerical simulations, the direction of **B** is varied in a discrete manner by cutting the surface of the first octant of the sphere of unit radius into smaller elements, each of which is associated with a direction of **u**. The resonance field corresponding to each direction is then calculated and a line attributed to it. In these simulations the spectrum is obtained by adding all the resonance lines, each weighted by its transition probability (see section 9.5).

Complement 4 – Expression giving the energy levels for any direction of B when the $\tilde{\mathbf{g}}$ and $\tilde{\mathbf{A}}$ matrices are anisotropic

We are interested in finding approximations of the eigenvalues of the Hamiltonian

$$\hat{H} = \hat{H}_{Zeeman} + \hat{H}_{hyperfine}$$

$$\hat{H}_{Zeeman} = \beta(g_x B_x \hat{S}_x + g_y B_y \hat{S}_y + g_z B_z \hat{S}_z)$$

$$\hat{H}_{hyperfine} = (A_x \hat{S}_x \hat{I}_x + A_y \hat{S}_y \hat{I}_y + A_z \hat{S}_z \hat{I}_z)$$

where $\hat{H}_{hyperfine}$ is a *perturbation* relative to $\hat{H}_{Zeeman}$. $\{x, y, z\}$ are the principal axes of the $\tilde{\mathbf{g}}$ and $\tilde{\mathbf{A}}$ matrices, and the direction of **B** is identified by the unit vector **u** which has the components (u_x, u_y, u_z).

▹ The eigenvalues and eigenvectors for the principal term $\hat{H}_{Zeeman}$ are obtained by writing it in the following form (section 3.3.2):

$$\hat{H}_{Zeeman} = g'\beta B\hat{S}_Z$$

Z is the direction of the unit vector **v** with components

$$v_x = (g_x/g')\,u_x,\ v_y = (g_y/g')\,u_y,\ v_z = (g_z/g')\,u_z \qquad [1]$$

and g' is given by:

$$g' = (g_x^2\,u_x^2 + g_y^2\,u_y^2 + g_z^2\,u_z^2)^{1/2} \qquad [2]$$

The eigenvectors of $\hat{H}_{Zeeman}$ take the form $\{|S, M_S\rangle|I, M_I\rangle\}$, where the $\{|S, M_S\rangle\}$ are the eigenvectors shared by $(\hat{\mathbf{S}}^2, \hat{S}_Z)$ and $\{|I, M_I\rangle\}$ is, for the moment, any basis in the $\mathcal{E}_I$ space. The eigenvalues

$$E(M_S) = g'\beta BM_S,\ M_S = -S, -S+1, \ldots S$$

are $(2I + 1)$-fold degenerate. Z is completed by two axes (X, Y) to produce a Cartesian reference frame.

▹ The second step consists, for each value of M_S, in constructing the matrix representing $\hat{H}_{hyperfine}$ in the sub-space $\{|S, M_S\rangle|I, M_I\rangle\}$. Its elements are written:

$$V_{M_S}(M_I, M_I') = \langle S, M_S|\langle I, M_I|\hat{H}_{hyperfine}|S, M_S\rangle|I, M_I'\rangle \qquad [3]$$

To calculate them, $\hat{H}_{hyperfine}$ must be expressed as a function of the operators $(\hat{S}_X, \hat{S}_Y, \hat{S}_Z)$ which have a known action on the kets $\{|S, M_S\rangle\}$. The change in reference frame is defined by:

$$\begin{bmatrix} S_x \\ S_y \\ S_z \end{bmatrix} = \begin{bmatrix} a_{11} & a_{12} & a_{13} \\ a_{21} & a_{22} & a_{23} \\ a_{31} & a_{32} & a_{33} \end{bmatrix} \begin{bmatrix} S_X \\ S_Y \\ S_Z \end{bmatrix} \quad [4]$$

The columns in this matrix are the components of the unit vectors of the $\{X, Y, Z\}$ reference frame in the $\{x, y, z\}$ reference frame. In particular, the last column represents the components of the unit vector along Z, which is none other than the vector $\mathbf{v}$ (equation [1]). We therefore have:

$$a_{13} = v_x,\ a_{23} = v_y,\ a_{33} = v_z \quad [5]$$

Using equations [4], $\hat{H}_{hyperfine}$ can be written:

$$\hat{H}_{hyperfine} = A_x (\ldots + v_x\hat{S}_Z)\,\hat{I}_x + A_y (\ldots + v_y\hat{S}_Z)\,\hat{I}_y + A_z (\ldots + v_z\hat{S}_Z)\hat{I}_z$$

We only specify the terms in $\hat{S}_Z$ as the ket $|S, M_S\rangle$ appears on either side of the matrix element [3], and only these terms give a non-null contribution. We therefore obtain:

$$V_{M_S}(M_I, M_I') = M_S \langle I, M_I|A_x\, v_x\, \hat{I}_x + A_y\, v_y\, \hat{I}_y + A_z\, v_z\, \hat{I}_z|I, M_I'\rangle$$

Using the method described in section 3.3.2, we can construct a basis $\{|I, M_I\rangle\}$ such that the matrix is diagonal. To do so, we must identify the unit vector $\mathbf{w}(w_x, w_y, w_z)$ such that

$$A_x\, v_x\, I_x + A_y\, v_y\, I_y + A_z\, v_z\, I_z = A'(w_x\, I_x + w_y\, I_y + w_z\, I_z) \quad [6]$$

By identification we deduce

$$A'w_x = A_x\, v_x,\ A'w_y = A_y\, v_y,\ A'w_z = A_z\, v_z$$

A' is determined by writing that the sums of the squares of the two sides of these equations are equal:

$$A'^2\,(w_x^2 + w_y^2 + w_z^2) = A_x^2\, v_x^2 + A_y^2\, v_y^2 + A_z^2\, v_z^2$$

As $\mathbf{w}$ is a unit vector, we deduce that:

$$A' = [A_x^2\,(g_x/g')^2\, u_x^2 + A_y^2\,(g_y/g')^2\, u_y^2 + A_z^2\,(g_z/g')^2\, u_z^2]^{1/2} \quad [7]$$

To do this, we used equations [1]. On the second side of equation [6] the value in brackets is the scalar product $\mathbf{w}\cdot\mathbf{I}$ which can be written $I_{Z'}$, where Z' is the direction of $\mathbf{w}$. $V_{M_S}(M_I, M_I')$ then becomes:

$$V_{M_S}(M_I, M_I') = M_S \langle I, M_I|A'\hat{I}_{Z'}|I, M_I'\rangle$$

By choosing the eigenvectors $\{|I, M_I\rangle\}$ shared by $(\hat{\mathbf{I}}^2, \hat{I}_{Z'})$ as the basis of $\mathcal{E}_I$, a diagonal matrix is obtained. The eigenvalues for this matrix are equal to $A' M_S M_I$.

To first order in perturbation theory, the energy levels are therefore written:

$$E(M_S, M_I) = g'\beta B M_S + A' M_S M_I$$

where g' and A' are respectively given by equations [2] and [7].

Complement 5 – An example of a study of a single crystal: identification of the site of Ti^{3+} fluorescence in $LaMgAl_{11}O_{19}$

Crystals of $LaMgAl_{11}O_{19}$ doped with transition ions or rare earth ions are good candidates for the production of tuneable lasers. Their lattice is hexagonal, with a 120 ° angle between the **a** and **b** axes, and a **c** axis perpendicular to the (**a**, **b**) plane. The complex has three Al^{3+} sites with deformed octahedral symmetry noted 2*a*, 4*f* and 12*k*, which can accommodate Ti^{3+} ($3d^1$) ions to form TiO_6 centres. Detailed experiments have shown that only one Ti^{3+} site is fluorescent, and the aim of this study was to identify it [Gourier *et al.*, 1988].

At 20 K, the EPR spectrum for a single crystal in which 1 % of the Al^{3+} ions were replaced by Ti^{3+} is composed of three signals denoted 1, 2, 3. Whereas signals 1 and 2 are visible up to room temperature, signal 3 becomes broader due to relaxation (section 5.4.3) from 100 K and disappears completely at around 130 K. We will see how the detailed study of these signals can allow their attribution to the 2*a*, 4*f* and 12*k* sites.

▹ The first step is to determine the principal values and the principal axes for the $\tilde{g}$ matrices of the three centres. When the direction of **B** is varied in a plane containing the **c** axis, the g' numbers which identify the positions of signals 1 and 3 vary with the angle $\theta = (\mathbf{c}, \mathbf{B})$ according to equation [4.5] (see figure 4.3), producing (figure 4.20a):

$$\text{signal 1: } g_{//} = 1.999, \ g_{\perp} = 1.958$$

$$\text{signal 3: } g_{//} = 1.9623, \ g_{\perp} = 1.7962$$

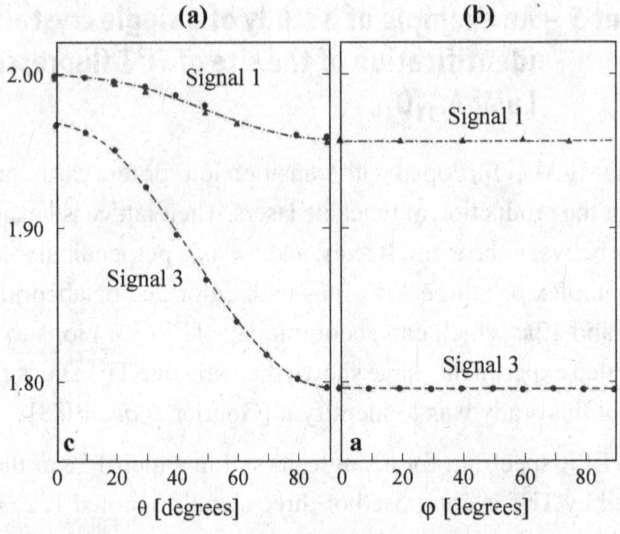

Figure 4.20 – Angular variation of g' for signals 1 and 3 for Ti^{3+} in $LaMgAl_{11}O_{19}$. Circles: 20 K, triangles: 295 K. **(a)** **B** is in a plane which passes through the **c** axis and $\theta = (\mathbf{c}, \mathbf{B})$. **(b)** **B** is in the **(a, b)** plane and $\varphi = (\mathbf{a}, \mathbf{B})$.
[Reproduced with permission from: Gourier D. *et al.*, *Journal of Applied Physics* **63**, 1144–1151 © 1969, American Institute of Physics]

When the direction of **B** varies in the (**a**, **b**) plane, the position of the lines is unaltered (figure 4.20b). The $\tilde{\mathbf{g}}$ matrices for the Ti^{3+} centres producing these signals are therefore axial with axis **c**. As the 3*d* subshell is less than half-full, the principal values are less than $g_e = 2.0023$ (section 4.2.2). Careful examination of the structures of the three sites reveals that only the 2*a* and 4*f* sites are deformed along a 3-fold axis parallel to **c**. We therefore deduce that they produce signals 1 and 3. The deformation of the 12*k* site is quasi axial along the **a** axis, in line with the angular dependence of signal 2. To avoid overloading the figure, the angular dependence of this signal is not represented in figure 4.20.

▹ Signals 1 and 3 remain to be attributed to sites 2*a* and 4*f*. According to ligand field theory, the small difference between the principal values $g_{//} = 1.999$ and $g_{\perp} = 1.958$ (signal 1) relative to $g_e = 2.0023$ suggests strong axial distortion, with a high-energy first excited level at 7000 cm^{-1} [Gourier *et al.*, 1988]. In contrast, the principal values $g_{//} = 1.9623$, $g_{\perp} = 1.7962$ (signal 3) can only be reproduced by assuming a mild axial distortion and a first excited level at around 1200 cm^{-1}. The disappearance of signal 3 due to broadening at low temperatures is in line with the existence of a relatively nearby excited state (section 5.4.3). The 4*f* site is strongly distorted, with three metal-ligand distances equal to 1.86 Å and three others equal to 1.97 Å, whereas the geometry of the 2*a* site is very close to that of a regular octahedron. From this information, the following attribution was proposed:

signal 1 → site 4*f*; signal 3 → site 2*a*

▹ Study of a series of single crystals demonstrates that the level of fluorescence of the Ti^{3+} centres correlates with the intensity of signal 3. This result indicates that only the "regular" 2*a* sites are responsible for the fluorescence.

References

AYANT Y. & BELORIZKY E. (2000) *Cours de mécanique quantique*, Dunod, Paris.

ATHERTON N.M. (1993) *Principles of Electron Spin Resonance*, Ellis Horwood PTR Prentice Hall, New York.

ATHERTON N.M. & HORSEWILL A.J. (1980) *Journal of the Chemical Society Faraday* II **76**: 660-666.

BANCI L. *et al.* (1981) *Inorganic Chemistry* **20**: 393-398.

BENDIX J., BORSON M. & SCHAFFER C.E. (1993) *Inorganic Chemistry* **32**: 2838-2849.

BERTRAND P. *et al.* (1983) *Chemical Physics Letters* **102:** 442-445.

CARRINGTON A. & MCLACHLAN A.D. (1979) *Introduction to Magnetic Resonance with Applications to Chemistry and Chemical Physics*, Chapman & Hall, London.

COSPER M.M. *et al.* (2005) *Inorganic Chemistry* **44**: 1290-1301.

DOLE F. *et al.* (1997) *Journal of the American Chemical Society* **119**: 11540-11541.

FRONCISZ W. & HYDE J.S. (1980) *Journal of Chemical Physics* **73**: 3123-3131.

GORDY W. (1980) *Theory and Applications of Electron Spin Resonance*, John Wileys and Sons, New York.

GOURIER D. *et al.*(1988) *Journal of Applied Physics* **63**: 1144-1151.

GRAY H.B. (1965) *Electrons and Chemical Bonding*, W.A. Benjamin Inc., New York.

HANSON G.R. *et al.* (1987) *Journal of the American Chemical Society* **109**: 2609-2616.

HUREAU C. *et al.* (2006) *Journal of Biological Inorganic Chemistry* **11**: 735-744.

KAHN O. (1993) *Molecular Magnetism*, VCH Publishers, New York.

KNEUBUHL F.K. (1960) *Journal of Chemical Physics* **33**: 1074-1078.

MARSHALL S.A. *et al.* (1964) *Molecular Physics* **8**: 225- 231.

MÖBIUS K. *et al.* (2005) *Magnetic Resonance in Chemistry* **43**: S4-S19.

MORE C., GAYDA J-P. & BERTRAND P. (1990) *Journal of Magnetic Resonance* **90**: 486-499.

MORTON J.R. (1964) *Chemical Review* **64**: 453-471.

OTTENWALDER X. *et al.* (2004) *Chemical Communications*: 504-505.

OVENALL D.W.& WHIFFEN D.H. (1961) *Molecular Physics* **4**: 135-144.

REUVENI A., LUZ Z. & SILVER B.L. (1970) *Journal of Chemical Physics* **53**: 4619-4623.

SCHRECKENBACH G. & ZIEGLER T. (1996) *Journal of Physical Chemistry* A **101**: 3388- 3399.

SCHWEIGER A. & JESCHKE G. (2001) *Principles of Pulse Electron Paramagnetic Resonance*, Oxford University Press, Oxford.

STONE A.J. (1963) *Molecular Physics* **6**: 509-515.

STONE A.J. (1964) *Molecular Physics* **7**: 311-316.

UN S., DORLET P. & RUTHERFORD A.W. (2001) *Applied Magnetic Resonance* **21**: 341-361.

VAN GASTEL M. *et al.* (2004) *Journal of the American Chemical Society* **126**: 2237-2246.

ZELDES H. & LIVINGSTON R. (1961) *Journal of Chemical Physics* **35**: 563-567.

Exercises

4.1. A paramagnetic centre is characterised by its spin S and its $\tilde{\mathbf{g}}$ matrix. Interactions between unpaired electrons and two *equivalent* nuclei of spin I_1 and I_2 are described by the operator:

$$\hat{H}_{hyperfine} = A\mathbf{S}\cdot\mathbf{I}_1 + A\mathbf{S}\cdot\mathbf{I}_2$$

By applying the approach presented in section 4.4.2, determine the energy of the allowed transitions for the canonical directions of **B**. Compare the result to the expression obtained in section 2.4 and deduce the shape of the spectrum when the nuclei are **a)** protons **b)** ^{14}N nuclei.

4.2. The spectrum in figure 4.10 was calculated at Q-band. What would the position of the resonance lines be at X-band (ν = 9.40 GHz) for the canonical directions of the field? Conclude.

4.3. Using the method from complement 3, demonstrate equation [4.15], which gives the density of the resonance lines $D_{M_I}(B)$ when the $\tilde{\mathbf{g}}$ matrix is isotropic and the $\tilde{\mathbf{A}}$ matrix is axial.

4.4. Justify equation [4.20] which gives the characteristic time δt when an anisotropic hyperfine interaction exists.

4.5. Determine the values of δt given in section 4.5.2 for three paramagnetic centres.

4.6. Numerical simulation of the spectrum for a frozen solution of nitroxide radicals recorded at high field shows that the $\tilde{\mathbf{A}}$ matrix due to interactions with the ^{14}N nucleus is axial with:

$$|A_{//}| = 98 \text{ MHz}, \; |A_{\perp}| = 14 \text{ MHz}$$

a) By comparing these numbers to $|A_{iso}|$ = 42 MHz measured in the isotropic regime, show that the principal values have the same sign as A_s.

b) The $\tilde{\mathbf{T}}$ matrix, which characterises the dipolar interaction between an electron in a $2p_z$ orbital from a nitrogen atom and the ^{14}N nucleus, is axial with $T_{//}°$ = 110 MHz, $T_{\perp}°$ = −55 MHz (appendix 3). Deduce the values of (A_s, $T_{//}$, $T_{\perp}$) and the spin population ρ_N in the $2p_z$ orbital of the nitrogen atom of the nitroxide radical. Compare it to the value of ρ_N obtained when using the method from exercise 2.3. The remaining spin density is mainly localised in the oxygen atom's $2p_z$ orbital. What can be said about the dipolar interaction between the electron occupying this orbital and the ^{14}N nucleus?

Answers to exercises

4.1. The difference compared to the case dealt with in section 4.4.2 relates to the kets, which now take the form $\{|S, M_S\rangle|I_1, M_1\rangle|I_2, M_2\rangle\}$, and to the expression for $\hat{H}_{hyperfine}$. For each value of M_S, we wish to determine the eigenvalues for the matrix for which the elements are written:

$$\langle S, M_S|\langle I_1, M_1|\langle I_2, M_2|A\mathbf{S}\cdot(\mathbf{I}_1 + \mathbf{I}_2)|S, M_S\rangle|I_1, M_1'\rangle|I_2, M_2'\rangle$$

Only the $A\hat{S}_z(\hat{I}_{1z} + \hat{I}_{2z})$ term contributes to the matrix element. The matrix is therefore diagonal and for **B** parallel to the axis i, the energy levels are given by the relations:

$$E(M_S, M_1, M_2) = g_i\beta B M_S + A M_S(M_1 + M_2) \qquad i = x, y, z$$

The selection rule becomes $\Delta M_S = \pm 1$, $\Delta M_1 = 0$, $\Delta M_2 = 0$, and the transitions have the energy

$$\Delta E(M_1, M_2) = g_i\beta B + A(M_1 + M_2) \qquad i = x, y, z$$

This expression is identical to equation [2.9]. For each canonical orientation, the hyperfine structure is similar to that shown in figure 2.9 for two protons, and that shown in figure 2.11 for two ^{14}N nuclei. Generalisation to n equivalent nuclei is straightforward.

4.2. The splittings are independent of the frequency and are equal to

$$A_x/g_x\beta = 6.33 \text{ mT}, A_y/g_y\beta = 5.0 \text{ mT}, A_z/g_z\beta = 10 \text{ mT}$$

For $\nu = 9.4$ GHz, the following resonance fields are obtained (in mT):

B parallel to x: 322, 329, 335, 341.

B parallel to y: 303, 308, 314, 319.

B parallel to z: 284, 294, 304, 314.

The hyperfine structures centred at B_{0x} and B_{0y} remain separate, but those centred at B_{0y} and B_{0z} overlap. Q-band is used to separate the patterns, which simplifies interpretation of the spectrum.

4.3. By using the same method as in complement 3, we obtain:

$D_{M_I}(B) = p(\theta)/(\mathrm{d}B/\mathrm{d}\theta)$. To calculate $\mathrm{d}B/\mathrm{d}\theta$, [4.13] must be differentiated after having squared its two sides; we use [4.14], which produces

$$\mathrm{d}B/\mathrm{d}\theta = (M_I/g\beta)\,(A_{//}^2 - A_\perp^2)\sin\theta\cos\theta/A'$$

$$D_{M_I}(B) = (g\beta/M_I)A'/(A_{//}^2 - A_\perp^2)\cos\theta$$

[4.13] and [4.14] are used once again to express A' and $\cos\theta$ as a function of B:

$$A' = g\beta(B_0 - B)/M_I;\ \cos^2\theta = (A'^2 - A_\perp^2)/(A_{//}^2 - A_\perp^2)$$

Which produces [4.15].

4.4. With hyperfine interaction, the energies ΔE_x and ΔE_y for the transitions for **B** parallel to x and y are $\Delta E_x(M_I) = g_x\beta B + A_x M_I$ and $\Delta E_y(M_I) = g_y\beta B + A_y M_I$ (section 4.4.2), which justifies [4.20].

4.5. See the text.

4.6. a) For a free radical, the principal values of the $\tilde{A}$ matrix take the form (equation [4.7]):

$A_{//} = A_s + T_{//}$, $A_\perp = A_s + T_\perp$. As the $\tilde{T}$ matrix has a null trace (equation [4.8]), we deduce $A_{iso} = (A_{//} + 2A_\perp)/3 = A_s$. The experimental values are such that $(|A_{//}| + 2\,|A_\perp|)/3 = 42$ MHz, which is the value of $|A_{iso}|$. We therefore deduce that $A_{//}$, $A_\perp$ and A_s have the same sign.

b) If the sign is positive: $T_{//} = A_{//} - A_s = 98 - 42 = 56$ MHz
$T_\perp = A_\perp - A_s = 14 - 42 = -28$ MHz

If the sign is negative: $T_{//} = A_{//} - A_s = -98 + 42 = -56$ MHz
$T_\perp = A_\perp - A_s = -14 + 42 = 28$ MHz

But, if ρ_N is the (positive) spin population of the $2p_z$ orbital for the nitrogen atom, we get $T_{//} = \rho_N T_{//}^\circ$, $T_\perp = \rho_N T_\perp^\circ$. The comparison with $T_{//}^\circ = 110$ MHz, $T_\perp^\circ = -55$ MHz shows that the first hypothesis is correct and that $\rho_N = 56/110 = 0.50$, in good agreement with the value $\rho_N \approx 0.52$ deduced from A_{iso} (exercise 2.3). The other half of the spin density is mainly localised in the $2p_z$ orbital of the oxygen atom. The dipolar coupling between the ^{14}N nucleus and an electron in this orbital is not axially symmetric. The fact that the experimental (^{14}N) $\tilde{A}$ matrix is axial shows that the effect of this interaction is negligible.

Chapter 5

Spectrum intensity, saturation, spin-lattice relaxation

5.1 – Introduction

In the previous chapter, we saw that the resonance lines are added together to form the EPR spectrum, and that the way they are added is determined by the anisotropy of the $\tilde{\mathbf{g}}$ and $\tilde{\mathbf{A}}$ matrices and by how the paramagnetic centres are organised in the sample. When the sample is a polycrystalline powder or a frozen solution, the spectrum presents characteristic features at specific positions which can be used to measure the *principal values* of these matrices. These data are very useful when seeking to identify the paramagnetic centres and to determine their structure. But the spectrum can also provide information on the *number N* of paramagnetic centres contained in the sample. To determine this number, the *intensity* of the spectrum must be measured, i.e., the *area* under the absorption signal. This intensity is proportional to N, which can be determined by comparison with the intensity of a spectrum produced by a reference sample containing a known number of paramagnetic centres. When performing this comparison, it is essential to consider the instrumental parameters and an "intensity factor" which depends on the principal values (g_x, g_y, g_z) of the paramagnetic centres.

However, comparison of the intensities of two spectra is only relevant if they are recorded in conditions where the paramagnetic centres are at *thermal equilibrium*. This is not necessarily the case in an EPR experiment. Indeed, thermal equilibrium is only maintained at resonance if the energy absorbed by the paramagnetic centres is dissipated rapidly enough towards the "lattice". In some conditions, the *relaxation processes* which allow this dissipation are not

P. Bertrand, *Electron Paramagnetic Resonance Spectroscopy*,
https://doi.org/10.1007/978-3-030-39663-3_5

sufficiently efficient and the system of paramagnetic centres is *not at thermal equilibrium*. As a result, the absorption signal is reduced and can even disappear due to a phenomenon known as *signal saturation*. It is important to be aware of this effect so as to select the appropriate experimental conditions to avoid it, or sometimes, to exploit it.

To quantitatively treat these phenomena, we must examine how the intensity of the *resonance lines* depends on the one hand on the interactions between the paramagnetic centres and the radiation, and on the other hand on the interactions between the centres and the "lattice" modes, which determine the sample's *temperature*. This chapter is the most "physical" part of this volume. To make it more readable, the homogeneous broadening of the lines is described using a simplified model which allows to define factors determining their characteristics (shape, intensity and width) and rapidly produces expressions which can be used in practice.

Although the phenomena described in this chapter relate to all types of paramagnetic centres, most of the expressions are derived for centres with spin $S = ½$. Their extension to centres with a spin greater than ½ is dealt with in chapter 6.

5.2 – Spectrum intensity at thermal equilibrium

In this section, we will consider the absorption signal resulting from interactions between the magnetic component $\mathbf{b}_1(t) = \mathbf{B}_1\cos 2\pi\nu t$ of the radiation and the magnetic moments of molecules *at thermal equilibrium*. The saturation phenomenon, which appears when centres are not at thermal equilibrium, will be studied in section 5.3.

5.2.1 – Absorption signal and intensity of a resonance line

When a paramagnetic centre of spin $S = ½$ without hyperfine interaction is placed in a magnetic field **B**, the splitting of the two energy levels is given by:

$$\Delta E = g'\beta B \quad [5.1]$$

with

$$g' = [g_x^2 u_x^2 + g_y^2 u_y^2 + g_z^2 u_z^2]^{½} \quad [5.2]$$

where (g_x, g_y, g_z) are the principal values of the $\tilde{\mathbf{g}}$ matrix and (u_x, u_y, u_z) are the components of the unit vector **u** in the direction of **B**, identified in the system of magnetic axes $\{x, y, z\}$ for the paramagnetic centre (section 3.3.2). The spin states associated with these levels are the eigenvectors $\{|½, -½\rangle_Z, |½, ½\rangle_Z\}$ shared by $(\hat{S}^2, \hat{S}_Z)$, where Z is the direction of the unit vector **v** with components (equation [3.18]):

$$v_x = u_x g_x/g',\ v_y = u_y g_y/g',\ v_z = u_z g_z/g'$$

The components (u_x, u_y, u_z) are linked to the angles $(\gamma_x, \gamma_y, \gamma_z)$ that **u** makes with the axes $\{x, y, z\}$ by

$$u_x = \cos\gamma_x,\ u_y = \cos\gamma_y,\ u_z = \cos\gamma_z$$

To define the direction of **B** relative to $\{x, y, z\}$, it is best to use *two independent parameters* such as the angles (θ, φ) of the spherical coordinates, which are such that (figure 5.1a):

$$u_x = \sin\theta\cos\varphi,\ u_y = \sin\theta\sin\varphi,\ u_z = \cos\theta$$

Equation [5.2] can then be written:

$$g'(\theta, \varphi) = [g_x^2 \sin^2\theta\cos^2\varphi + g_y^2 \sin^2\theta\sin^2\varphi + g_z^2 \cos^2\theta]^{½} \qquad [5.3]$$

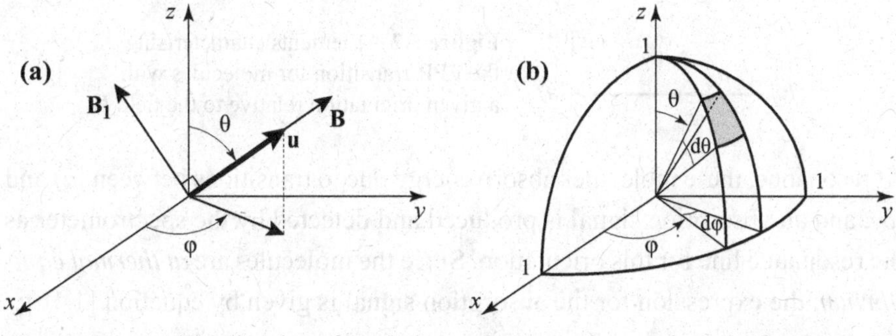

Figure 5.1 – **(a)** The direction of **B** relative to the magnetic axes $\{x,y,z\}$ of molecule M is identified by (θ, φ). In the resonance cavity, the directions of **B** and $\mathbf{B}_1$ are fixed, with $\mathbf{B}_1 \perp \mathbf{B}$, and these angles vary as a function of the orientation of the molecule. **(b)** Surface element defined by $(\theta, d\theta)$ and $(\varphi, d\varphi)$ on the sphere of unit radius.

◇ *Transition and resonance line for molecules characterised by* (θ, φ)

Consider a molecule M with magnetic axes $\{x, y, z\}$ with an orientation such that the direction of **B** is identified by (θ, φ) (figure 5.1a). If the $\{x,y,z\}$ reference frame is rotated around **B**, a new system $\{x', y', z'\}$ is obtained such that $\gamma_{x'} = \gamma_x$, $\gamma_{y'} = \gamma_y$, $\gamma_{z'} = \gamma_z$. The direction of **B** is therefore identified in this new system by (θ', φ') with $\theta' = \theta$ and $\varphi' = \varphi$. Therefore, all the molecules deduced from M by rotation around **B** are characterised by the same (θ, φ) pair. When all the orientations of the molecules are possible and equally probable in the sample, the expression of the number d*N* of molecules characterised by spherical coordinates in the interval (θ,θ +dθ) and (φ,φ +dφ) is readily obtained. Indeed, the ratio of d*N* (hereafter denoted *n* for simplicity) to the total number of molecules *N* is equal to the ratio of the surface element represented in figure 5.1b, equal to sinθdθdφ, to the total surface, 4π, of the sphere of unit radius:

$$n/N = \sin\theta \, d\theta \, d\varphi / 4\pi \qquad [5.4]$$

All these molecules share the same value of g' given by equation [5.3] and the same spin states $\{|½, -½\rangle_Z, |½, ½\rangle_Z\}$ defined below. They are noted $\{|a\rangle, |b\rangle\}$ for simplicity, and n_a and n_b are their populations which verify:

$$n_a + n_b = n$$

The states $\{|a\rangle, |b\rangle\}$, their splitting ΔE and their populations define the *transition* for the direction defined by (θ, φ) (figure 5.2).

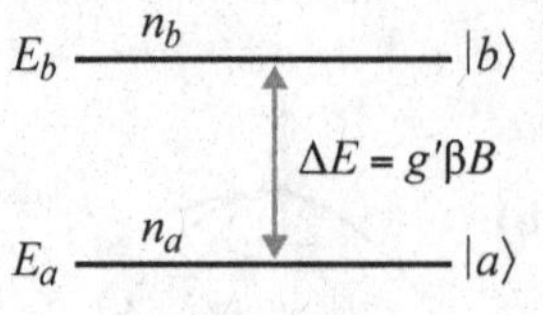

Figure 5.2 – Elements characterising the EPR transition for molecules with a given orientation relative to the field **B**.

At resonance, these molecules absorb energy due to transitions between $|a\rangle$ and $|b\rangle$, and an absorption signal is produced and detected by the spectrometer as the resonance line for this orientation. Since the molecules are *at thermal equilibrium*, the expression for the absorption signal is given by equation [1.19]:

$$s = \alpha \, W \, n \tanh(h\nu/2k_B T)/B_1 \qquad [5.5]$$

T is the temperature, B_1 the amplitude of the magnetic component of the radiation, W the transition probability per second between the two spin states, and α is a scale factor which we assume hereafter to be equal to 1. To use equation [5.5], we must first develop the expression of W.

◇ *Expression of the transition probability per second for a (θ, φ) orientation*

A transition probability per unit time can be defined for transitions from $|a\rangle$ to $|b\rangle$ when the difference $(E_b - E_a)$ can take a set of values that form a quasi continuum. This distribution is due to weak interactions between the paramagnetic molecules; the nature of these interactions is detailed in chapter 7. In a simplified description, it can be considered that these interactions add a small offset, ε, to the splitting produced by **B**, such that equation [5.1] becomes:

$$E_b - E_a = g'\beta B + \varepsilon$$

The value of ε varies from one molecule to another and its distribution is described by a *probability density* $\rho(\varepsilon)$. $\rho(\varepsilon_0)d\varepsilon$ is the fraction of molecules for which $\varepsilon_0 < \varepsilon < \varepsilon_0 + d\varepsilon$. This probability density is *normalised*:

$$\int_{-\infty}^{+\infty} \rho(\varepsilon)\, d\varepsilon = 1 \qquad [5.6]$$

When the molecules are randomly oriented and diluted in the sample, its shape is *Lorentzian* and its half width at half maximum is equal to $\hbar/T_2$ (figure 5.3a), where T_2 is the "spin-spin relaxation time" characterising the rate of internal equilibration of the paramagnetic molecules [Abragam, 1983]: this is the time required for their energy levels to be "populated". This relation between T_2 and $\rho(\varepsilon)$ is due to the fact that the same intercentre interactions are responsible for the internal equilibration of the system of paramagnetic centres and the offset ε.

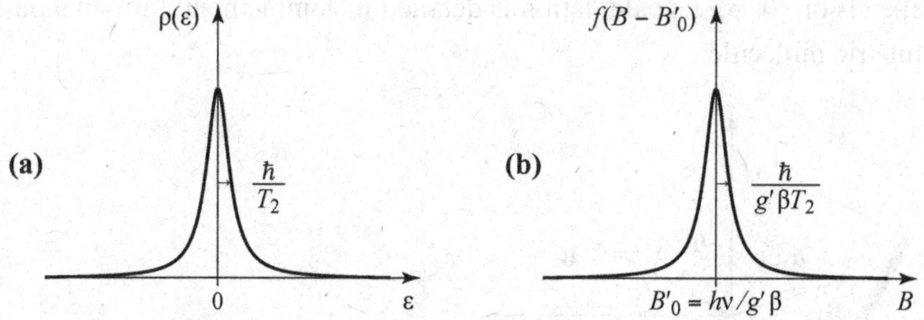

Figure 5.3 – **(a)** Lorentzian form of the probability density ρ(ε). **(b)** Function describing the lineshape.

Among all the molecules characterised by (θ, φ), we consider those for which the offset has a given value of ε. For these molecules, resonance takes place when the value B of the field satisfies:

$$g'\beta B + \varepsilon = h\nu \qquad [5.7]$$

Interaction between the magnetic moment of the molecules and the magnetic component $\mathbf{b}_1(t) = \mathbf{B}_1\cos 2\pi\nu t$ of the radiation causes the spin states to vary, and we can define a *transition probability per second* from $|a\rangle$ to $|b\rangle$ which is given by "Fermi's golden rule" (complement 1):

$$W_{ab} = \frac{\pi}{2\hbar} |V_{ab}|^2 \rho(\varepsilon) \qquad [5.8]$$

V_{ab} is the matrix element defined in section 3.4:

$$V_{ab} = \langle b|\hat{H}_1|a\rangle; \ \hat{H}_1 = \beta \, \mathbf{B}_1 \cdot \tilde{\mathbf{g}} \cdot \mathbf{S} \qquad [5.9]$$

However, there is a complication as the value of V_{ab} is not the same for all the molecules characterised by (θ, φ). Indeed, when the axes $\{x, y, z\}$ are rotated around **B** (figure 5.1a), the angles (θ, φ) remain unchanged, but the direction of $\mathbf{B}_1$ (fixed and perpendicular to **B** in the cavity) relative to the axes varies, and the value of V_{ab} is modified (equation [5.9]). In equation [5.8], we must therefore replace $|V_{ab}|^2$ by its *average value* $|V_{ab}|^2_{av}$ when $\{x, y, z\}$ rotate around **B**. To determine this mean value, it is simpler to consider the axes fixed and to rotate $\mathbf{B}_1$ in the plane perpendicular to **B**. If the $\mathbf{B}_1$ direction is identified in this plane by an angle α defined from an arbitrary direction (dotted line in figure 5.4), the mean of $|V_{ab}(\alpha)|^2$ is written:

$$|V_{ab}|^2_{av} = \frac{1}{2\pi} \int_0^{2\pi} |V_{ab}(\alpha)|^2 d\alpha$$

It depends on (θ, φ). Its calculation is detailed in complement 2 for an axially symmetric molecule.

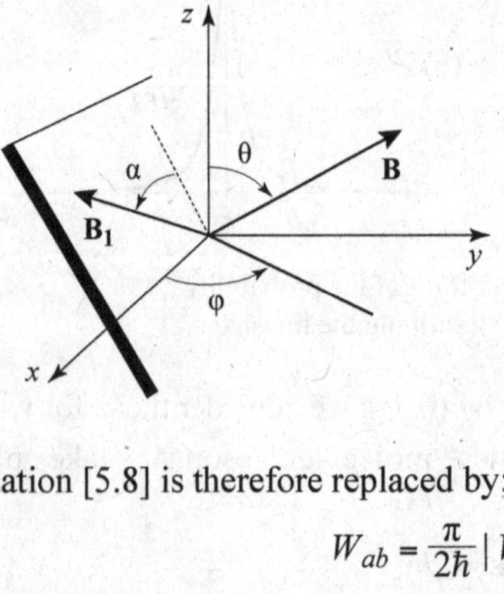

Figure 5.4 – Identifying the direction of $\mathbf{B}_1$ in the plane perpendicular to **B**.

Equation [5.8] is therefore replaced by:

$$W_{ab} = \frac{\pi}{2\hbar} |V_{ab}|^2_{av} \rho(\varepsilon)$$

The transition probability per second for the transition from $|b\rangle$ to $|a\rangle$ is obtained by replacing V_{ab} by V_{ba}. As the operator $\hat{H}_1$ is *Hermitian*, the matrix elements V_{ab} (equation [5.9]) and V_{ba} are complex conjugates (section 3.2.3), and thus:

$$W_{ab} = W_{ba}$$

◇ *Expression for the absorption signal and intensity of the line. The intensity factor*

The existence of a relation between the offset ε and the resonance field B (equation [5.7]) indicates that the transition probability depends on B and that the resonance field is itself distributed. The normalised function describing this distribution is written (exercise 5.1):

$$f(B - B_0') = g'\beta\, \rho(h\nu - g'\beta B) \qquad [5.10]$$

where $B_0' = h\nu/g'\beta$. The half width at half maximum for this function, δB, is linked to the spin-spin relaxation time T_2 by (figure 5.3b):

$$\delta B = \hbar/g'\beta T_2 \qquad [5.11]$$

Note – In the presence of hyperfine interaction (section 5.2.4), the transition energy is not *proportional* to B as in equation [5.1], rather it takes the form $\Delta E = g'\beta B + C$. In this case, we define $f(B - B_0')$ using an equation similar to [5.10] in which $h\nu$ is replaced by $(h\nu - C)$ (exercise 5.1).

With these notations, the transition probability becomes:

$$W_{ab}(B) = \frac{\pi}{2\hbar}\frac{|V_{ab}|^2{}_{av}}{g'\beta} f(B - B_0')$$

The value $|V_{ab}|^2{}_{av}$ is homogeneous to the square of an energy and it is proportional to $(B_1)^2$ (equation [5.9]). We can therefore write:

$$|V_{ab}|^2{}_{av} = (g_1\beta B_1)^2$$

where g_1 is a dimensionless number with an order of magnitude of one which depends on (θ, φ), like g'. The transition probability can therefore be written:

$$W_{ab}(B) = \frac{\pi\beta}{2\hbar}\frac{g_1{}^2}{g'} B_1{}^2 f(B - B_0') \qquad [5.12]$$

It is proportional to the square of the amplitude B_1 of the magnetic component and therefore to the power P of the radiation (section 1.5.2).

The absorption signal for the orientation defined by (θ, φ) is obtained by substituting this expression into equation [5.5]:

$$s(B,\theta,\varphi) = \frac{\pi\beta}{2\hbar} g_P(\theta,\varphi) B_1 n(\theta,\varphi) \tanh\left(\frac{h\nu}{2k_B T}\right) f(B - B_0') \qquad [5.13]$$

$$g_P(\theta, \varphi) = (g_1{}^2/g')$$

where $n(\theta, \varphi)$ is given by equation [5.4]. The resonance line is centred at $B_0' = h\nu/g'\beta$ and its shape reproduces that of the density $\rho(\varepsilon)$ (figure 5.3). The *dimensionless quantity* $g_P(\theta, \varphi)$ which weights the absorption signal depending on the orientation of the molecules, is the *intensity factor* for the line. When the $\tilde{\mathbf{g}}$ matrix is *axial*, it is given by (complement 2):

$$g_P(\theta) = (g_\perp)^2 \,[1 + (g_{//}/g')^2]/8g' \qquad [5.14]$$

where θ is the angle between the symmetry axis for the molecule and **B**, and g' is given by (equation [4.5]):

$$(g')^2 = (g_{//})^2 \cos^2\theta + (g_\perp)^2 \sin^2\theta$$

In the most general case, the expression for the intensity factor is as follows [Aasa and Vänngård, 1975]:

$$g_P = [g_x^2 g_y^2 (1-u_z^2) + g_y^2 g_z^2 (1-u_x^2) + g_z^2 g_x^2 (1-u_y^2)]/(8g'^3) \qquad [5.15]$$

where g' is defined by equation [5.2].

The *intensity* $I(\theta, \varphi)$ of the line is the *area* under the absorption signal. As the function $f(B - B_0')$ is normalised, $s(B, \theta, \varphi)$ (equation [5.13]) can be directly integrated:

$$I(\theta,\varphi) = \frac{\pi\beta}{2\hbar}\, g_P(\theta,\varphi) B_1 \, n(\theta,\varphi) \tanh\left(\frac{h\nu}{2k_B T}\right) \qquad [5.16]$$

In equations [5.13] and [5.16] the quantities depending on (θ, φ) were made explicit. The dependence of n on (θ, φ) (equation [5.4]) is the result of the definition of the surface element represented in figure 5.1b and is not linked to the anisotropy of a physical quantity. In contrast, the anisotropy of the *intensity factor* is due to the anisotropy of the $\tilde{\mathbf{g}}$ matrix (equation [5.15]). To illustrate the importance of this factor, in table 5.1 we have presented the values of g_P calculated for the "canonical" directions of the field for three very different centres:

- ▷ a nitroxide radical for which the three principal values (g_x, g_y, g_z) are very close to g_e (figure 4.13).
- ▷ a Cu^{2+} ion complex with square planar coordination for which the anisotropy of the principal values is moderate (figure 4.11).
- ▷ a Fe^{3+} ion complex in a low-spin situation with strong anisotropy of its principal values (figure 4.8).

Table 5.1 – Value of the intensity factor g_P for three paramagnetic centres

Paramagnetic centre		$\mathbf{B}//x$ [a]	$\mathbf{B}//y$ [a]	$\mathbf{B}//z$ [a]	$(g_P)_{av}$ [b]
nitroxide radical	$g_x = 2.0089$ $g_y = 2.0064$ $g_z = 2.0027$	0.500	0.501	0.503	0.502
Cu^{2+} complex	$g_x = 2.050$ $g_y = 2.050$ $g_z = 2.200$	0.563	0.563	0.467	0.528
Fe^{3+} complex	$g_x = 1.45$ $g_y = 2.24$ $g_z = 3.01$	1.21	0.628	0.292	0.572

a: intensity factor g_P calculated from equation [5.15]
b: factor $(g_P)_{av}$ calculated from equation [5.19]

The anisotropy of g_P is nearly two-fold larger than that of g': it is negligible for the nitroxide radical, around 20 % for the Cu^{2+} complex, and reaches a factor of around 4 for the Fe^{3+} complex. It is therefore essential to take this parameter into account when simulating the EPR spectrum for a transition ion complex.

5.2.2 – Expressions for the absorption signal and the intensity of the spectrum for a powder or a frozen solution

The EPR signal produced by a sample is the sum of the $s(B)$ signals (equation [5.13]) for all possible values of (θ, φ). In the case of a polycrystalline sample or a frozen solution, where all the orientations of the molecules relative to $\mathbf{B}$ are possible and equally probable, this sum can be written:

$$S(B)=\frac{\pi\beta}{2\hbar}\left[\int_{\theta=0}^{\pi}\int_{\varphi=0}^{2\pi} g_P(\theta,\varphi) f\left(B-\frac{h\nu}{\beta g'(\theta,\varphi)}\right)\sin\theta\, d\theta \frac{d\varphi}{4\pi}\right] B_1 N \tanh\left(\frac{h\nu}{2k_B T}\right)$$

In this expression, we have replaced B_0' by $h\nu/g'\beta$ and $n(\theta, \varphi)$ by its expression (equation [5.4]). The *shape* of the spectrum is determined by the integral in brackets. This integral is different from the *density of the resonance lines* $D(B)$ defined in chapter 4, which can be written:

$$D(B) = \int_{\theta=0}^{\pi}\int_{\varphi=0}^{2\pi} \delta\left(B-\frac{h\nu}{\beta g'(\theta,\varphi)}\right)\sin\theta\, d\theta \frac{d\varphi}{4\pi}$$

where $\delta(B - B_0')$ is a Dirac function centred at B_0' (section 1.4.4). The differences between the two expressions relate to two points:

1. The *shape* and *width* of the resonance lines appear explicitly in the expression for $S(B)$ whereas the lines are assumed to be infinitely narrow in the expression for $D(B)$.
2. In the expression for $S(B)$, each line is weighted by its *intensity factor* $g_P(\theta, \varphi)$.

We will now consider the *total intensity* of the spectrum. By adding the intensities $I(\theta, \varphi)$ of the resonance lines (equation [5.16]) for all possible values of (θ, φ), we obtain:

$$I = \frac{\pi\beta}{2\hbar}(g_P)_{av}\, B_1 N \tanh\left(\frac{h\nu}{2k_\mathrm{B}T}\right) \qquad [5.17]$$

where:

$$(g_P)_{av} = \int_{\theta=0}^{\pi}\int_{\varphi=0}^{2\pi} g_P(\theta,\varphi)\sin\theta\, \mathrm{d}\theta \frac{\mathrm{d}\varphi}{4\pi} \qquad [5.18]$$

$g_P(\theta, \varphi)$ can be deduced from equation [5.15] by expressing (u_x, u_y, u_z) in spherical coordinates (section 5.2.1). Unlike the absorption signal, the total intensity depends neither on the *shape* nor the *width* of the resonance lines, and it can be used to determine the number N of paramagnetic molecules. There is no closed-form expression for $(g_P)_{av}$, but a detailed numerical study showed that in the vast majority of cases, its value is adequately reproduced by the following expression [Aasa and Vänngård, 1975]:

$$(g_P)_{av} = \frac{\frac{2}{3}\left[\left(g_x^2 + g_y^2 + g_z^2\right)/3\right]^{1/2} + \frac{1}{3}\left[(g_x + g_y + g_z)/3\right]}{4} \qquad [5.19]$$

This is the expression that is used in practice. The "instrumental" parameters included in equation [5.17] are the frequency ν of the spectrometer, the temperature T of the sample and the value of B_1 which is proportional to $\sqrt{P}$, where P is the power of the radiation (section 1.5.2). Equations [5.17] and [5.19] are the basis from which the number N of paramagnetic centres can be determined from the intensity of the spectrum. It should be remembered that $\tanh(h\nu/2k_\mathrm{B}T)$ can be replaced by $h\nu/2k_\mathrm{B}T$ with less than 2 % error when $k_\mathrm{B}T \geq 2h\nu$.

In the original paper by [Aasa and Vänngård, 1975], the intensity factor is defined such that the factor ¼ does not appear in equation [5.19]. As equation [5.17] only gives the intensity to within a multiplicative factor, it is sufficient to use the same definition when comparing two intensities.

In the last column of table 5.1 we indicated the value of $(g_P)_{av}$ calculated from equation [5.19] for the three paramagnetic centres. Although the intensity factor g_P of the *resonance lines* can be very anisotropic, the number $(g_P)_{av}$ which characterises *the whole of the spectrum* is not very sensitive to the anisotropy of the principal values (g_x, g_y, g_z) and it remains close to 0.5 in all cases.

◇ Case of a solution of molecules in the isotropic regime

We saw in section 4.5.2 that in the isotropic regime, the principal values (g_x, g_y, g_z) are replaced by their *mean* $g_{iso} = (g_x + g_y + g_z)/3$. The absorption signal is then written:

$$S(B) = \frac{\pi\beta}{2\hbar}\, g_P B_1 N \tanh\left(\frac{h\nu}{2k_B T}\right) f(B - B_0')$$

where $B_0 = h\nu/g_{iso}\beta$ and $g_P = g_{iso}/4$ according to equation [5.15]. The total intensity of the signal is therefore given by:

$$I = \frac{\pi\beta}{2\hbar}\, g_P B_1 N \tanh\left(\frac{h\nu}{2k_B T}\right)$$

5.2.3 – Intensity of the spectrum produced by a single crystal

As equivalent molecules in the crystal have *parallel* principal axes, they are characterised by the same (θ, φ) pair and the same angle α (figure 5.4). We must therefore replace n by N in equation [5.13] and use the expression [5.8] for the transition probability without averaging on α. If we write:

$$|V_{ab}(\alpha)|^2 = [g_1(\theta, \varphi, \alpha)\, \beta B_1]^2$$

we obtain the following expressions for the absorption signal and its intensity:

$$s(B,\theta,\varphi,\alpha) = \frac{\pi\beta}{2\hbar}\, g_P(\theta,\varphi,\alpha) B_1 N \tanh\left(\frac{h\nu}{2k_B T}\right) f(B - B_0')$$

$$I(\theta,\varphi,\alpha) = \frac{\pi\beta}{2\hbar}\, g_P(\theta,\varphi,\alpha) B_1 N \tanh\left(\frac{h\nu}{2k_B T}\right)$$

$$g_P(\theta, \varphi, \alpha) = (g_1(\theta, \varphi, \alpha))^2/g'(\theta, \varphi)$$

Given their triple angular dependence on (θ, φ, α), the amplitude and intensity of the lines produced by a single crystal vary in a very complex manner depending on the molecules' orientation relative to **B** and $\mathbf{B}_1$.

5.2.4 – Intensity of the resonance lines and of the spectrum in the presence of hyperfine interactions

When a hyperfine interaction with a nucleus of spin I exists, each of the energy levels E_a and E_b in figure 5.2 splits to give $(2I + 1)$ equidistant levels separated by $A'/2$, the value of which depends on the orientation of the molecule relative to **B** (equation [4.11]). Each value of M_I has a corresponding transition of energy (figure 5.5)

$$\Delta E(M_I) = g'\beta B + A'M_I$$

which gives a line centred at $B_0'(M_I) = B_0' - (A'/g'\beta)M_I$, where $B_0' = h\nu/g'\beta$ (equation [4.12]). To determine the intensity of this line, we will return to the calculation performed in section 5.2.1. Based on the note following equation [5.11], expression [5.12] can be used to determine the transition probability, on the condition that we replace $h\nu$ by $(h\nu - A'M_I)$, i.e., replace B_0' by the field $B_0'(M_I)$. In addition, we saw in section 3.5.2 that the matrix element involved in the transition probability is not modified in the presence of hyperfine interaction. The intensity of the line (M_I) is therefore given by an expression similar to equation [5.16]:

$$I(M_I,\theta,\varphi) = \frac{\pi\beta}{2\hbar} g_P(\theta,\varphi) B_1 n(M_I) \tanh\left(\frac{h\nu}{2k_B T}\right) \qquad [5.20]$$

where $n(M_I) = n_a(M_I) + n_b(M_I)$ is the sum of the populations of energy levels between which transitions take place (figure 5.5). When the intensities of the $(2I + 1)$ lines for the hyperfine structure are added, the sum $\sum_{M_I=-I}^{I} n(M_I)$ emerges, which is equal to $n = n_a + n_b$ and we find the intensity of the line in the absence of interaction.

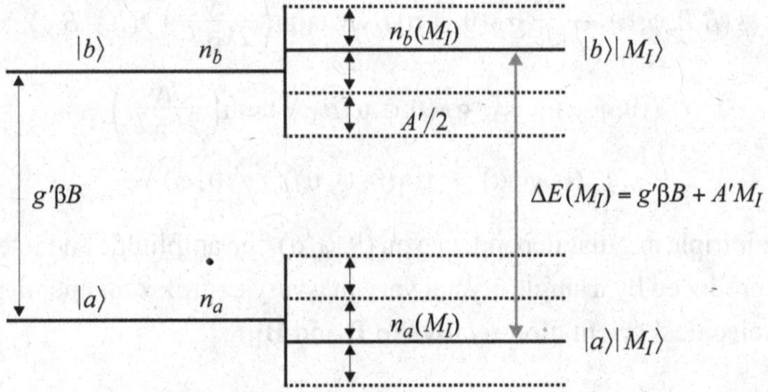

Figure 5.5 – Definition of the $n_a(M_I)$ and $n_b(M_I)$ populations of the energy levels involved in the transition with energy $\Delta E(M_I)$.

This result also applies to the *total intensity* of the spectrum obtained by integrating over all the values of (θ, φ): the hyperfine interaction modifies the *shape* of the spectrum but does not alter its *intensity*. This property, which can readily be generalised if interactions exist with several nuclei, allows two successive integrations of the experimental spectrum to be performed without worrying about the hyperfine patterns, whether they are resolved or not.

We will now consider the *relative intensities* of the different lines in the hyperfine pattern. At thermal equilibrium, the populations of the neighbouring levels in figure 5.5 verify:

$$n_a(M_I)/n_a(M_I+1) = \exp(-A'/2k_BT)$$
$$n_b(M_I)/n_b(M_I+1) = \exp(A'/2k_BT)$$

With the usual values of A', these ratios are practically equal to 1 when the temperature exceeds a fraction of a kelvin (exercise 5.2). The populations $n_a(M_I)$ and $n_b(M_I)$, and consequently the values $n(M_I)$ in equation [5.20], are therefore the same for all transitions: all the hyperfine lines in the pattern have *the same intensity*—this property can be verified in figures 2.3, 2.6a, 2.7, 4.10, 4.11—but their *widths* (and therefore their *amplitudes*) can sometimes be different. The reason for this effect will become clear in section 5.4.3 where we discuss spectra for liquid samples.

5.3 – Signal saturation

In the expressions for the absorption signal (equation [5.13]) and line intensity (equation [5.16]), the population difference *at thermal equilibrium* was used (equation [1.16]):

$$\Delta n_e = n \tanh(h\nu/2k_BT) \qquad [5.21]$$

However, in some conditions that we will define, the paramagnetic centres are *not at thermal equilibrium* and these expressions must be modified.

5.3.1 – Saturation of an EPR transition

Consider once again the transition defined by the molecules with an orientation characterised by a given (θ, φ) pair (figure 5.2). The probabilities per second for the $|a\rangle \rightarrow |b\rangle$ and $|b\rangle \rightarrow |a\rangle$ transitions, which take place at resonance, are *equal*: $W_{ab} = W_{ba}$ (section 5.2.1). Due to these "resonant" transitions, the populations n_a and n_b change as predicted by equations (figure 5.6a):

$$\mathrm{d}n_a/\mathrm{d}t = -\mathrm{d}n_b/\mathrm{d}t = W_{ab}(n_b - n_a)$$

We deduce that the *population difference* $\Delta n = n_a - n_b$ verifies the differential equation:

$$\mathrm{d}(\Delta n)/\mathrm{d}t = -2W_{ab}\Delta n$$

which is solved as (figure 5.6a):

$$\Delta n(t) = \Delta n(0)\exp(-2W_{ab}t)$$

where $\Delta n(0)$ is the initial value of Δn. The effect of the resonant transitions is to cancel Δn and, as a consequence, the intensity of the resonance line proportional to this quantity (section 1.4.4). If this line exists, it is thanks to "spin-lattice relaxation" processes which tend to maintain the population difference at its value Δn_e at *thermal equilibrium* (equation [5.21]). Whereas resonant transitions are induced by interaction with the radiation, relaxation transitions occur spontaneously and continuously in the sample, like all "thermalisation" processes. To clarify their effect, we will assume that the paramagnetic molecules are initially *not at thermal equilibrium* and that only relaxation transitions take place. If we denote w_{ab} and w_{ba} their probability per second, evolution of the populations is determined by the following rate equations (figure 5.6b):

$$\mathrm{d}n_a/\mathrm{d}t = -\mathrm{d}n_b/\mathrm{d}t = w_{ba}n_b - w_{ab}n_a$$

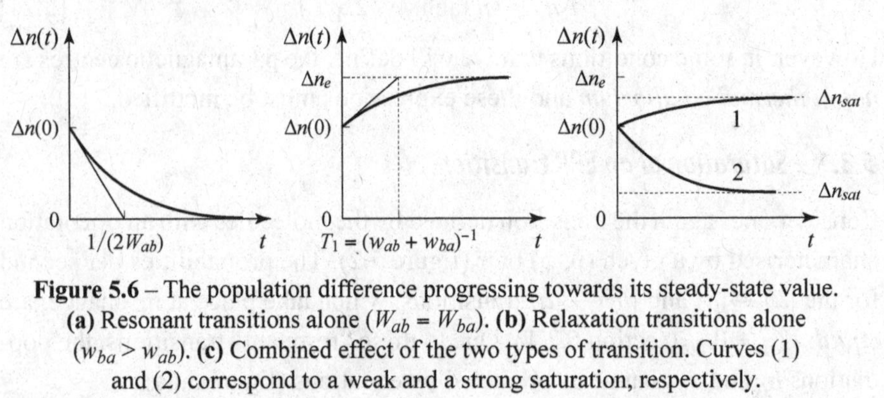

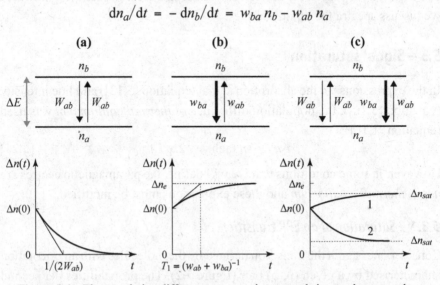

Figure 5.6 – The population difference progressing towards its steady-state value. **(a)** Resonant transitions alone ($W_{ab} = W_{ba}$). **(b)** Relaxation transitions alone ($w_{ba} > w_{ab}$). **(c)** Combined effect of the two types of transition. Curves (1) and (2) correspond to a weak and a strong saturation, respectively.

By introducing $n = n_a + n_b$, we deduce that the population difference Δn verifies the differential equation

$$\mathrm{d}(\Delta n)/\mathrm{d}t = -(w_{ab} + w_{ba})\,\Delta n + n\,(w_{ba} - w_{ab})$$

for which the solution is:

$$\Delta n(t) = (\Delta n(0) - \Delta n_e)\exp(-t/T_1) + \Delta n_e$$

We set:

$$\Delta n_e = n\,(w_{ba} - w_{ab})/(w_{ab} + w_{ba});\;\; T_1 = (w_{ab} + w_{ba})^{-1}$$

Under the influence of the relaxation transitions, $\Delta n(t)$ progresses towards the value Δn_e with the time constant T_1 (figure 5.6b). These transitions also take place at thermal equilibrium, but they do not alter the populations because their probability verifies the relation (exercise 5.3):

$$w_{ab}/w_{ba} = \exp(-\Delta E/k_\mathrm{B}T)$$

T_1 is the *spin-lattice relaxation time*, which is shorter when the relaxation processes are more efficient. Its value depends on the ($|a\rangle$, $|b\rangle$) states and consequently on the orientation of the molecules relative to **B**. We will see in section 5.4 that its value is highly dependent on the nature of the molecules and that it decreases rapidly as the temperature increases.

We are now ready to calculate the value of Δn *at resonance* under the combined effects

- ▷ of the resonant transitions with a probability per second W_{ab}.
- ▷ of the relaxation transitions with probabilities per second w_{ab} and w_{ba}.

The time course of the populations is determined by the following equations (figure 5.6c):

$$\mathrm{d}n_a/\mathrm{d}t = -\mathrm{d}n_b/\mathrm{d}t = n_b(w_{ba} + W_{ab}) - n_a(w_{ab} + W_{ab}) \qquad [5.22]$$

The population difference therefore verifies:

$$\mathrm{d}(\Delta n)/\mathrm{d}t = -(2W_{ab} + 1/T_1)\,\Delta n + \Delta n_e/T_1$$

We have used the expressions for T_1 and Δn_e given above. The solution is (figure 5.6c):

$$\Delta n(t) = (\Delta n(0) - \Delta n_{sat})\exp(-(2W_{ab} + 1/T_1)\,t) + \Delta n_{sat}$$

The quantity

$$\Delta n_{sat} = \Delta n_e/(1 + 2W_{ab}T_1) \qquad [5.23]$$

is the steady-state value of Δn in the *saturated regime*. It results from *competition* between the resonant transitions which tend to cancel Δn and the relaxation transitions which tend to maintain Δn_e (figure 5.7):

- ▹ When W_{ab} is very small relative to $1/T_1$, Δn_{sat} is equal to Δn_e and thermal equilibrium is maintained. This is the unsaturated regime dealt with in section 5.2.
- ▹ When W_{ab} is comparable to $1/T_1$, Δn_{sat} is smaller than Δn_e (equation [5.23]). The population difference, and as a consequence the absorption signal and its intensity, have lower values than at thermal equilibrium. The transition is said to be *partially saturated*.
- ▹ When W_{ab} is much greater than $1/T_1$, $\Delta n_{sat} \ll \Delta n_e$ the transition is *strongly saturated*; as a result the signal is very weak and can even disappear. We will see in section 5.4 that this situation occurs at low temperatures where T_1 is long.

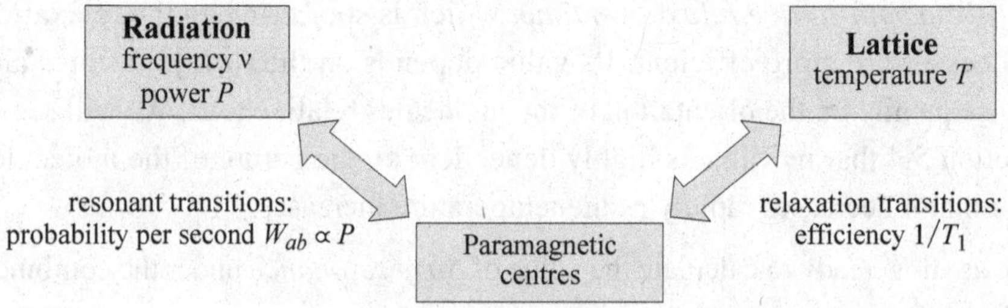

Figure 5.7 – Energy exchanges between paramagnetic centres and the radiation, or the lattice. The population difference in the steady-state depends on W_{ab} and T_1.

The spectrum is generally recorded in "slow passage" conditions where the steady-state is always reached. To do so, the field sweep must not be too rapid and the modulation frequency must not be too high (see section 9.2). In these conditions, the power absorbed by the molecules is entirely transmitted to the "lattice" thanks to relaxation processes, and the absorption signal is proportional to the value Δn_{sat} (equation [5.23]). This power can readily be calculated using equations [5.22] and [5.23] (exercise 5.4).

5.3.2 – Expression for the absorption signal in the saturated regime

The expression for the absorption signal in the steady-state can be obtained by replacing Δn_e (equation [5.21]) by Δn_{sat} (equation [5.23]) in equation [5.13]. Using expression [5.12] for W_{ab} and the definition $g_P = g_1{}^2/g'$ for the intensity factor, the following expression emerges:

$$s(B) = \frac{\pi\beta}{2\hbar} g_P \Delta n_e \frac{B_1 f(B - B_0')}{1 + \frac{\pi\beta}{\hbar} g_P B_1^2 T_1 f(B - B_0')} \qquad [5.24]$$

The variation of $s(B)$ as a function of B_1 is illustrated in figure 5.8a. It should be remembered that B_1 is proportional to $\sqrt{P}$, where P is the power of the incident radiation on the sample (section 1.5.2).

- ▹ When B_1 is small enough for the denominator in the second part of the equation to be practically equal to 1, equation [5.24] gives equation [5.13]. In this *unsaturated regime*, the signal is proportional to $\sqrt{P}$.
- ▹ Saturation begins to appear when the second term in the denominator is no longer negligible, and it is observed as curving of the line, as illustrated in figure 5.8a.
- ▹ In the limit of strong saturation, $s(B)$ tends towards zero. Naturally, this limit is not always reached with the maximum power available on the spectrometer.

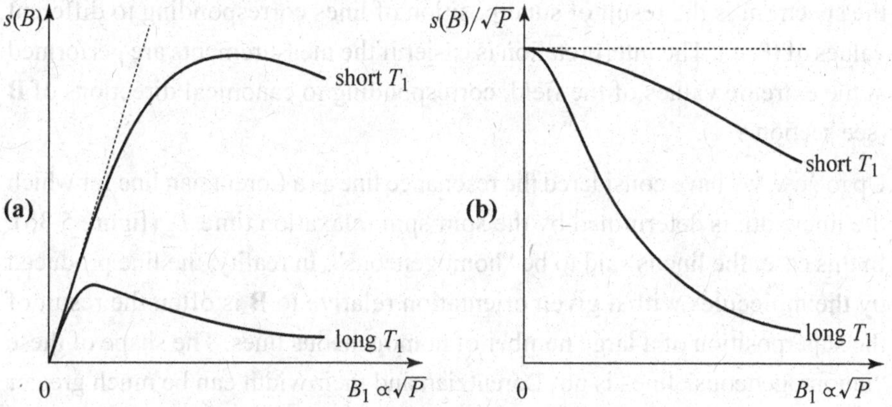

Figure 5.8 – **(a)** Variation in the amplitude of the absorption signal $s(B)$ as a function of B_1. **(b)** Variation of the ratio $s(B)/\sqrt{P}$. The dashed lines correspond to the unsaturated regime.

Expression [5.24] shows that the value of B_1 from which saturation appears is smaller with long T_1 (figure 5.8a) and large g_P values. It also shows that saturation is stronger when $f(B - B_0)$ is large: the centre of the line is more easily saturated than its wings; as a result the resonance line *broadens* under strong saturation conditions. Another representation is sometimes used, where the ratio $s(B)/\sqrt{P}$ is plotted as a function of $\sqrt{P}$ (figure 5.8b): this ratio is constant as long as the signal is not saturated; when saturated it decreases, tending towards zero at strong saturation. In the two representations, the precise shape of these "saturation curves" depends on T_1, the intensity factor g_P, and the shape of the line described by $f(B - B_0')$.

All of the above relates to the resonance line produced by molecules for which the orientation relative to **B** is defined by a given (θ, φ) pair. In equation [5.24], the quantities that depend on this orientation are the position, B_0', of the line, its intensity factor g_P, and T_1. It is obvious that the EPR spectrum, which results from the superposition of the lines produced by all the molecules in the sample, will have a complex behaviour in the saturated regime. To describe this behaviour, we can construct an experimental saturation curve representing the amplitude variation at a point of the spectrum as a function of $\sqrt{P}$, and the "resistance to saturation" at this point can be characterised by the power $P_{1/2}$ at which the (amplitude/$\sqrt{P}$) ratio decreases 2-fold. Although the shapes of these curves are similar to those presented in figure 5.8, their quantitative analysis using equation [5.24] is far from simple as the amplitude at a point in the spectrum is the result of superposition of lines corresponding to different values of (θ, φ). The interpretation is easier if the measurements are performed at the extreme values of the field, corresponding to canonical directions of **B** (see section 9.4).

Up to now, we have considered the resonance line as a Lorentzian line for which the linewidth is determined by the spin-spin relaxation time T_2 (figure 5.3b). In this case, the line is said to be "homogeneous". In reality, the line produced by the molecules with a given orientation relative to **B** is often the result of the superposition of a large number of homogeneous lines. The shape of these "inhomogeneous" lines is not Lorentzian and their width can be much greater than that determined by T_2. The various causes of inhomogeneity of a line and their consequences on its intensity and its saturation behaviour are addressed in complement 3.

5.3.3 – Significance of the saturation phenomenon

At cryogenic temperatures (liquid nitrogen or helium), the spin-lattice relaxation time T_1 for all paramagnetic centres is long and the EPR spectrum can become saturated if it is recorded with a high microwave power. The saturation phenomenon can therefore be observed for all types of centres, but to various extents: for those which "relax rapidly", saturation only occurs at very low temperatures, and it is enough to reduce the power P slightly or to decrease T_1 by slightly raising the temperature to "desaturate" the signal; for other centres, saturation is observed even at high temperatures.

◇ ***Should saturation of the EPR spectrum be avoided?***

In the unsaturated regime, the spectrum amplitude is proportional to $\sqrt{P}$ (figure 5.8a) and the power can be increased to produce an adequate signal-to-noise ratio. When the signal is weak, it can be tempting to use a slightly saturating power to take advantage of the increased signal at the start of saturation (figure 5.8a). This approach entails the risk that the spectrum may become deformed if the different parts do not saturate in the same way. However, when we wish to compare the intensities of two spectra, it is essential to record them with a *non-saturating* power. Indeed, the area under the line described by equation [5.24] is smaller than that of the unsaturated line (equation [5.13]) by an unknown factor which depends on T_1, the shape of the line and its width. The problem is even more complex for the spectrum, which is the result of superposition of lines which may not be saturated in the same way.

◇ ***Can the saturation phenomenon be exploited?***

In some applications, the sample being studied may contain radical impurities which produce large-amplitude EPR signals which superpose on the spectrum in the $g \approx 2$ region. If these impurities are characterised by a relatively long spin-lattice relaxation time, their signals can be attenuated by specifically saturating them, without causing saturation of the remainder of the spectrum.

This "selective saturation" procedure is also very useful when analysing the complex spectra produced by samples containing several types of paramagnetic centres producing overlapping signals. If the relaxation time T_1 for these centres is sufficiently different for their signals to start to saturate at different powers, it is possible to distinguish between them by proceeding as follows: start by recording the spectrum with a non-saturating power, then progressively increase the power, which preferentially saturates the signals for centres which "relax slowly"—i.e., which have a larger value of T_1—and amplifies the signals for centres which "relax rapidly". This procedure can be repeated at several temperatures to enhance the differences between the relaxation times.

We will see some other applications of saturation below.

5.4 – Spin-lattice relaxation

In an EPR experiment, the "sample temperature" in fact characterises the degrees of freedom of the "lattice", i.e., all the modes of translation, rotation, and vibration which are linked to molecular *movements*. All these modes are strongly coupled with the surrounding medium, composed of the low-temperature cryogenic liquid (nitrogen or helium) or the room-temperature resonance cavity, and they can be considered to be always at thermal equilibrium. Therefore, the paramagnetic centres arrive at thermal equilibrium through energy exchange with the lattice thanks to relaxation transitions. In this section, we describe the processes producing these transitions. Knowledge of these processes will allow us to better understand why relaxation phenomena play such an important role in EPR spectroscopy.

5.4.1 – The various spin-lattice relaxation processes

In order for relaxation transitions to take place between the $|a\rangle$ and $|b\rangle$ states of a paramagnetic centre, there must exist a *time-dependent interaction* between the centre and one of the lattice modes which verifies certain conditions. Any interaction which verifies these conditions defines a *relaxation process*, and the contributions of the different processes to the relaxation rate $1/T_1$ are additive:

$$\frac{1}{T_1} = \Sigma_k \left(\frac{1}{T_1}\right)_k$$

When the sample temperature is raised, the population of the lattice modes increases and all relaxation processes become more efficient. However, as the temperature-dependence of $(1/T_1)$ is not the same for all processes, at any given temperature there may exist a process which is more efficient than the others; this process determines the value of T_1. Hereafter, we describe the main relaxation processes by distinguishing between samples in solid and liquid media.

◇ In solid medium

Numerous studies performed on transition ions and cations of rare earth elements substituted into crystals [Orbach and Stapleton, 1972] and on frozen solutions of paramagnetic molecules [Bertrand *et al.*, 1982; Stapleton *et al.*, 1980; Zhou *et al.*, 1999] have allowed characterisation of the most effective relaxation processes, which involve *vibrational modes*. In solids, these modes cover a wide range of frequencies:

▹ The angular frequencies of *collective modes* known as "acoustic phonons" form a quasi continuum from very low frequencies corresponding to elastic waves, up to a maximum value ω_{max} which is higher when the interactions between molecules are strong and their masses are small. At low frequency, the density $g(\omega)$ of the acoustic phonons is proportional to ω^2 (figure 5.9). In crystalline solids, "optical phonons" also exist, with an angular frequency greater than ω_{max}.

▹ The internal vibrations of the molecules have a more *localised* nature, and their frequencies are generally higher than those of acoustic phonons.

These vibrations modulate the geometry of the paramagnetic centres, causing the potential energy of the unpaired electrons to vary. This produces a time-dependent interaction which can induce transitions between their spin states. Several relaxation processes can be defined which are experimentally distinguished based on the *temperature-dependence* of $1/T_1$.

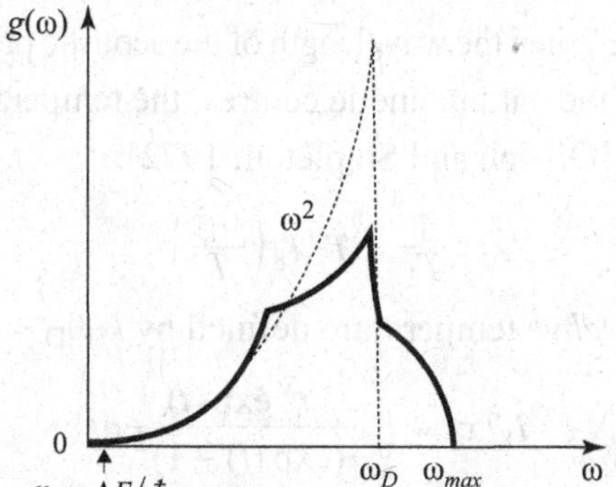

Figure 5.9 – Profile of the density $g(\omega)$ of the acoustic phonons. ω_r is the angular frequency of the resonant modes involved in the direct process. The dashed curve represents the Debye approximation, used to calculate T_1 in the Raman process.

1. The *direct process* involves "resonant phonons" for which the energy $\hbar\omega_r$ is equal to the splitting ΔE between the $|a\rangle$ and $|b\rangle$ states. Based on the value of ΔE (0.3 cm^{-1} at X-band), these are very low-frequency acoustic phonons (figure 5.9). In this process, transition between two energy levels for the paramagnetic centre is associated with a reverse transition between two energy levels for the resonant mode of angular frequency ω_r (figure 5.10a). This process, which has a $1/T_1 \propto T$ temperature-dependence (appendix 6), generally dominates at very low temperatures (liquid helium) where only the lowest frequency modes are populated. It is sometimes masked by the

"phonon bottleneck" phenomenon which occurs when the energy received by the resonant phonons is not dissipated quickly enough into the helium bath. In this case, the return to thermal equilibrium of the paramagnetic centres is not only dictated by T_1, and its rate often varies as T^2.

2. The *Raman process* involves two phonons, for which the angular frequencies ω_1 and ω_2, verify the following relation (figure 5.10b):

$$\hbar\omega_1 - \hbar\omega_2 = \Delta E \qquad [5.25]$$

As the angular frequency $\Delta E/\hbar$ is much smaller than the $[0, \omega_{max}]$ range for acoustic phonons (figure 5.9), the number of modes participating in this process increases rapidly with the temperature. The relaxation rate $1/T_1$ increases very quickly as long as $k_B T < \hbar\omega_{max}$, then less rapidly when all the modes in the $[0, \omega_{max}]$ range are populated. To calculate the temperature-dependence, the phonon density is modelled by extending the ω^2 dependence up to a value ω_D such that the area under the curve is equal to the area under $g(\omega)$ (the Debye model, figure 5.9). When the wavelength of the acoustic phonons by far exceeds the dimension of the paramagnetic centres, the temperature-dependence of $1/T_1$ is given by [Orbach and Stapleton, 1972]:

$$\frac{1}{T_1} \propto T^9 I_8\left(\frac{\theta_D}{T}\right)$$

where θ_D is the *Debye* temperature defined by $k_B\theta_D = \hbar\omega_D$, and

$$I_8(x) = \int_0^x \frac{t^8 \exp(t)}{(\exp(t) - 1)^2} dt$$

This expression leads to $1/T_1 \propto T^9$ for $T \ll \theta_D$ and to $1/T_1 \propto T^2$ for $T \gg \theta_D$. Another Raman process, which is generally less efficient for centres with an odd number of unpaired electrons, produces the following temperature-dependence [Orbach and Stapleton, 1972]:

$$\frac{1}{T_1} \propto T^7 I_6\left(\frac{\theta_D}{T}\right)$$

The validity of the hypotheses from which these authors determined the temperature-dependences is discussed in [Bertrand, 1986].

3. The *Orbach process* is a two-phonon resonance process which occurs when the paramagnetic centre has an excited state of energy Δ and there exists a vibrational mode such that $\hbar\omega = \Delta$ (figure 5.10c). The following temperature-dependence is observed:

$$\frac{1}{T_1} \propto \frac{1}{\exp(\Delta/k_B T) - 1} \approx \exp(-\Delta/k_B T) \quad \text{when } k_B T \ll \Delta$$

This process is particularly efficient in polynuclear complexes of transition ions and complexes of rare earth elements which often have low energy excited states (sections 7.5.2 and 8.3.5).

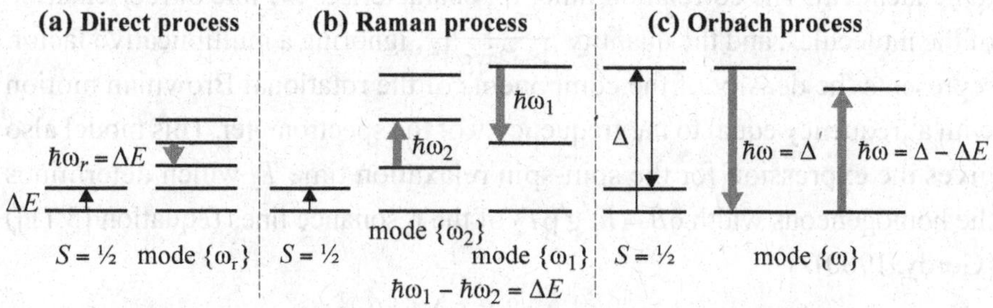

Figure 5.10 – Relaxation processes involving concerted transitions with energy conservation between the energy levels for the paramagnetic centre (black arrows) and the vibrational energy levels (grey arrows).

◇ *In liquid solution*

In liquid solution, the orientation of the molecules relative to **B** changes continuously due to the effect of *rotational Brownian motion*, which causes their Hamiltonian $\hat{H}(t)$ to vary due to the anisotropy of the $\tilde{\mathbf{g}}$ and $\tilde{\mathbf{A}}$ matrices (section 4.5.2). If the reorientation movements are rapid enough, and if they allow all the directions in space to be explored with equal probability, the EPR spectrum is determined by an *average Hamiltonian* $\hat{H}_{av}$ characterised by the parameters

$$g_{iso} = (g_x + g_y + g_z)/3; \quad A_{iso} = (A_x + A_y + A_z)/3$$

This is the "isotropic regime", which was discussed in detail in chapter 2. But these movements also have another effect: the fluctuations of $\hat{H}(t)$ relative to $\hat{H}_{av}$ constitute a random time-dependent interaction, which is stronger when the anisotropy of the $\tilde{\mathbf{g}}$ and $\tilde{\mathbf{A}}$ matrices is large. If the Fourier transform of these fluctuations has components at the frequency of the spectrometer, transitions can be induced between the spin states of the paramagnetic molecules.

The model describing this mechanism is similar to the one used in NMR [Bloembergen *et al.*, 1948]. In the presence of hyperfine interaction with a nucleus of spin I, a simplified model of the spin-lattice relaxation time for the (M_I) transition gives the following expression [Gordy, 1980]:

$$\frac{1}{T_1} = \frac{2}{15}\left[\frac{\Delta g \beta B + \Delta A M_I}{\hbar}\right]^2 \frac{\tau_C}{1+(2\pi\nu\tau_C)^2} \qquad [5.26]$$

In this expression, Δg and ΔA represent the greatest differences between the principal values of the $\tilde{\mathbf{g}}$ and $\tilde{\mathbf{A}}$ matrices, for which the principal axes are assumed to be identical. The correlation time, τ_C, characterises the rate of reorientation of the molecules, and the quantity $\frac{\tau_C}{1+(2\pi\nu\tau_C)^2}$, ignoring a multiplicative factor, represents the density of the components of the rotational Brownian motion with a frequency equal to the frequency ν of the spectrometer. This model also gives the expression for the spin-spin relaxation time T_2 which determines the homogeneous width $\delta B = \hbar / g'\beta T_2$ of the resonance lines (equation [5.11]) [Gordy, 1980]:

$$\frac{1}{T_2} = \frac{16}{45}\left[\frac{\Delta g \beta B + \Delta A M_I}{\hbar}\right]^2 \tau_C \qquad [5.27]$$

The linewidth decreasing with τ_C is the "motional narrowing" phenomenon which produces the narrow NMR lines observed in liquid solutions. In equations [5.26] and [5.27], the dimension of the quantity in square brackets is $(\text{time})^{-1}$. This time is in fact the duration δt determining the emergence of the isotropic regime (equation [4.20]). This is not surprising, since this duration is also linked to the anisotropy of the $\tilde{\mathbf{g}}$ and $\tilde{\mathbf{A}}$ matrices. More elaborate models are described in [Carrington and McLachlan, 1979; Atherton, 1993].

5.4.2 – How can the spin-lattice relaxation time T_1 be measured?

In some EPR applications, the value of T_1 is required. When T_1 is long enough for the signal to be saturated with the microwave power available with the spectrometer, the saturation phenomenon is used. The most direct method consists in strongly saturating a transition with a high-power microwave pulse, and then using a non-saturating power to observe the exponential recovery of the signal with a time constant T_1 (figure 5.6b). In this "saturating pulse" method, the power of the radiation must be changed very rapidly while simultaneously protecting the detection circuit. This can be achieved using complex commutation devices which are not installed on conventional continuous wave spectrometers. Pulsed spectrometers are, in contrast, ideally suited to performing this type of experiment [Schweiger and Jeschke, 2001]. To overcome the material limitations in continuous wave EPR, another method called progressive saturation is used instead. This method consists in monitoring a *saturation curve* by studying how the amplitude at a point in the spectrum varies as a function of the microwave power; this curve will have a profile similar to those in figure 5.8. By simulating this curve using a model which takes the inhomogeneous nature of the resonance lines into account (complement 3), T_1 can be determined (see section 9.4).

The previous methods can only be applied if T_1 is long enough to allow signal saturation. When this is not possible, T_1 can be deduced from the *spectral broadening*, which sometimes appears as the temperature increases. Indeed, the width of the resonance lines is often temperature-dependent for the following reasons:

- At very low temperature, T_1 is longer than the spin-spin relaxation time T_2 and the *homogeneous* width of the resonance line is determined by T_2 (equation [5.11]).
- When the temperature increases, T_1 decreases and T_2 remains fairly constant. If the temperature is high enough for T_1 to be shorter than T_2, T_1 will determine the *homogeneous* width of the resonance line, which becomes:

$$\delta B = \hbar / g' \beta T_1 \qquad [5.28]$$

Shortening of T_1 leads to *broadening* of the line and thus of the spectrum when the temperature increases, and T_1 can be determined by simulating this broadening using an appropriate model (complement 3 to chapter 9).

To illustrate these different methods, we have represented the temperature-dependence of the spin-lattice relaxation time T_1 for the $S = ½$ ground state of the $[2Fe - 2S]^{1+}$ centre in a protein (section 7.4.3) in figure 5.11a. The T_1 values measured at X-band between 1.2 K and 130 K cover 9 orders of magnitude. They were determined by the saturating pulse technique between 1.2 K and 13 K, the progressive saturation technique between 8 K and 30 K and the spectral broadening technique between 50 K and 130 K. At very low temperature, the data reveal a resonant phonon bottleneck phenomenon. At temperatures greater than around 5 K, a Raman process characterised by $\theta_D = 60$ K becomes more efficient, and an Orbach process involving an excited state of energy $\Delta = 250\ cm^{-1}$ leads to broadening of the spectrum above 40 K, followed by its disappearance [Bertrand *et al.*, 1982]. The temperature-dependence of the relaxation time for the ground state $S = ½$ of a $[4Fe - 4S]^{1+}$ centre presents similar characteristics (figure 5.11b).

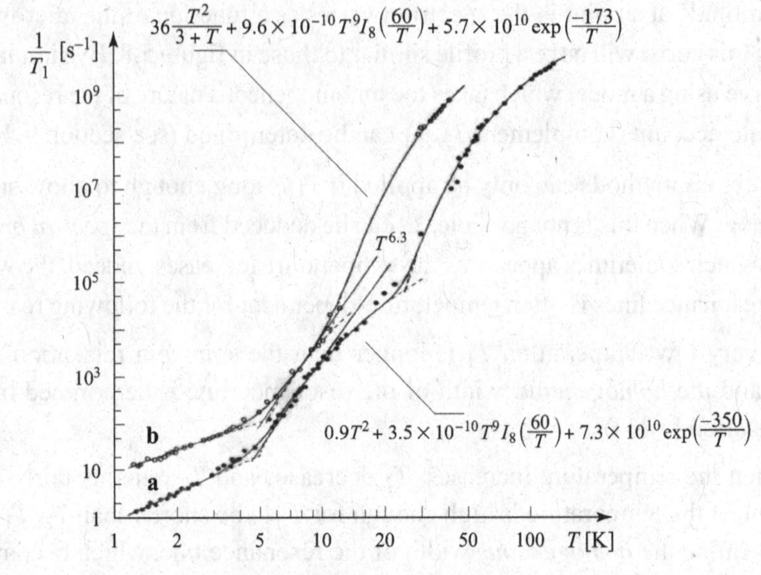

Figure 5.11 – Temperature-dependence of $1/T_1$ for iron-sulfur centres:
(a) $[2Fe-2S]^+$ centre for a protein from the cyanobacterium *Spirulina maxima*.
(b) $[4Fe-4S]^+$ centre for a protein from the bacterium *Bacillus stearothermophilus*.
The $T^{6.3}$ temperature-dependence is that predicted by a "fractal" model [Stapleton *et al.*, 1980].

[Reproduced with permission from: Bertrand P. *et al.*, *Journal of Chemical Physics*, **76**, 4715–4719 © 1969, American Institute of Physics]

5.4.3 – Relaxation phenomena and EPR spectroscopy in practice

In any spectroscopic technique, the molecules are brought to thermal equilibrium by "thermalisation" processes. In optical or vibrational spectroscopies, we are interested in transitions between states which are strongly coupled to the lattice due to molecular collisions (in liquid solution) or deformations (in solid medium). Thermalisation processes are very efficient, and thermal equilibrium is continuously maintained. In addition, these processes contribute little to the linewidth which is determined by other causes. Apart from in exceptional cases, thermalisation processes therefore do not alter the spectrum.

The situation is very different in EPR, and more generally in magnetic spectroscopies, where paramagnetic centres are brought to thermal equilibrium by the very specific processes of spin-lattice relaxation.

◇ *In solid samples*

In solid samples, vibrational movements modulate the geometry of paramagnetic centres, and therefore act on their *orbital states*. This generally affects their *spin states* indirectly, thanks to *spin-orbit coupling*. The interaction responsible for transitions between spin states is therefore weak: at very low temperatures, where the vibrational modes are weakly populated, T_1 is long and the absorption signal is readily saturated. When the temperature increases, two-phonon relaxation processes become efficient and lead to rapid shortening of T_1. In some cases, this shortening is extensive enough to cause spectral *broadening* and thus reduce the amplitude, finally leading to disappearance of the signal. This effect is particularly common with complexes of rare earth elements and polynuclear transition ion complexes in which low-lying excited states promote Orbach-type relaxation processes (figure 5.11).

◇ *In liquid samples*

In liquid samples, we saw that spin-lattice relaxation is determined by rotational Brownian motion of the molecules and that it is more effective when the anisotropy of the $\tilde{\mathbf{g}}$ and $\tilde{\mathbf{A}}$ matrices is large. We will use equations [5.26] and [5.27] to assess the relaxation times (T_1, T_2) and the homogeneous width δB, for the three types of paramagnetic centre considered in section 4.5.2. At X-band ($\nu = 9 \times 10^9$ Hz) and for an aqueous solution at 20 °C, the following results are obtained:

1. for a nitroxide radical: with the values $\delta t = 2 \times 10^{-9}$ s and $\tau_C = 10^{-10}$ s, obtained in section 4.5.2, we calculate $T_1 = 10^{-5}$ s, $T_2 = 1.1 \times 10^{-7}$ s and $\delta B = 5 \times 10^{-5}$ T.
2. for a Cu^{2+} complex with square planar geometry: with $\delta t = 2 \times 10^{-10}$ s and $\tau_C = 2 \times 10^{-11}$ s, we calculate $T_1 = 4 \times 10^{-8}$ s, $T_2 = 4.5 \times 10^{-9}$ s and $\delta B = 1.2 \times 10^{-3}$ T.
3. for a Fe^{3+} complex in the low-spin situation: with $\delta t = 2.4 \times 10^{-11}$ s and $\tau_C = 2 \times 10^{-11}$ s, we calculate $T_1 = 5.3 \times 10^{-10}$ s, $T_2 = 6.7 \times 10^{-11}$ s and $\delta B = 9 \times 10^{-2}$ T.

Although these values were obtained using simplified models, they qualitatively correlate with the experimental observations:

▹ Thanks to the motional narrowing mechanism, the EPR spectra for free radicals are characterised by narrow linewidths, even if broadening is sometimes induced by causes other than the homogeneous T_2-related broadening. The dependence of $1/T_2$ on M_I (equation [5.27]) explains the differences in *linewidth of hyperfine lines* that are often observed on these spectra. In some favourable cases, analysis of these widths can be used to attribute a value to M_I for each line, and thus to determine the *sign* of the hyperfine constant A_{iso} [Atherton, 1993]. In addition, the relatively short relaxation time T_1 explains why the spectrum for free radicals is not readily saturated at room temperature.

▹ For transition ion complexes of spin ½ with moderately anisotropic $\tilde{\mathbf{g}}$ and $\tilde{\mathbf{A}}$ matrices, such as V^{4+}, Cu^{2+} and Mo^{5+}, relaxation broadening is weak enough for the spectrum to be observed at room temperature, but the hyperfine pattern is not always resolved. The spin-lattice relaxation time is very short, and the signals are not saturated.

▹ In contrast, for complexes characterised by a strongly anisotropic $\tilde{\mathbf{g}}$ matrix, such as Fe^{3+} in a weak spin situation, the broadening of the resonance lines is too extensive for the spectrum to be observable.

We have described the Orbach process for solid samples. When the process involves vibrational modes internal to the paramagnetic molecules, this process is also very efficient in liquid samples. This explains why the spectra produced by complexes of rare earths and polynuclear complexes of transition ions are generally not observed at room temperature.

5.5 – Points to consider in applications

We have already underlined the importance of the phenomena dealt with in this chapter for practical EPR spectroscopy. Here, we will review a few points and refine others.

5.5.1 – Intensity of the resonance lines and the spectrum

▹ The *anisotropy of the magnetic moment* of molecules appears in two ways on the resonance line:

- The interaction of the moment with the field **B** causes anisotropy of g' and thus of the *position* for the line.
- Interaction with the $\mathbf{b}_1(t)$ component of the radiation results in an anisotropic intensity factor g_P, and thus an anisotropic *intensity* of the line.

- ▷ The *sensitivity* of a spectroscopic technique is largely determined by the intensity of the spectral lines. In EPR, the transition probability is proportional to β^2 (equations [5.8] and [5.9]) whereas in NMR it is proportional to β_N^2 which is six orders of magnitude smaller; this difference favours EPR spectroscopy. As interactions involving the electric dipoles are greater than magnetic interactions, UV-visible absorption spectroscopy and vibrational spectrometry techniques are more sensitive than magnetic spectroscopies.
- ▷ When the Zeeman term is greater than the other terms in the Hamiltonian, the *shape* of the EPR spectrum depends on these terms, but its *intensity* does not. By measuring the intensity by two successive integrations of the spectrum and comparing it to the intensity for a reference sample, the number N of centres contained in the sample can be determined. This information is very important in numerous applications and could be used more routinely, for example to verify that the intensity effectively corresponds to the concentration of molecules being investigated.
- ▷ The intensity of the EPR spectrum provides very useful information when studying the temperature-dependence of the magnetisation of magnetic materials, such as metal complexes, polyradicals, nanomaterials, etc. Indeed, when the magnetic susceptibility of a sample is measured, the magnetisation is the result of contributions from all the energy levels, each of which is weighted by an appropriate Boltzmann factor [Kahn, 1993]. The temperature-dependence of the susceptibility is often too smooth to unambiguously deduce the energy level diagram. As the EPR spectrum only involves a small number of energy levels, the temperature-dependence of its intensity is more pronounced and easier to interpret.

5.5.2 – Use of spin-lattice relaxation

The saturation and relaxation broadening phenomena restrict the conditions (power, temperature) in which a spectrum can be recorded. But these phenomena can also be exploited in a multitude of ways:

- ▷ T_1 *measurements* have numerous applications, e.g. to determine the Debye temperature for a solid, to determine the energy of the excited states of a paramagnetic molecule, or to study the dynamic effects of intercentre interactions (section 7.5.2).
- ▷ When a sample contains several types of paramagnetic centre, their signals can be "sorted" using their relaxation properties.

▹ When a nitroxide radical is "attached" onto a macromolecule in solution, the shortening of the relaxation times T_1 and T_2 due to interaction with a "relaxing agent" (e.g Ni^{2+} or Cr^{3+} complex) added to the solution can be used to determine the solvent accessibility of the binding site (section 7.6.3). Similarly, in NMR imaging, complexes of rare earth elements are used to accelerate the relaxation of protons in water molecules so as to increase contrast.

▹ Saturation of a transition is also used in techniques other than EPR spectroscopy. To obtain an ENDOR signal, an EPR transition is saturated and its desaturation caused by NMR transitions is observed. Similarly, by saturating a transition we create the population inversion required to amplify microwave radiation in three-level solid state *masers* (masers have a similar principle to lasers).

Complement 1 – Fermi's golden rule

We provide some information on the origins of equation [5.8], readers interested in the details can consult a reference work on quantum mechanics [Ayant and Belorizky, 2000; Cohen-Tannoudji *et al.*, 2015].

Consider a molecule which has two possible states represented by the kets $|a\rangle$ and $|b\rangle$ with energies E_a and E_b (figure 5.12a), which is initially in the $|a\rangle$ state. At time $t = 0$, the molecule is exposed to the interaction $u(t) = U\cos 2\pi\nu t$. In the space of states, U is an *operator* $\hat{U}$ such that the matrix element U_{ab} defined by

$$U_{ab} = \langle b|\hat{U}|a\rangle$$

is non-null. Under the influence of the interaction $u(t)$, the molecule's state evolves and we are interested in the probability $p_{ab}(t)$ that this state is $|b\rangle$ at time t. Calculations demonstrate that $p_{ab}(t)$ is a *sinusoidal function* of time with an amplitude proportional to $|U_{ab}|^2$ and a frequency equal to $|\Delta E/h - \nu|$.

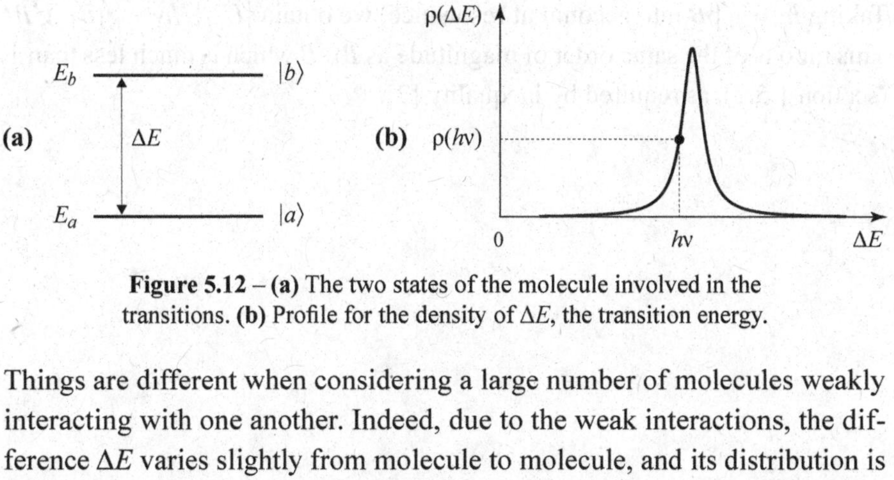

Figure 5.12 – **(a)** The two states of the molecule involved in the transitions. **(b)** Profile for the density of ΔE, the transition energy.

Things are different when considering a large number of molecules weakly interacting with one another. Indeed, due to the weak interactions, the difference ΔE varies slightly from molecule to molecule, and its distribution is characterised by a *probability density* $\rho(\Delta E)$ (figure 5.12b); $\rho(\Delta E)\mathrm{d}(\Delta E)$ is the fraction of molecules for which $(E_b - E_a)$ is between ΔE and $\Delta E + \mathrm{d}(\Delta E)$. This density is normalised:

$$\int_0^\infty \rho(\Delta E)\,\mathrm{d}(\Delta E) = 1$$

Because of this distribution, the *mean value* $\langle p_{ab}(t)\rangle$ calculated over all the values of ΔE must now be considered. Under a specific set of conditions, this mean value very rapidly becomes *proportional to t*:

$$\langle p_{ab}(t)\rangle \propto W_{ab}\, t$$

W_{ab} is the *probability per second* of transition from $|a\rangle$ towards $|b\rangle$, given by:

$$W_{ab} = \frac{\pi}{2\hbar} |U_{ab}|^2 \rho(h\nu) \qquad [1]$$

where $\rho(h\nu)$ is the density for $\Delta E = h\nu$ (figure 5.12b). This expression shows that the transition from $|a\rangle$ towards $|b\rangle$ only takes place if $\rho(h\nu)$ is different from zero, and that "energy conservation" is only verified to within the width of $\rho(\Delta E)$. For equation [1] to be valid, the following condition in particular must be satisfied:

$$|U_{ab}| \ll h\nu \qquad [2]$$

If these results are transposed to EPR transitions, the difference $(E_b - E_a)$ takes the form (section 5.2.1):

$$E_b - E_a = g'\beta B + \varepsilon$$

The distribution of $(E_b - E_a)$ is due to that of ε and the density of ΔE is replaced by that of ε. In addition, we have set $|U_{ab}| = g_1\beta B_1$ (section 5.2.1). Taking $h\nu = g'\beta B$ into account at resonance, we obtain $|U_{ab}|/h\nu = g_1 B_1/g'B$. This ratio is of the same order of magnitude as B_1/B which is much less than 1 (section 1.5.2), as required by inequality [2].

Complement 2 – Expression of the intensity factor for an axially symmetric centre of spin ½

The matrix element involved in the transition probability takes the following expression (equation [5.9]):

$$V_{ab} = {}_Z\langle ½, ½|\hat{H}_1|½, -½\rangle_Z \qquad [1]$$

$$\hat{H}_1 = \beta(B_{1x}\, g_x\, \hat{S}_x + B_{1y}\, g_y\, \hat{S}_y + B_{1z}\, g_z\, \hat{S}_z) \qquad [2]$$

$\{x, y, z\}$ are the principal axes of the $\tilde{\mathbf{g}}$ matrix, the kets $\{|½, ½\rangle_Z, |½, -½\rangle_Z\}$ are the eigenvectors shared by $(\hat{\mathbf{S}}^2, \hat{S}_Z)$, Z is the direction of the vector **v** with components:

$$v_x = u_x\, g_x/g',\ v_y = u_y\, g_y/g',\ v_z = u_z\, g_z/g' \qquad [3]$$

and (u_x, u_y, u_z) are the components of the unit vector **u** in the direction of **B** (equation [3.18]).

To calculate V_{ab}, $\hat{H}_1$ must be expressed as a function of the operators $(\hat{S}_X, \hat{S}_Y, \hat{S}_Z)$ for which the action is known on the kets $\{|½, ½\rangle_Z, |½, -½\rangle_Z\}$. In axial symmetry, $g_x = g_y = g_\perp$, $g_z = g_{//}$, and x can be taken in the $(z, \mathbf{B})$ plane (figure 5.13). If we set $\psi = (z, Z)$ and X is placed in the (z, Z) plane, we shift from the $\{x, y, z\}$ reference frame to the $\{X, Y, Z\}$ reference frame by rotation through an angle ψ around y (figure 5.13 and exercises 3.6 and 3.7). As a result:

$$S_x = \cos\psi\, S_X + \sin\psi\, S_Z,\ S_y = S_Y,\ S_z = -\sin\psi\, S_X + \cos\psi\, S_Z \qquad [4]$$

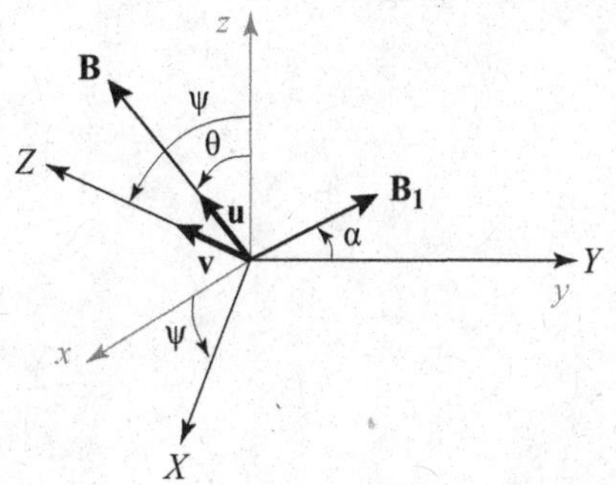

Figure 5.13 – The axes and angles used to calculate the intensity factor in axial symmetry. The x, X, Z axes are in the $(z, \mathbf{B})$ plane and $\mathbf{B}_1$ is perpendicular to **B**.

The angle ψ is linked to $\theta = (z, \mathbf{B})$. Indeed:

$$u_x = \sin\theta,\ u_y = 0,\ u_z = \cos\theta \qquad [5]$$

which, according to equation [3], produces:

$$\tan\psi = v_x/v_z = (g_\perp/g_{//})\tan\theta \qquad [6]$$

In addition, as $\mathbf{B}_1$ is perpendicular to $\mathbf{u}$ defined by equations [5], its components can be written:

$$B_{1x}' = B_1 \cos\theta \sin\alpha, \; B_{1y} = B_1 \cos\alpha, \; B_{1z} = -B_1 \sin\theta \sin\alpha$$

where $\alpha = (y, \mathbf{B}_1)$ (figure 5.13). Using these relations and equations [4], $\hat{H}_1$ can be expressed (equation [2]) as a function of $(\hat{S}_X, \hat{S}_Y, \hat{S}_Z)$. Only the operators $\hat{S}_X$ and $\hat{S}_Y$ contribute to V_{ab} to produce:

$$V_{ab}(\alpha) = \frac{\beta B_1}{2}[\sin\alpha(\cos\theta\cos\psi\, g_\perp + \sin\theta\sin\psi\, g_{//}) - \mathrm{i}\cos\alpha\, g_\perp]$$

$$|V_{ab}(\alpha)|^2 = \frac{(\beta B_1)^2}{4}[\sin^2\alpha(\cos\theta\cos\psi\, g_\perp + \sin\theta\sin\psi\, g_{//})^2 + \cos^2\alpha\,(g_\perp)^2]$$

As the angle α can take any value between 0 and 2π with equal probability, the mean value of $\sin^2\alpha$ and $\cos^2\alpha$ is ½. Using equation [6], the mean value of $|V_{ab}(\alpha)|^2$ is written:

$$|V_{ab}|^2_{av} = \frac{(\beta B_1)^2 (g_\perp)^2 [1 + (g_{//}/g')^2]}{8}$$

where g' is defined by equation [4.5]. The dimensionless number g_1 defined by $|V_{ab}|^2_{av} = (g_1\, \beta B_1)^2$ is therefore equal to:

$$g_1^2 = (g_\perp)^2\,[1 + (g_{//}/g')^2]/8$$

From this, we deduce the *intensity factor* $g_P = g_1^2/g'$ for the line:

$$g_P(\theta) = (g_\perp)^2\,[1 + (g_{//}/g')^2]/8g' \qquad [7]$$

Complement 3 – Homogeneous and inhomogeneous lines

The *width* of the resonance lines is an important feature of the spectrum. At equal intensity, narrow lines have a greater amplitude than broad lines, and they produce better-resolved features. Here, we briefly describe the main mechanisms causing EPR lines to broaden.

◇ *Homogeneous line*

A line is said to be *homogeneous* when its width is determined by the lifetime of the states between which the transition takes place. In EPR spectroscopy, dipolar interactions between the paramagnetic centres in a dilute sample produce homogeneous lines with a Lorentzian shape, for which the half width at half maximum, δB, is linked to the spin-spin relaxation time T_2 by equation [5.11] (figure 5.14a):

$$\delta B = \hbar / g'\beta T_2$$

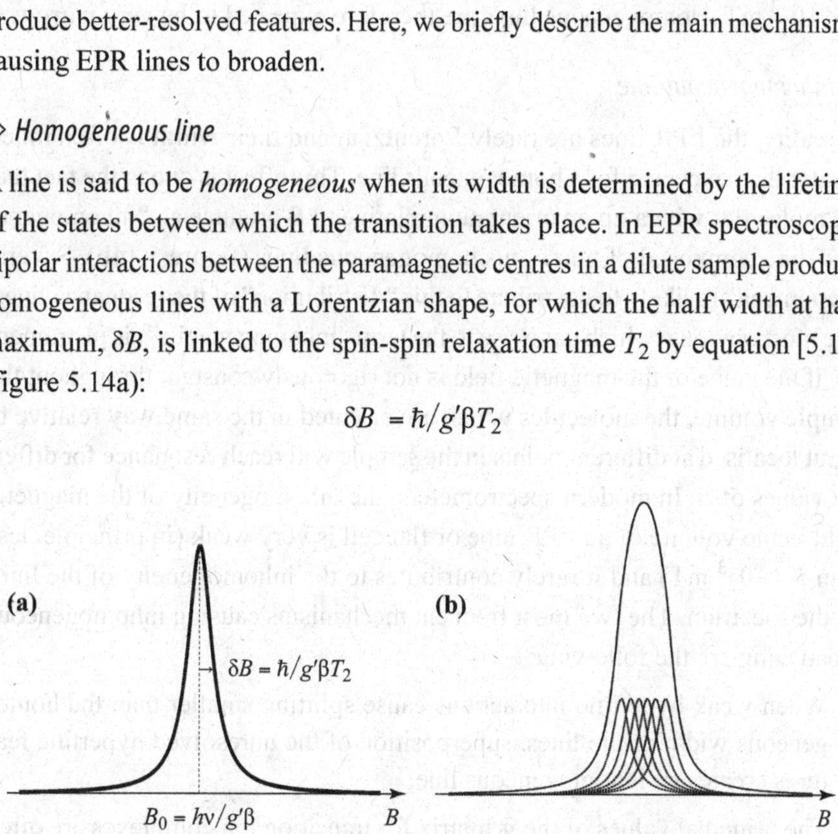

Figure 5.14 – Homogeneous and inhomogeneous lines **(a)** Homogeneous line: its width is determined by the spin-spin relaxation time T_2. **(b)** Inhomogeneous line resulting from the superposition of a set of homogeneous lines.

Dipolar interactions between two paramagnetic entities vary as the inverse cube of the distance separating them ($1/r^3$) (appendix 3). When N centres are uniformly distributed in a volume V, the cube of the mean distances between them is approximately equal to V/N. δB is therefore proportional to N/V, i.e., to the *concentration*. By adapting the expression obtained for nuclear spins [Abragam, 1983] to paramagnetic centres of spin ½ with an isotropic $\tilde{\mathbf{g}}$ matrix, we obtain

$$\delta B \approx (2\pi^2/3\sqrt{3})\, g\beta (N/V)$$

which can be written numerically in the form:

$$\delta B[\text{mT}] = 2.1 \times 10^{-3}\, g\, [\text{concentration in mM}]$$

For a 1 mM concentration and $g = 2.0$, this relation predicts a linewidth of 4×10^{-3} mT. Homogeneous lines are therefore expected to be very narrow.

◇ *Inhomogeneous line*

In reality, the EPR lines are rarely Lorentzian and their width is often much greater than expected for a homogeneous line. This effect is due to the fact that the molecules with a given orientation relative to **B** produce an "inhomogeneous" line composed of numerous homogeneous lines (figure 5.14b). Several phenomena are likely to contribute to this "distribution" of the resonance lines. The most obvious (which gave rise to the term "inhomogeneous") is instrumental: if the value of the magnetic field is not rigorously constant throughout the sample volume, the molecules which are oriented in the same way relative to **B** but localised at different points in the sample will reach resonance for different values of B. In modern spectrometers, the inhomogeneity of the magnetic field in the volume of an EPR tube or flat cell is very weak (in principle, less than 5×10^{-3} mT) and it rarely contributes to the inhomogeneity of the lines on the spectrum. The two most frequent mechanisms causing inhomogeneous broadening are the following:

- ▹ When weak hyperfine interactions cause splitting smaller than the homogeneous width of the lines, superposition of the unresolved hyperfine features creates an inhomogeneous line.
- ▹ The principal values of the $\tilde{\mathbf{g}}$ matrix for transition ion complexes are often very sensitive to the molecular structure, which can vary slightly from one molecule to the next. To demonstrate how this "*g*-strain" mechanism affects spectra, we need only observe that a relative variation $\delta g/g = 10^{-3}$ around $g = 2$ causes the resonance field to vary by around 0.3 mT at X-band.

The inhomogeneous broadening produced by these two mechanisms behave differently when the spectrometer *frequency* is varied:

- ▹ As the $A'/g'\beta$ splitting between lines in a hyperfine pattern is independent of the frequency (figure 5.15a), the same is true for the linewidth due to unresolved hyperfine features.
- ▹ We now consider paramagnetic centres affected by the "*g*-strain" phenomenon. Among the centres with a given orientation relative to **B**, those

characterised by $(g'+\delta g'/2)$ and $(g'-\delta g'/2)$ produce lines separated by (figure 5.15b):

$$\Delta B = (h\nu/\beta)\,[1/(g'-\delta g'/2) - 1/(g'+\delta g'/2)] \approx (h\nu/g'\beta)\,(\delta g'/g')$$

where ν is the spectrometer frequency. Broadening due to "g-strain" is thus proportional to the frequency.

The relative contributions of these two mechanisms can therefore be determined by studying how the linewidth is affected by the spectrometer's frequency [Hearshen *et al.*, 1986].

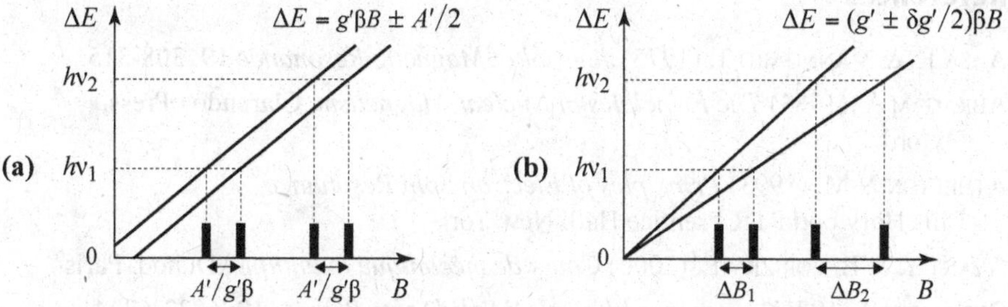

Figure 5.15 – Effect of the spectrometer frequency, ν, on the splitting between two lines. $\Delta E(B)$ is the transition energy. **(a)** The splitting between two hyperfine lines is $A'/g'\beta$. **(b)** The splitting between two lines characterised by $g'+\delta g'/2$ and $g'-\delta g'/2$ is $\Delta B = (h\nu/\beta g')(\delta g'/g')$.

◇ *Consequences of the inhomogeneous nature of a line on the intensity and saturation of the spectrum*

The expressions presented in this chapter were obtained by assuming the line to be homogeneous (section 5.2.1). How are they altered in the very frequent situation where the line is inhomogeneous?

▷ The *shape* of the spectrum, which is the result of superposition of the lines produced by molecules with the same orientation relative to **B** is, naturally, modified, but what about its *intensity*? Since an inhomogeneous line is due to splitting of a unique homogeneous line into several components, it might be imagined that its intensity would be the same as that of the unique homogeneous line in the absence of splitting. In section 5.2.4 we showed that this was indeed the case when the splitting is caused by hyperfine interactions. With the g-strain phenomenon, each component of the inhomogeneous line is characterised by a (g_x, g_y, g_z) triplet and thus a specific *intensity factor* g_P (equation [5.15]). But these variations are generally negligible and we can safely use expressions [5.17] and [5.19] when the spectrum is made of inhomogeneous lines.

▹ Things are different when one considers the *saturation* phenomenon. Indeed, saturation of a homogeneous line is described by equation [5.24]. In contrast, at any point on an inhomogeneous line, only the components contributing to the amplitude at this point will be saturated. Inhomogeneous lines are therefore *more difficult* to saturate than homogeneous lines (section 9.4).

References

AASA R. & VÄNNGARD T. (1975) *Journal of Magnetic Resonance* **19**: 308-315

ABRAGAM A. (1983) *The Principles of Nuclear Magnetism*, Clarendon Press, Oxford.

ATHERTON N.M. (1993) *Principles of Electron Spin Resonance*, Ellis Horwood PTR Prentice Hall, New York.

AYANT Y. & BELORIZKY E. (2000) *Cours de mécanique quantique*, Dunod, Paris

BERTRAND P. (1986) *Journal of Physics C: Solid State Physics* **19**: 6833-6844.

BERTRAND P., GAYDA J.-P. & RAO K.K.(1982) *Journal of Chemical Physics* **76**: 4715-4719.

BLOEMBERGEN N., PURCELL E.M. & POUND R.V. (1948) *Physical Review* **73**: 679-692.

CARRINGTON A. & MCLACHLAN A.D. (1979) *Introduction to Magnetic Resonance with Applications to Chemistry and Chemical Physics*, Chapman & Hall, London.

COHEN-TANNOUDJI C., DIU B. & LALOË F. (2015) *Quantum Mechanics*, Wiley-VcH, New York.

GORDY W. (1980) *Theory and Applications of Electron Spin Resonance*, John Wileys and Sons, New York.

HEARSHEN D.O. *et al.* (1986) *Journal of Magnetic Resonance* **69**: 440- 459.

KAHN O. (1993) *Molecular Magnetism*, Wiley-VCH Publishers, New York.

ORBACH R. & STAPLETON H.J. (1972) "Electron Spin-Lattice Relaxation" in *Electron Paramagnetic Resonance*, GESCHWIND S. ed. Plenum Press, New York.

SCHWEIGER A. & JESCHKE G. (2001) *Principles of Pulse Electron Paramagnetic Resonance*, Oxford University Press, Oxford.

STAPLETON H.J. *et al.* (1980) *Physical Review Letters* **45**: 1456-1459.

ZHOU Y. *et al.* (1999) *Journal of Magnetic Resonance* **139**: 165-174.

Exercises

5.1. **a)** Show that the function $f(B - B_0')$ defined by equation [5.10] is normalised.

b) We assume that the energy of the transition takes the form $\Delta E = g'\beta B + C$. What is the field B_0' of the resonance line? Show that the function $f(B - B_0')$ defined by $f(B - B_0') = g'\beta\rho(h\nu - C - g'\beta B)$ is normalised.

5.2. Calculate the quantity $\exp(A/2k_BT)$ for $A = 1000$ MHz, $T = 1$ K. What can be deduced about the intensity of the lines in a hyperfine pattern (section 5.2.4)?

5.3. In section 5.3, the population difference at thermal equilibrium appears in two forms:

(1) $\Delta n_e/n = \tanh(\Delta E/2k_BT)$: equation [5.21] deduced from the Boltzmann relation (section 1.4.4).

(2) $\Delta n_e/n = (w_{ba} - w_{ab})/(w_{ba} + w_{ab})$ deduced from the kinetic model illustrated in figure 5.6b.

Deduce that w_{ba} and w_{ab} are linked by $w_{ab}/w_{ba} = \exp(-\Delta E/k_BT)$.

5.4. Consider the paramagnetic centres that are subjected to the transitions represented in figure 5.6c.

a) Show that they absorb a power $\Delta n W_{ab} \Delta E$ from the radiation, where $\Delta n = n_a - n_b$. What power do they emit towards the lattice? Show that these two powers are equal in the steady-state and give the expression for their shared value, P_{stat}. Represent P_{stat} as a function of the power P of the radiation.

b) The "spin temperature" T_S for a system of paramagnetic centres is defined by $n_b/n_a = \exp(-\Delta E/k_BT_S)$. Determine its expression as a function of T, ΔE, $W_{ab}T_1$. Examine the two following limit cases:

- unsaturated regime: $W_{ab}T_1 \ll 1$.
- strongly saturated regime: $W_{ab}T_1 \gg 1$.

Note: $[1 - \tanh(x/2)]/[1 + \tanh(x/2)] = \exp(-x)$

5.5. This exercise aims to numerically evaluate the transition probability and the radiation power absorbed at resonance by a sample in the unsaturated regime. The sample is a solution of N paramagnetic molecules with an *isotropic* $\tilde{g}$ matrix. We assume the resonance line to be Lorentzian:

$$f(B-B_0)=\frac{1}{\pi}\frac{\delta B}{(B-B_0)^2+\delta B^2}$$

The cavity is such that the amplitude B_1 of the magnetic component of the radiation is B_1[tesla] = $10^{-4}\sqrt{P}$, where P[watt] is the power of the incident radiation.

a) Using equation [5.12], give the expression for the transition probability at the centre of the line as a function of P. *Numerical application*: $g = 2$, $\delta B = 10^{-1}$ mT.

b) Express the power P_a absorbed at resonance as a function of P.

Numerical application: the sample contains 150 μL of a 1 mM solution of molecules. What is the number N of molecules in the sample? Given that the frequency of the spectrometer is $\nu = 9$ GHz, determine the P_a/P ratio for $T = 4$ K, and for $T = 300$ K.

Answers to exercises

5.1. a) Let $I = \int_0^{+\infty} f(B - h\nu/g'\beta)\,\mathrm{d}B = g'\beta\int_0^{+\infty} \rho(h\nu - g'\beta B)\,\mathrm{d}B$. The change of variable $x = h\nu - g'\beta B$ leads to $I = \int_{-\infty}^{h\nu} \rho(x)\,\mathrm{d}x$. The upper bound can be considered to be $+\infty$. As the density $\rho(x)$ is normalised, we effectively have $I = 1$.

b) By writing $\Delta E = h\nu$, $B_0' = (h\nu - C)/g'\beta$ is obtained. We set $I' = g'\beta\int_0^{+\infty} \rho(h\nu - C - g'\beta B)\,\mathrm{d}B$. As a result of the change of variable $x = h\nu - C - g'\beta B$, we have $I' = \int_{-\infty}^{h\nu - C} \rho(x)\,\mathrm{d}x$, which is equal to 1 for the same reason as previously.

5.2. The calculation produces $A/2k_BT = 2.4 \times 10^{-2}$, $\exp(A/2k_BT) = 1.024$. In general, the hyperfine constants are less than 1000 MHz, the temperatures exceed 1 K, and this ratio is even closer to 1. The lines in a hyperfine structure can therefore be considered to have the same intensity.

5.3. The form (2) is written:

$$\Delta n_e/n = (1 - w_{ab}/w_{ba})/(1 + w_{ab}/w_{ba}) = [1 - \exp(-x)]/[1 + \exp(-x)]$$

with $\exp(-x) = w_{ab}/w_{ba}$. By factorising $\exp(-x/2)$ in both the numerator and the denominator, $\Delta n_e/n = \tanh(x/2)$ is obtained. Comparison with form (1) shows that $x = \Delta E/k_BT$.

5.4. a) Power absorbed from the radiation: $\Delta n\, W_{ab}\Delta E$, power emitted towards the lattice: $(w_{ba}\, n_b - w_{ab}\, n_a)\Delta E$. In the steady state, the first side of [5.22] is null and these two powers are *equal*. $P_{stat} = \Delta n_{sat} W_{ab}\Delta E$ is obtained where Δn_{sat} is given by [5.23]. From this result, we deduce that $P_{stat} = \Delta n_e \Delta E\, W_{ab}/(1 + 2W_{ab}T_1)$. As W_{ab} is proportional to P, the variation of P_{stat} as a function of P is the following:

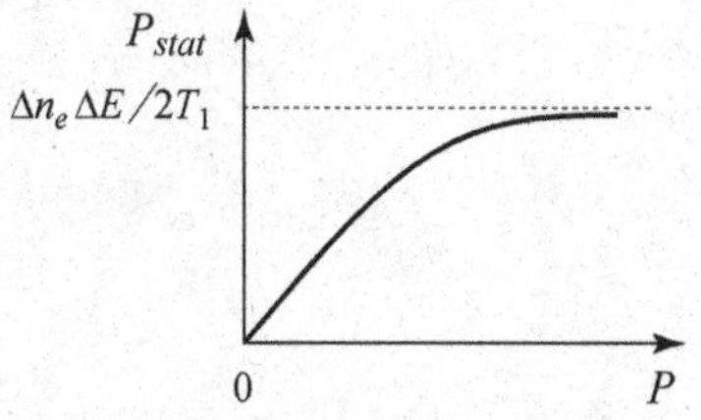

b) From $\Delta n = n_a - n_b$ and $n = n_a + n_b$, we deduce

$n_b/n_a = (1 - \Delta n/n)/(1 + \Delta n/n)$. By replacing Δn by Δn_{sat}, we obtain:

$\exp(-\Delta E/k_B T_S) = n_b/n_a$

$= [1+2W_{ab}T_1 - \tanh(\Delta E/2k_B T)]/[1+2W_{ab}T_1 + \tanh(\Delta E/2k_B T)]$

- In the absence of saturation: $2W_{ab}T_1 \ll 1$ and $\exp(-\Delta E/k_B T_S) = \exp(-\Delta E/k_B T)$: we have $T_S = T$.
- Upon very strong saturation $2W_{ab}T_1 \gg 1$: $\exp(-\Delta E/k_B T_S) \rightarrow 1$ and $T_S \rightarrow +\infty$.

5.5. a) $W_{ab}(B_0) = (\beta/2\hbar)\, g_P\, B_1^2/\delta B$ with $g_P = g/4$ (section 5.2.2). Numerically, $W_{ab}[\mathrm{s}^{-1}] = 2.2 \times 10^6\, P$.

b) $P_a = \Delta n W_{ab} \Delta E = 2.2 \times 10^6\, N \tanh(h\nu/2k_B T)\, h\nu P$.
We find $N = 9 \times 10^{16}$ molecules, $P_a/P = 1.2 \tanh(h\nu/2k_B T) = 6 \times 10^{-2}$ at $T = 4$ K and 9×10^{-4} at $T = 300$ K.

Even in this favourable case (narrow line, high concentration), only a small part of the power of the incident radiation is absorbed by the sample at resonance.

Chapter 6

The zero-field splitting term. EPR spectrum for paramagnetic centres of spin greater than ½

6.1 – Introduction

Up to now, we have been interested in paramagnetic molecules, free radicals and transition ion complexes, where the unpaired electrons are subjected to anisotropic interactions due to the applied magnetic field and the surrounding paramagnetic nuclei. From the eigenvalues and eigenvectors of the Hamiltonian describing how these interactions affect the ground state, we were able to determine how the position (chapter 4) and intensity (chapter 5) of the resonance lines varied depending on the orientation of the molecules relative to the field, and we deduced the *shape* and *intensity* of the EPR spectrum for different types of sample. However, several times we pointed out that these results only related to centres of spin ½. Indeed, when the spin is greater than ½, which is the case of organic molecules in a triplet state and many transition ion complexes, the position and intensity of the resonance lines also depend on another term in the Hamiltonian, the "zero-field splitting" term. The importance of this term relative to the Zeeman term varies considerably depending on the nature of the molecules and the magnitude of the magnetic field, such that paramagnetic centres with a spin greater than ½ can produce a great diversity of spectral forms. These spectra are generally more difficult to analyse than those for centres of spin ½. Nevertheless, the spectrum is simplified in some situations that are frequently encountered in applications. These specific cases will be dealt with in detail for situations where the calculations are simple, then the results obtained will be generalised.

P. Bertrand, *Electron Paramagnetic Resonance Spectroscopy*,
https://doi.org/10.1007/978-3-030-39663-3_6

In this chapter, the matrix elements of operators constructed from $(\hat{S}_x, \hat{S}_y, \hat{S}_z)$ will frequently be referred to, these elements are tabulated in appendix 7.

6.2 – The zero-field splitting term

6.2.1 – The $\tilde{\mathbf{D}}$ matrix

Up to now, we have ignored the *zero-field splitting* term in the spin Hamiltonian. This term is larger than the hyperfine terms and sometimes even than the Zeeman term. It is written

$$\hat{H}_{ZFS} = \mathbf{S}\cdot\tilde{\mathbf{D}}\cdot\mathbf{S}$$

$\tilde{\mathbf{D}}$ is a *symmetric* matrix which includes contributions of very different origins:

- ▹ Dipolar interactions between the magnetic moments of the unpaired electrons.
- ▹ Spin-orbit coupling (see appendix 2).

The relative weight of these contributions depends on the nature of the paramagnetic centre: the first is the most significant in organic molecules in a triplet state, whereas the second generally dominates in transition ion complexes. In the latter case, the principal values for the $\tilde{\mathbf{g}}$ and $\tilde{\mathbf{D}}$ matrices are linked by a simple, but approximate, relation [Abragam and Bleaney, 1986]. The $\tilde{\mathbf{D}}$ matrix is *diagonal* in a Cartesian system of molecular axes $\{X, Y, Z\}$ where the $\hat{H}_{ZFS}$ operator is written:

$$\hat{H}_{ZFS} = D_X \hat{S}_X^2 + D_Y \hat{S}_Y^2 + D_Z \hat{S}_Z^2 \qquad [6.1]$$

The quantities (D_X, D_Y, D_Z) are the *principal values* of the $\tilde{\mathbf{D}}$ matrix.

◇ *A remarkable property of the $\hat{H}_{ZFS}$ operator*

Consider the $\hat{H}'_{ZFS}$ operator defined by:

$$\begin{aligned}\hat{H}'_{ZFS} &= (D_X + C)\,\hat{S}_X^2 + (D_Y + C)\,\hat{S}_Y^2 + (D_Z + C)\hat{S}_Z^2 \\ &= \hat{H}_{ZFS} + C\,\hat{\mathbf{S}}^2\end{aligned}$$

The action of $\hat{\mathbf{S}}^2 = \hat{S}_X^2 + \hat{S}_Y^2 + \hat{S}_Z^2$ on any ket $|u\rangle$ in the $\mathcal{E}_S$ space is such that (section 3.2.1):

$$\hat{\mathbf{S}}^2\,|u\rangle = S(S+1)\,|u\rangle \qquad [6.2]$$

As a result, if $|u\rangle$ is an *eigenvector* of $\hat{H}_{ZFS}$, the equation

$$\hat{H}_{ZFS}\,|u\rangle = E\,|u\rangle$$

implies that $$\hat{H}'_{ZFS}\,|u\rangle = [E + C\,S(S+1)]\,|u\rangle$$

The eigenvectors of $\hat{H}'_{ZFS}$ are therefore identical to those of $\hat{H}_{ZFS}$ and its eigenvalues are offset by $C\,S(S+1)$. If we are only interested in the eigenvectors and the *differences* between the eigenvalues, the operators $\hat{H}_{ZFS}$ and $\hat{H}'_{ZFS}$ are *equivalent*. It is convenient to select:

$$C = -(D_X + D_Y + D_Z)/3$$

so as to *nullify the trace* of the $\tilde{\mathbf{D}}$ matrix. Hereafter, we will write the zero field splitting term in the form given by expression [6.1], assuming that the parameters are linked by

$$D_X + D_Y + D_Z = 0 \qquad [6.3]$$

As the mean of the principal values is null, the $\hat{H}_{ZFS}$ term is "purely" *anisotropic*. For example, it has no effect on the shape of the EPR spectrum in the "isotropic regime" described in section 4.5. When relation [6.3] is verified, the trace of the $\hat{H}_{ZFS}$ *operator* is *null*. Indeed, as the traces of the $(\hat{S}_X^2, \hat{S}_Y^2, \hat{S}_Z^2)$ operators are equal (section 3.2.5), we can write:

$$\text{trace}\,(\hat{H}_{ZFS}) = (D_X + D_Y + D_Z)\,\text{trace}\,(\hat{S}_X^2) = 0$$

◇ The specific case of centres of spin $S = ½$

The matrix representing the $\hat{H}_{ZFS}$ operator is constructed in the basis made of the kets $\{|½, ½\rangle, |½, -½\rangle\}$, the eigenvectors shared by $(\hat{\mathbf{S}}^2, \hat{S}_Z)$. We saw in section 3.2.5 that $(\hat{S}_X^2, \hat{S}_Y^2, \hat{S}_Z^2)$ are represented in this basis by the same diagonal matrix:

	$|½, ½\rangle$	$|½, -½\rangle$
$\langle ½, ½|$	¼	0
$\langle ½, -½|$	0	¼

Therefore, the $\hat{H}_{ZFS}$ operator defined by equation [6.1] is represented by a diagonal matrix in which the diagonal elements $(D_X + D_Y + D_Z)/4$ are null according to equation [6.3]: for centres of spin $S = ½$ (and only them, see exercise 6.2), the $\hat{H}_{ZFS}$ term does not contribute to the splitting between energy levels and can be omitted from the Hamiltonian, as done in chapters 4 and 5. The present chapter therefore relates to organic molecules in a triplet state ($S = 1$) and transition ion complexes with a spin greater than ½.

◇ *How the molecule's symmetry properties affect the* $\tilde{\mathbf{D}}$ *matrix*

The symmetry properties of the paramagnetic centre are manifested in the *direction* of the principal axes and the *eigenvalues* of the $\tilde{\mathbf{D}}$ matrix, like in the case of the $\tilde{\mathbf{g}}$ matrix:

- When a molecule has a symmetry axis of order greater than or equal to 2, it is a principal axis for the $\tilde{\mathbf{g}}$ and $\tilde{\mathbf{D}}$ matrices.
- If this axis z is of order greater than or equal to 3, we have $g_x = g_y$ and $D_x = D_y$. The molecule is said to have *axial symmetry* and we can arbitrarily choose x and y in the plane perpendicular to z.
- When three perpendicular directions (x, y, z) are equivalent, the molecule is said to have *cubic symmetry*. We thus have:

$$g_x = g_y = g_z; \; D_x = D_y = D_z$$

  and equation [6.3] gives $D_x = D_y = D_z = 0$: the zero-field splitting term is absent from the Hamiltonian. This specific situation occurs in transition ion complexes with octahedral (O_h) or tetrahedral (T_d) symmetry.

Note – To interpret some spectra, it is sometimes necessary to add 4th degree terms in $(\hat{S}_x, \hat{S}_y, \hat{S}_z)$ to the spin Hamiltonian. In particular, these terms are required in cubic symmetry where the terms in $(\hat{S}_x^2, \hat{S}_y^2, \hat{S}_z^2)$ are null.

6.2.2 – The D and E parameters

In general, the $\hat{H}_{ZFS}$ operator depends on two *independent parameters* which are usually defined by favouring the principal axis of the $\tilde{\mathbf{D}}$ matrix which corresponds to the principal value with the largest absolute value. If Z is the principal axis such that

$$|D_Z| \geq |D_X|, |D_Y| \qquad [6.4]$$

expression [6.1] is transformed as follows:

$$\begin{aligned}\hat{H}_{ZFS} &= D_Z \hat{S}_Z^2 + [(D_X + D_Y)/2\,](\hat{S}_X^2 + \hat{S}_Y^2) + [(D_X - D_Y)/2](\hat{S}_X^2 - \hat{S}_Y^2)\\ &= [D_Z - (D_X + D_Y)/2]\hat{S}_Z^2 + [(D_X + D_Y)/2]\hat{\mathbf{S}}^2 + [(D_X - D_Y)/2](\hat{S}_X^2 - \hat{S}_Y^2)\end{aligned}$$

Using equations [6.2] and [6.3], this expression can be written:

$$\hat{H}_{ZFS} = D\,[\hat{S}_Z^2 - S(S+1)/3] + E\,(\hat{S}_X^2 - \hat{S}_Y^2) \qquad [6.5]$$

by setting

$$D = 3D_Z/2; \; E = (D_X - D_Y)/2 \qquad [6.6]$$

This is the form that is generally used. Strictly speaking, we should write $[S(S+1)/3]\hat{I}$ in the first term on the second side of equation [6.5], where $\hat{I}$ is *the identity operator* represented by the *unit matrix* $\tilde{1}$ in any basis. As a result of inequalities [6.4], the E/D ratio satisfies:

$$|E/D| \leq ⅓ \qquad [6.7]$$

Indeed, using equations [6.3] and [6.6], we can write:

$$\frac{E}{D} = \frac{D_X - D_Y}{3D_Z} = \frac{D_X - D_Y}{-3(D_X + D_Y)}$$

Equation [6.3] and inequality [6.4] imply that D_X and D_Y have the same sign, which leads to

$$|D_X - D_Y| \leq |D_X + D_Y|$$

and demonstrates inequality [6.7]. This inequality suggests three limit situations:

- ▹ $E = 0$, which happens when $D_X = D_Y$. This is the case for molecules with *axial symmetry* along the Z axis, as mentioned above. In mononuclear transition ion complexes with axial symmetry, the value and the sign of D depend on the nature of the ligands and their geometry. For example, in a complex with O_h ($D = 0$) symmetry, stretching of a metal-ligand bond along an 4th-order axis leads to $D > 0$, whereas its compression produces $D < 0$ [Neese, 2006].
- ▹ $E/D = ⅓$ when $D_X = 0$, which implies $D_Y = -D_Z$.
- ▹ $E/D = -⅓$ when $D_Y = 0$, which implies $D_X = -D_Z$.

Unlike the case $E = 0$, the $|E/D| = ⅓$ limit does not correspond to a specific symmetry property of the paramagnetic centre. The terminology "completely rhombic symmetry" is often used to designate this situation, even though it has no *geometric* meaning.

Note – If the principal axes are annotated such that $|D_Y| \geq |D_X|$, the expression given above shows that $E/D \geq 0$, and as a result $0 \leq E/D \leq ⅓$.

6.3 – Definition and general characteristics of "high-field" and "low-field" situations

Consider a paramagnetic centre of spin $S > ½$ characterised by its $\tilde{\mathbf{g}}$ and $\tilde{\mathbf{D}}$ matrices. When this centre is placed in a magnetic field **B**, its lowest energy levels are described by the spin Hamiltonian:

$$\hat{H}_S = \hat{H}_{Zeeman} + \hat{H}_{ZFS} \quad [6.8]$$

$$\hat{H}_{Zeeman} = \beta\,(g_x B_x \hat{S}_x + g_y B_y \hat{S}_y + g_z B_z \hat{S}_z)$$

$$\hat{H}_{ZFS} = D\,[\hat{S}_Z^2 - S(S+1)/3] + E(\hat{S}_X^2 - \hat{S}_Y^2)$$

- ▹ In $\hat{H}_{Zeeman}$, $\{x,y,z\}$ are the principal axes of the $\tilde{\mathbf{g}}$ matrix. This term can vary between 0 and around 0.3 cm^{-1} when using an X-band spectrometer, and between 0 and 12 cm^{-1} when using a spectrometer with a frequency of 360 GHz.
- ▹ In $\hat{H}_{ZFS}$, $\{X,Y,Z\}$ are the principal axes of the $\tilde{\mathbf{D}}$ matrix. The D and E parameters are closely linked to the nature of the paramagnetic centre. Their absolute value varies between 0.01 and around 1 cm^{-1} in organic molecules in a triplet state [Gordy, 1980], and between 0.1 and 5 cm^{-1} in transition ion complexes [Abragam and Bleaney, 1986]. Higher values, of around 10 cm^{-1}, have been measured for some iron-containing complexes of spin $S = 5/2$ [Scholes, 1971].

The precise expression for the eigenvalues and eigenvectors for the Hamiltonian [6.8] cannot be determined in the most general case, but approximate expressions can be found by applying perturbation theory when one of the terms is much larger than the other. These two limit situations will be studied in detail in the following sections. We will see that the shape of the spectrum is very different in the two cases and that its examination provides information on the order of magnitude of the D parameter. In this section, we highlight the main differences between these two situations by considering the example of a centre with *axial symmetry* for which the energy levels can be readily calculated for the canonical directions of the magnetic field.

6.3.1 – The energy levels for a centre with axial symmetry for the canonical directions of the magnetic field

For a centre with axial symmetry, the Hamiltonian [6.8] becomes:

$$\hat{H}_S = \beta(g_\perp B_x \hat{S}_x + g_\perp B_y \hat{S}_y + g_{//} B_z \hat{S}_z) + D\,[\hat{S}_z^2 - S(S+1)/3] \quad [6.9]$$

▷ When **B** is *parallel* to the symmetry axis, z, this equation is written:

$$\hat{H}_S = g_{//}\,\beta B \hat{S}_z + D\,[\hat{S}_z^2 - S(S+1)/3]$$

This operator is represented by a *diagonal* matrix in the $\{|S, M_S\rangle_z, M_S = -S, -S+1, \ldots S\}$ basis of eigenvectors shared by $(\hat{\mathbf{S}}^2, \hat{S}_z)$, and its eigenvalues are equal to:

$$E(M_S) = g_{//}\,\beta B M_S + D\,[M_S^2 - S(S+1)/3] \quad [6.10]$$

Figures 6.1a and 6.2a show the variations of the $E(M_S)/D$ ratios as a function of $r = g_{//}\,\beta B/D$ for $S = 1$ and $S = ³⁄₂$.

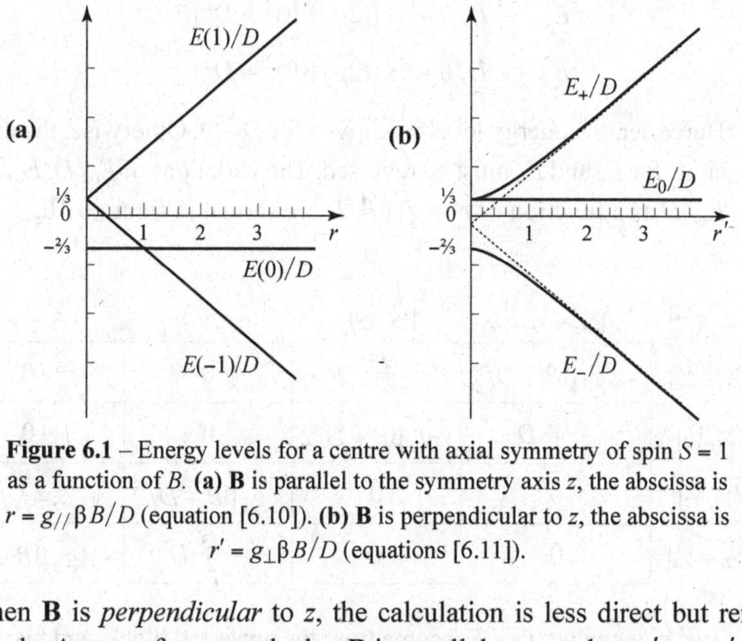

Figure 6.1 – Energy levels for a centre with axial symmetry of spin $S = 1$ as a function of B. **(a) B** is parallel to the symmetry axis z, the abscissa is $r = g_{//}\,\beta B/D$ (equation [6.10]). **(b) B** is perpendicular to z, the abscissa is $r' = g_\perp \beta B/D$ (equations [6.11]).

▷ When **B** is *perpendicular* to z, the calculation is less direct but remains simple. Assume for example that **B** is parallel to x, and set $z' = x$, $y' = z$ so as to return to a familiar notation. The Hamiltonian [6.9] is written:

$$\hat{H}_S = g_\perp \beta B \hat{S}_{z'} + D\,[\hat{S}_{y'}^2 - S(S+1)/3]$$

To determine its eigenvalues, the matrix representing it in the $\{|S, M_S\rangle_{z'}\}$ basis of eigenvectors shared by $(\hat{\mathbf{S}}^2, \hat{S}_{z'})$ is constructed using the matrix elements from appendix 7. We perform the calculation for $S = 1$ and $S = ³⁄₂$, ordering the kets so that "blocks" emerge in the matrix:

- $S = 1$

	$\|1,-1\rangle_{z'}$	$\|1,1\rangle_{z'}$	$\|1,0\rangle_{z'}$
$_{z'}\langle 1,-1\|$	$-g_\perp\beta B - D/6$	$-D/2$	0
$_{z'}\langle 1,1\|$	$-D/2$	$g_\perp\beta B - D/6$	0
$_{z'}\langle 1,0\|$	0	0	$D/3$

The eigenvalue $E_0 = D/3$ is associated with the eigenvector $|1,0\rangle_{z'}$ and by resolving a second degree equation, we can obtain the two other eigenvalues:

$$E_+ = -D/6 + \tfrac{1}{2}\,[(2g_\perp\beta B)^2 + D^2]^{1/2}$$

$$E_- = -D/6 - \tfrac{1}{2}\,[(2g_\perp\beta B)^2 + D^2]^{1/2} \qquad [6.11]$$

Hereafter, the energy levels are given for $D > 0$. Otherwise, the expressions for E_+ and E_- must be reversed. The variations of E_0/D, E_+/D and E_-/D as a function of $r' = g_\perp\beta B/D$ are shown in figure 6.1b.

- $S = 3/2$

	$\|3/2,-3/2\rangle_{z'}$	$\|3/2,1/2\rangle_{z'}$	$\|3/2,3/2\rangle_{z'}$	$\|3/2,-1/2\rangle_{z'}$
$_{z'}\langle 3/2,-3/2\|$	$-3/2 g_\perp\beta B - D/2$	$-\frac{\sqrt{3}}{2}D$	0	0
$_{z'}\langle 3/2,1/2\|$	$-\frac{\sqrt{3}}{2}D$	$1/2 g_\perp\beta B + D/2$	0	0
$_{z'}\langle 3/2,3/2\|$	0	0	$3/2 g_\perp\beta B - D/2$	$-\frac{\sqrt{3}}{2}D$
$_{z'}\langle 3/2,-1/2\|$	0	0	$-\frac{\sqrt{3}}{2}D$	$-1/2 g_\perp\beta B + D/2$

Two eigenvalues (E_+, E_-) come from the upper left block, and two others (E_+', E_-') from the lower right block:

$$E_+ = -\tfrac{1}{2}g_\perp\beta B + [(g_\perp\beta B)^2 + g_\perp\beta BD + D^2]^{1/2}$$

$$E_- = -\tfrac{1}{2}g_\perp\beta B - [(g_\perp\beta B)^2 + g_\perp\beta BD + D^2]^{1/2}$$

$$E_+' = \tfrac{1}{2}g_\perp\beta B + [(g_\perp\beta B)^2 - g_\perp\beta BD + D^2]^{1/2} \qquad [6.12]$$

$$E_-' = \tfrac{1}{2}g_\perp\beta B - [(g_\perp\beta B)^2 - g_\perp\beta BD + D^2]^{1/2}$$

The variations of $E_+/D, E_-/D, E_+'/D, E_-'/D$ as a function of r' are illustrated in figure 6.2b.

6.3.2 – "High-field" and "low-field" situations

Some properties of the curves in figures 6.1 and 6.2 are general. For example, the energy levels are only partially degenerate for $B = 0$: the term $\hat{H}_{ZFS}$ produces a zero-field splitting, as its name indicates. As the trace of the $\hat{H}_S$ Hamiltonian is null, the mean of the eigenvalues is null whatever the value of B, as verified using equations [6.11] and [6.12]. Unlike for centres of spin ½, the energies *do not vary linearly* with B (the linearity observed in figures 6.1a and 6.2a only occurs for centres with axial symmetry when **B** is parallel to z). Examination of figures 6.1b and 6.2b reveals other properties:

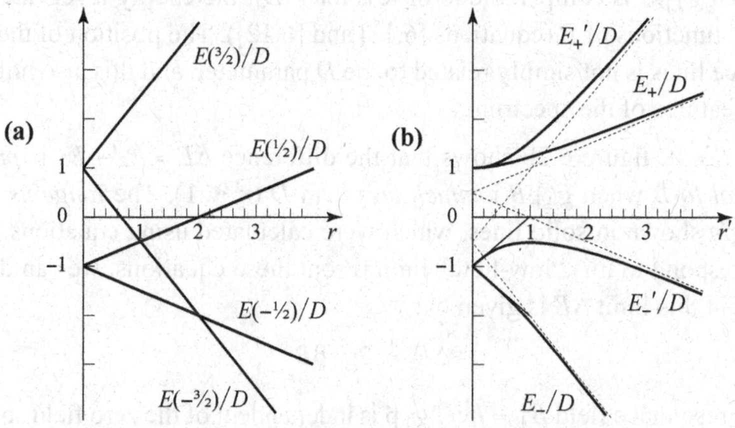

Figure 6.2 – Energy levels for a centre with axial symmetry of spin $S = 3/2$ as a function of B. **(a)** **B** is parallel to the symmetry axis z, the abscissa is $r = g_{//}\beta B/D$ (equation [6.10]). **(b)** **B** is perpendicular to z, the abscissa is $r' = g_\perp \beta B/D$ (equations [6.12]).

▷ When $g_\perp \beta B$ is much greater than $|D|$ ($r' \gg 1$), the Zeeman term dominates in the Hamiltonian and the energy levels vary linearly with B. The dashed lines in figures 6.1b and 6.2b represent the *asymptotes* deduced from equations [6.11] and [6.12], which correspond to this "high-field" limit. In this limit, we will see that only transitions between adjacent levels are allowed. According to figures 6.1b and 6.2b, the energy of these transitions takes the form

$$\Delta E = g_\perp \beta B + d \qquad [6.13]$$

where d is equal to $(-D/2, D/2)$ for $S = 1$ and to $(-D, 0, D)$ for $S = 3/2$. As resonance occurs when $\Delta E = h\nu$, the resonance field is given by

$$B_{\perp} = B_{\perp 0} - d/g_{\perp}\beta$$

where $B_{\perp 0} = h\nu/g_{\perp}\beta$ is much greater than $d/g_{\perp}\beta$. The lines are centred at $B_{\perp 0}$, which would be their position for $D = 0$, and their separation is proportional to D. This situation is similar to that for centres of spin ½ with hyperfine interaction (e.g. compare to equation [4.12]): just as the remarkable features in the powder spectra for centres of spin ½ can be used to determine the principal values of the $\tilde{\mathbf{g}}$ and $\tilde{\mathbf{A}}$ matrices, they can also be used to determine the principal values of the $\tilde{\mathbf{g}}$ matrix and the zero-field splitting parameters for centres with spin greater than ½ in the high-field situation (section 6.5).

- When $g_{\perp}\beta B$ is comparable to or less than $|D|$, the energy levels are complex functions of B (equations [6.11] and [6.12]). The position of the resonance lines is not simply related to the D parameter, and this also holds for all features of the spectrum.
- For $S = 3/2$, figure 6.2b shows that the difference $\Delta E = E_-' - E_-$ is *proportional to B* when $g_{\perp}\beta B$ is *much less* than D ($r' \ll 1$). The *tangents at the origin* shown in solid lines, which were calculated using equations [6.12] correspond to this "low-field" limit. From these equations, we can deduce that in this limit ΔE is given by:

$$\Delta E = 2g_{\perp}\beta B \qquad [6.14]$$

  The resonance field $B_{\perp} = h\nu/2g_{\perp}\beta$ is independent of the zero field splitting terms and it is *half* the value of $B_{\perp 0}$. Resonance lines at positions that are very different from the positions observed when $D = 0$ are characteristic of *half-integer spin* complexes in the "low-field" situation (section 6.6). We will see that the features of the spectrum can be identified by "*effective g values*", which can be very different from $g_e = 2.0023$.

Hereafter, we will focus on the "high-field" and "low-field" situations that can be dealt with *for any direction* of **B** relative to the molecule using perturbation theory. This treatment will allow us to determine the principal characteristics of the spectrum.

6.4 – General properties of the spectrum in the high-field situation

In the case of organic molecules in a triplet state, D is generally less than 0.5 cm^{-1} and the high-field situation can be achieved at Q-band, and sometimes even at X-band. But in the case of transition ion complexes, this situation can only be achieved if D is small enough, and high magnetic fields, which are only available in specialised laboratories, are required. In this situation, we can deduce some general characteristics from the Hamiltonian $\hat{H}_S$ (equation [6.8]), such as the number and intensity of the resonance lines for a given orientation of the field, as well as the total intensity of the spectrum.

6.4.1 – Energy levels and allowed transitions

In the high-field situation, the principal term in $\hat{H}_S$ is the Zeeman term. For a given direction of **B** relative to the principal axes of the $\tilde{\mathbf{g}}$ matrix, this term is written (section 3.3.2):

$$\hat{H}_{Zeeman} = g'\beta B\hat{S}_{z'}$$

where g' is defined by equation [3.19] and z' is the direction of the unit vector **v** for which the components along the principal axes of the $\tilde{\mathbf{g}}$ matrix are given by equations [3.18]. The eigenvectors of $\hat{H}_{Zeeman}$ are those shared by $(\hat{\mathbf{S}}^2, \hat{S}_{z'})$, and its eigenvalues are equal to $g'\beta BM_S$. As these eigenvalues are *non-degenerate*, the eigenvalues and eigenvectors of the Hamiltonian $\hat{H}_S$ are given to first order in perturbation theory by:

$$E(M_S) = g'\beta BM_S + {}_{z'}\langle S, M_S|\hat{H}_{ZFS}|S, M_S\rangle_{z'}; \ |S, M_S\rangle_{z'}, M_S = -S, -S+1, \dots S \quad [6.15]$$

The allowed transitions can be determined without further knowledge of the matrix elements:

- ▹ These matrix elements are much smaller than $g'\beta B$, therefore the $(2S+1)$ energy levels are ranked in increasing order of the values of M_S.
- ▹ The probability per second for the transition $|S, M_S\rangle_{z'} \leftrightarrow |S, M_S'\rangle_{z'}$ is proportional to the square of the modulus of the matrix element V defined by (equation [5.9]):

$$V = {}_{z'}\langle S, M_S|\hat{H}_1|S, M_S'\rangle_{z'} \quad [6.16]$$

with: $$\hat{H}_1 = \beta(B_{1x}\, g_x\, \hat{S}_x + B_{1y}\, g_y\, \hat{S}_y + B_{1z}\, g_z\, \hat{S}_z) \quad [6.17]$$

where $\{x, y, z\}$ are the principal axes of the $\tilde{\mathbf{g}}$ matrix. The components (S_x, S_y, S_z) are linear combinations of the components $(S_{x'}, S_{y'}, S_{z'})$ in any $\{x', y', z'\}$ reference frame defined by z'. The $\hat{H}_1$ operator can therefore be written in the form:

$$\hat{H}_1 = \beta B_1 (c_{x'}\hat{S}_{x'} + c_{y'}\hat{S}_{y'} + c_{z'}\hat{S}_{z'}) \qquad [6.18]$$

The coefficients $(c_{x'}, c_{y'}, c_{z'})$ depend on the principal values of the $\tilde{\mathbf{g}}$ matrix, the angles (θ, φ) which define the direction of **B**, and the angle α which identifies the direction of $\mathbf{B}_1$ in the plane perpendicular to **B** (figure 5.4). As equations [6.16] and [6.18] are similar to those used in section 3.4, they produce the same selection rule $\Delta M_S = \pm 1$: only the $2S$ transitions between adjacent levels are allowed. Since these transitions have different energies, $2S$ resonance lines will exist for a given orientation of **B**.

When the sample is a single crystal, the spectrum produced by the molecules oriented in the same way relative to **B** is composed of a pattern of $2S$ lines. This pattern is sometimes termed the *fine structure* of the spectrum, as the *hyperfine* qualifier is used for interactions with paramagnetic nuclei. This explains why the zero-field splitting terms are called "fine structure terms" by some authors [Abragam and Bleaney, 1986].

6.4.2 – Intensity of the resonance lines and the spectrum

Consider a polycrystalline powder or frozen solution in which N molecules are trapped in random orientations, and note $n(\theta, \varphi)$ the number of molecules which have a given orientation relative to **B** to within $d\theta$ and $d\varphi$ (section 5.2.1). For these molecules, the energy of the $|S, M_S\rangle_{z'} \leftrightarrow |S, M_S + 1\rangle_{z'}$ transition is given by (equation [6.15]):

$$\begin{aligned}\Delta E(M_S) &= E(M_S + 1) - E(M_S)\\ &= g'\beta B + ({}_{z'}\langle S, M_S + 1|\hat{H}_{ZFS}|S, M_S + 1\rangle_{z'} - {}_{z'}\langle S, M_S|\hat{H}_{ZFS}|S, M_S\rangle_{z'})\end{aligned}$$

ΔE is a *linear function* of B, and expression [5.16] can be used to determine the intensity of the line which, in this case, is written:

$$I_{M_S}(\theta, \varphi) = \frac{\pi\beta}{2\hbar} g_P(M_S) B_1 \Delta n(M_S) \qquad [6.19]$$

▹ $g_P(M_S)$ is the *intensity factor* determined by equation [6.16] with $M_S' = M_S + 1$. In complement 1, we show that it can be written:

$$g_P(M_S) = g_P\,[S(S + 1) - M_S\,(M_S + 1)] \qquad [6.20]$$

where g_P is the intensity factor *for a paramagnetic centre of spin ½* (equation [5.15]).

▹ $\Delta n(M_S)$ is the *population difference* between the $E(M_S)$ and $E(M_S + 1)$ levels. In the "high temperature" limit, this difference is the same for all transitions (complement 1):

$$\Delta n = \frac{n(\theta,\varphi)h\nu}{(2S+1)k_B T} \quad [6.21]$$

Now consider *the pattern* of $2S$ resonance lines which appears for the (θ, φ) orientation. Its intensity $I(\theta, \varphi)$ is the sum of the intensities of the $2S$ lines (equations [6.19 – 6.21]):

$$I(\theta,\varphi) = \frac{\pi\beta}{2\hbar} g_P(\theta,\varphi) B_1 \Delta n \sum_{M_S=-S}^{S-1} [S(S+1) - M_S(M_S+1)]$$

The sum is equal to $2S(S + 1)(2S + 1)/3$ (exercise 6.3). Using equation [6.21], we obtain

$$I(\theta,\varphi) = \frac{\pi\beta}{2\hbar} g_P(\theta,\varphi) B_1 n(\theta,\varphi) \frac{2S(S+1)}{3} \frac{h\nu}{k_B T}$$

The total intensity I of the spectrum is obtained by replacing $n(\theta, \varphi)$ by its expression (equation [5.4]) and by integrating $I(\theta, \varphi)$ over all orientations relative to **B**, this produces:

$$I = \frac{\pi\beta}{2\hbar} (g_P)_{av} B_1 N \frac{4S(S+1)}{3} \frac{h\nu}{2k_B T} \quad [6.22]$$

The number $(g_P)_{av}$ is defined by equation [5.18] and can be evaluated using equation [5.19]. The total intensity is independent of the zero field splitting parameters which are much less than $g'\beta B$ in the high-field limit. The situation is similar to that in section 5.2.4, where we saw that the presence of *hyperfine interactions* modifies the *shape* of the spectrum without changing its *intensity*.

All other things being equal, the intensity of the spectrum produced by a centre of spin S in the high-field situation and in the high temperature limit is greater than that for a centre of spin ½, by a factor equal to $4S(S + 1)/3$. When the high temperature limit is not reached, the general expression for $\Delta n(M_S)$ (see complement 2) must be replaced in equation [6.19] and the total intensity cannot be written in the compact form of equation [6.22].

6.5 – Shape of the powder spectrum in the high-field situation

The shape of the powder spectrum can be deduced from the *density* $D(B)$ of the resonance lines. Hereafter, we calculate its expression for a centre with *axial* symmetry. In the case of "rhombic" symmetry, the calculation of the resonance fields will be limited to the canonical directions of **B**, and the positions of the notable features of the spectrum will be deduced.

6.5.1 – Expression for the resonance field in axial symmetry

To calculate the density of the resonance lines, the expression for the resonance field must be known, and therefore the matrix element $_{z'}\langle S, M_S|\hat{H}_{ZFS}|S, M_S\rangle_{z'}$ in equation [6.15] must be explicitly determined. To do so, $\hat{H}_{ZFS} = D[\hat{S}_z^2 - S(S+1)/3]$ must be expressed as a function of the $(\hat{S}_{x'}, \hat{S}_{y'}, \hat{S}_{z'})$ operators with known action on the kets $\{|S, M_S\rangle_{z'}\}$. θ is the angle between the symmetry axis z for the paramagnetic centre and **B**, and the x and x' axes are placed in the $(z, \mathbf{B})$ plane (figure 6.3). The components in the $\{x, y, z\}$ reference frame for the unit vector **v**, which defines z', are given by (equations [3.18]):

$$v_x = (g_\perp/g') \sin\theta, \; v_y = 0, \; v_z = (g_{//}/g') \cos\theta \qquad [6.23]$$

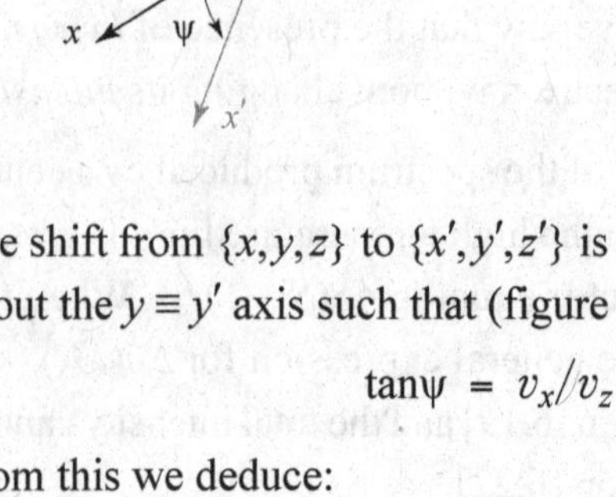

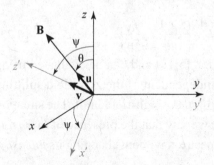

Figure 6.3 – Definition of the angles θ and ψ used to calculate the resonance field for a centre with axial symmetry.

The shift from $\{x, y, z\}$ to $\{x', y', z'\}$ is achieved by rotation through an angle ψ about the $y \equiv y'$ axis such that (figure 6.3):

$$\tan\psi = v_x/v_z = (g_\perp/g_{//}) \tan\theta \qquad [6.24]$$

From this we deduce:

$$\hat{S}_z = -\sin\psi \, \hat{S}_{x'} + \cos\psi \, \hat{S}_{z'}$$

Taking care to distinguish between the products $\hat{S}_{x'}\hat{S}_{z'}$ and $\hat{S}_{z'}\hat{S}_{x'}$, we can therefore write:

$$\hat{H}_{ZFS} = D\,[\cos^2\psi\,\hat{S}_{z'}^2 + \sin^2\psi\,\hat{S}_{x'}^2 - \sin\psi\cos\psi\,(\hat{S}_{x'}\hat{S}_{z'} + \hat{S}_{z'}\hat{S}_{x'}) - S(S+1)/3]$$

By substituting into equation [6.15] and using the matrix elements given in appendix 7, the energy levels to first order in perturbation theory are obtained:

$$E(M_S) = g'\beta B M_S + D/2\,(3\cos^2\psi - 1)\,[\,M_S^2 - S(S+1)/3\,] \qquad [6.25]$$

g' is given by equation [4.5] and ψ is linked to θ by equation [6.24]. It is interesting to compare this expression to the *exact* results obtained in section 6.3.1 for the canonical directions of **B**.

- ▹ When **B** is parallel to z ($\theta = 0$), $g' = g_{//}$, $\psi = 0$, equation [6.25] is identical to the exact expression given by equation [6.10].
- ▹ When **B** is perpendicular to z ($\theta = \pi/2$), $g' = g_\perp$, $\psi = \pi/2$, and for $S = 1$ and $S = 3/2$ we find the *asymptotes* represented by the dashed lines in figures 6.1b and 6.2b. These asymptotes therefore correspond to the approximation given by first order perturbation theory when $g_\perp\beta B$ is much greater than $|D|$, and the distance between the curves and the asymptotes represents the error associated with this approximation. For $S = 1$ and $S = 3/2$, this error can be seen to be small when $|g_\perp\beta B/D|$ exceeds a few units.

According to equation [6.25], the energy for the $|S, M_S\rangle \leftrightarrow |S, M_S + 1\rangle$ transition is equal to:

$$\Delta E(M_S) = g'\beta B + D(3\cos^2\psi - 1)\,(M_S + ½) \qquad [6.26]$$

The resonance field is determined by writing $\Delta E(M_S) = h\nu$:

$$\begin{aligned} B(M_S) &= B_0' - (D/g'\beta)(3\cos^2\psi - 1)\,(M_S + ½) \\ M_S &= -S, -S+1, \ldots S-1 \end{aligned} \qquad [6.27]$$

with $B_0' = h\nu/g'\beta$. For a given value of θ, there exist $2S$ *equidistant* lines separated by $|3\cos^2\psi - 1||D/g'\beta|$, for which the intensity factors may be *different* (equation [6.20]).

6.5.2 – Shape of the spectrum in axial symmetry

Calculation of the density of resonance lines defined by equation [6.27] is difficult in the most general case, but it is simple in the frequent situation where the anisotropy of the Zeeman term is weak relative to the zero-field splitting term. If we set $g_{//} = g_\perp = g$, we have $\psi = \theta$ (equation [6.24]) and equation [6.27] becomes

$$B(M_S) = B_0 - (D/g\beta)(3\cos^2\theta - 1)(M_S + ½) \qquad [6.28]$$
$$M_S = -S, -S+1, \dots S-1$$

with $B_0 = h\nu/g\beta$. The density of the lines for the transition $|S, M_S\rangle \leftrightarrow |S, M_S + 1\rangle$ is given by (exercise 6.4):

$$D(B) = ½\,[(B_{//} - B_\perp)(B - B_\perp)]^{-½} \qquad [6.29]$$

$$B_{//} = B_0 + 2\Delta B(M_S);\ B_\perp = B_0 - \Delta B(M_S);\ \Delta B(M_S) = -D(M_S + ½)/g\beta$$

$B_{//}$ and $B_\perp$ are the resonance fields when **B** is parallel and perpendicular to the molecular axis z, and we have assumed that $\Delta B > 0$. The density $D(B)$ is illustrated in figure 6.4. Its profile is similar to that of figure 4.6 which corresponds to an axial $\tilde{\mathbf{g}}$ matrix, but in the present case there are $2S$ densities which must be added to obtain the total density, each of which is weighted by its intensity factor (equation [6.20]).

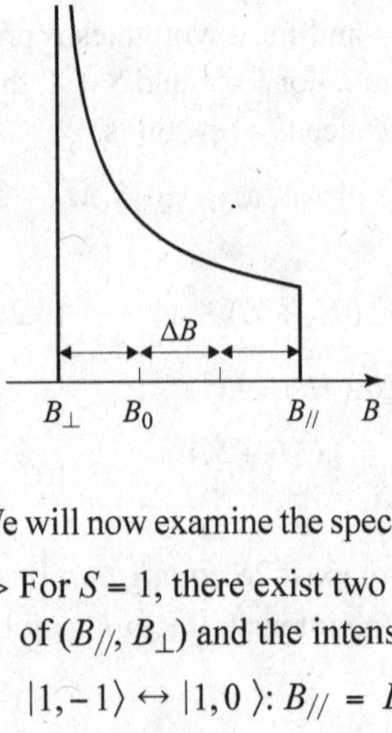

Figure 6.4 – Density of the resonance lines for a transition $|S, M_S\rangle \leftrightarrow |S, M_S+1\rangle$ for a centre with axial symmetry with an isotropic $\tilde{\mathbf{g}}$ matrix in the high-field situation (equation [6.29]). B_0 is the resonance field in the absence of zero-field splitting and $\Delta B = -D(M_S+½)/g\beta$. The figure is drawn for $\Delta B > 0$.

We will now examine the specific cases $S = 1$ and $S = 3/2$ considered in section 6.3:

▹ For $S = 1$, there exist two transitions characterised by the following values of $(B_{//}, B_\perp)$ and the intensity factors:

$$|1,-1\rangle \leftrightarrow |1,0\rangle\text{: } B_{//} = B_0 + D/g\beta;\ B_\perp = B_0 - D/2g\beta;\ g_P(-1) = 2g_P$$

$$|1,0\rangle \leftrightarrow |1,1\rangle\text{: } B_{//} = B_0 - D/g\beta;\ B_\perp = B_0 + D/2g\beta;\ g_P(0) = 2g_P$$

The total density, the absorption signal and the spectrum are shown in figure 6.5.

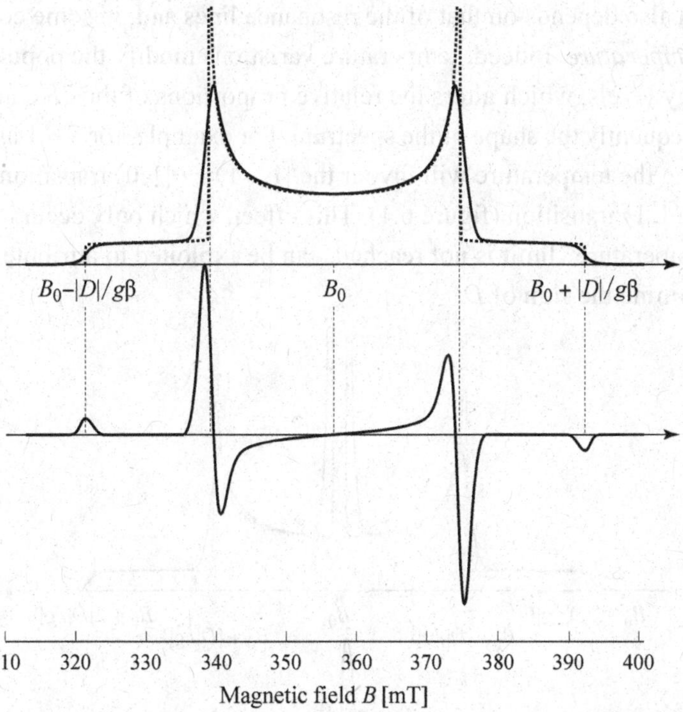

Figure 6.5 – Density (dashed line) and shape of the spectrum for a centre with axial symmetry of spin $S = 1$ in the high-field situation. The spectrum was calculated for the frequency $\nu = 10$ GHz with $g = 2.00$ and $|D| = 0.033$ cm^{-1} (1000 MHz).

▷ For $S = 3/2$, three transitions exist:

$$|3/2, -3/2\rangle \leftrightarrow |3/2, -1/2\rangle: B_{//} = B_0 + 2D/g\beta;\ B_{\perp} = B_0 - D/g\beta;\ g_P(-3/2) = 3g_P$$

$$|3/2, -1/2\rangle \leftrightarrow |3/2, 1/2\rangle: B_{//} = B_{\perp} = B_0;\qquad g_P(-1/2) = 4g_P$$

$$|3/2, 1/2\rangle \leftrightarrow |3/2, 3/2\rangle: B_{//} = B_0 - 2D/g\beta;\ B_{\perp} = B_0 + D/g\beta;\ g_P(1/2) = 3g_P$$

The density, the absorption signal and the spectrum are shown in figure 6.6. The peaks at the extremities of the spectrum at $B = B_0 \pm 2D/g\beta$ are not always easy to observe. To first order in perturbation theory, the $|3/2, -1/2\rangle \leftrightarrow |3/2, 1/2\rangle$ transition produces a line at B_0 whatever the value of θ. The higher order terms produce a small splitting of this line which is not visible in figure 6.6, but would appear for $|D| \geq 0.033$ cm^{-1} (1000 MHz).

As foreseen in section 6.3.2, the characteristic features of the spectrum recorded in the high-field situation can be used to determine the spin S, the value of g and the *absolute value* of the D parameter (figures 6.5 and 6.6). The shape of the spectrum also depends on that of the resonance lines and, in some conditions, on the *temperature*. Indeed, temperature variations modify the populations of the energy levels, which alters the relative proportions of the $2S$ components and consequently the shape of the spectrum. For example, for $S = 1$ and $D > 0$, decreasing the temperature will favour the $|1,-1\rangle \leftrightarrow |1,0\rangle$ transition over the $|1,0\rangle \leftrightarrow |1,1\rangle$ transition (figure 6.1). This effect, which only occurs when the "high temperature" limit is not reached, can be exploited to attribute the lines and determine the sign of D.

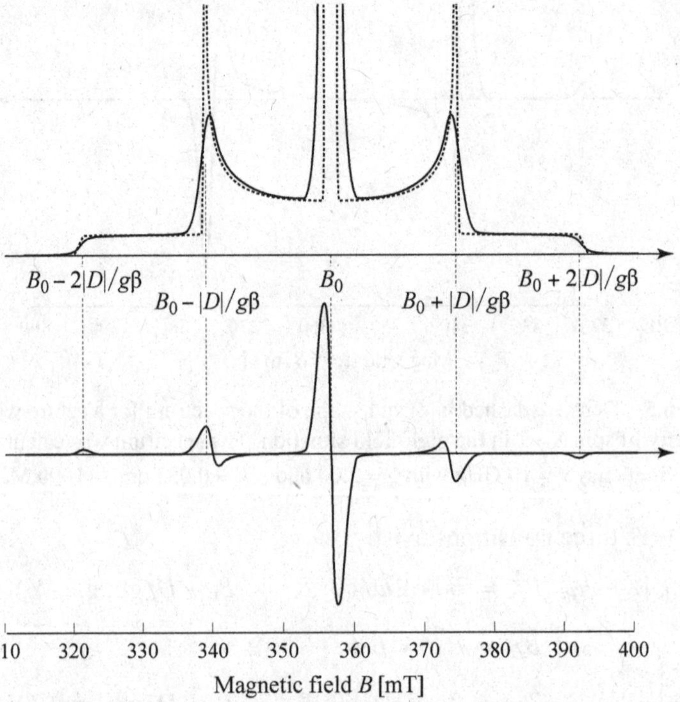

Figure 6.6 – Density (dashed line) and shape of the spectrum for a centre with axial symmetry of spin $S = 3/2$ in the high-field situation. The spectrum is calculated for the frequency $\nu = 10$ GHz with $g = 2.00$ and $|D| = 0.017$ cm^{-1} (500 MHz).

Note – The value of the spin S is absent from expression [6.27] for the resonance field. We can therefore determine the total density corresponding to the spin $(S + 1)$ by adding the two densities due to the new transitions to the $2S$ densities corresponding to S. For example, to determine the density corresponding to $S = 5/2$, the densities of the $|5/2,-5/2\rangle \leftrightarrow |5/2,-3/2\rangle$ and $|5/2,3/2\rangle \leftrightarrow |5/2,5/2\rangle$ transitions are simply added to that illustrated in figure 6.6. However, it should be noted that the value of S is involved in the transition probability (equation [6.20]).

6.5.3 – The "half-field" line for $S = 1$

Until now, we considered transitions verifying the selection rule $\Delta M_S = \pm 1$, which are the only ones allowed in the high-field limit where the eigenvectors for the Hamiltonian are the "pure" kets $\{|S, M_S\rangle_{z'}\}$. But when the EPR spectrum is recorded over a broad field range, B takes values at which the high-field limit is not reached. For these values, the eigenvectors are *mixtures* of the kets $\{|S, M_S\rangle_{z'}\}$, and "weakly allowed" transitions produce additional low-intensity lines when the resonance condition is met. For example, consider the case $S = 1$. In figure 6.7 we have reproduced the energy level diagrams from figure 6.1 assuming, for simplicity, that $g_{//} = g_{\perp} = g$. The transitions $\Delta M_S = \pm 1$ produce two lines centred at $B_0 = h\nu/g\beta$, the position of which is highly dependent on the direction of **B** relative to the symmetry axis z.

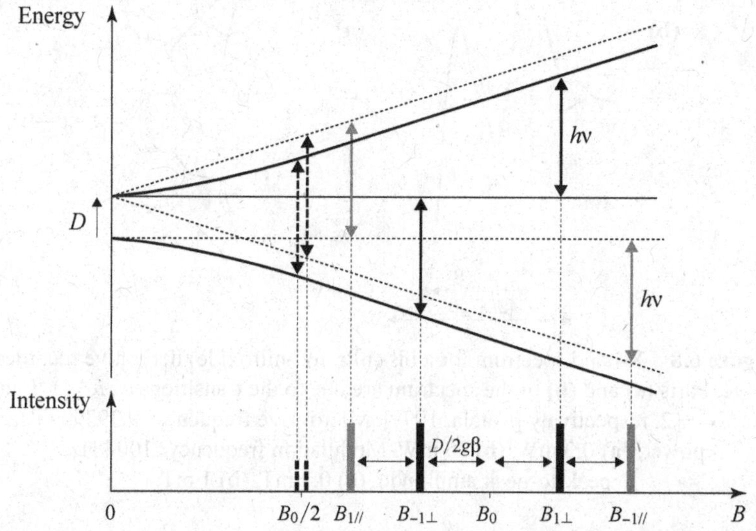

Figure 6.7 – Energy levels for a centre with axial symmetry of spin $S = 1$ for the canonical directions of the field and $D > 0$. The transitions are shown in grey for **B** parallel to z, in black for **B** perpendicular to z. The indices – 1 and 1 correspond to the $|1,-1\rangle \leftrightarrow |1,0\rangle$ and $|1,1\rangle \leftrightarrow |1,0\rangle$ transitions. The half-field transitions are shown by the dashed arrows and the lines near $B = B_0/2$ are represented with a lower intensity.

Moreover, a transition which would be qualified as "$\Delta M_S = \pm 2$" in the high-field situation is observed to verify the resonance condition for $B \approx B_0/2$. Its probability is null when **B** is parallel or perpendicular to z, but it is different from zero for the other directions. The intensity of this "half-field line" is much lower than that of the $\Delta M_S = \pm 1$ lines, by a factor of around $(D/h\nu)^2$, but its position is not very dependent on the direction of **B**. It is therefore often visible on spectra produced by organic molecules in a triplet state.

Figure 6.8 shows the X-band spectrum for a bis-(nitronyl-nitroxide)-type biradical in a polystyrene matrix, recorded at low temperature. To make the spectrum more readable, the parts of the spectrum produced by the $\Delta M_S = \pm 1$ and "$\Delta M_S = \pm 2$" transitions are separated and the second is considerably amplified. From the spectrum, one can deduce $g = 2.007$ and $D = 9.3 \times 10^{-3}$ cm^{-1}. The spectrum produced by the same molecules in the isotropic regime is represented in figure 7.10.

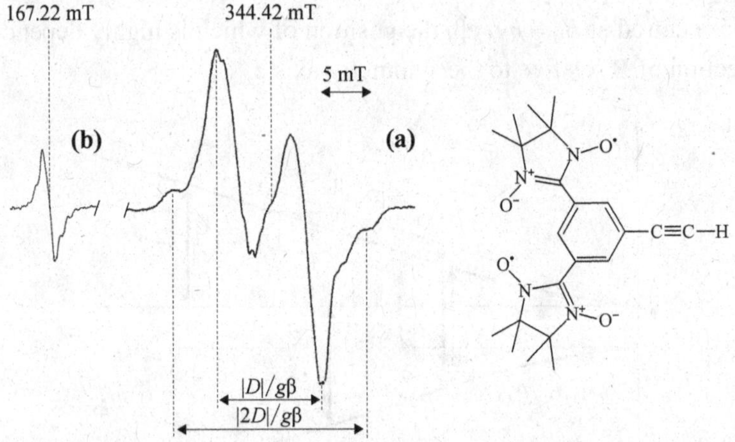

Figure 6.8 – X-band spectrum for a bis-(nitronyl-nitroxide) derivative recorded at 4 K. Parts **(a)** and **(b)** of the spectrum are due to the transitions $\Delta M_S = \pm 1$ and $\Delta M_S = \pm 2$, respectively [Catala, 1999]. Microwave frequency: 9.39392 GHz, power: **(a)** 0.2 mW, **(b)** 20 mW. Modulation frequency: 100 kHz, peak-to-peak amplitude: **(a)** 0.1 mT, **(b)** 1 mT.

To facilitate detection of the half-field line, the spectrum can be recorded with the magnetic component $\mathbf{b}_1(t)$ oriented *parallel* to **B** (see section 6.7). The expression for the $\hat{H}_1$ operator is then modified and the probability of "$\Delta M_S = \pm 2$" transitions is enhanced. When S is greater than 1, "$\Delta M_S = \pm 3$" etc. transitions can also produce low-field lines, but their position is more difficult to analyse.

6.5.4 – Shape of the spectrum in "rhombic" symmetry

In the case of molecules of spin ½, lowering the symmetry causes new features to appear on the powder spectrum (compare figures 4.5 and 4.7). A similar phenomenon occurs for molecules of spin greater than ½ when the symmetry is lowered from axial ($E = 0$) to "rhombic" ($E \neq 0$). To demonstrate this effect, we will calculate the resonance fields (B_X, B_Y, B_Z) for the canonical directions of **B** which determine the positions of the features on the spectrum. For simplicity, we will assume that the anisotropy of the $\tilde{\mathbf{g}}$ matrix is negligible and we set $g_x = g_y = g_z = g$. In this case, the quantization axis z' is the direction of **B** (section 3.3.1) and equation [6.15] can then be written

$$\begin{aligned} E(M_S) &= g\beta B M_S + {}_{z'}\langle S, M_S|\hat{H}_{ZFS}|S, M_S\rangle_{z'} \\ |S, M_S\rangle_{z'},\ M_S &= -S, -S+1, \ldots S \end{aligned} \qquad [6.30]$$

where $\hat{H}_{ZFS}$ is defined by equation [6.8]. Calculation of the position of the $2S$ resonance lines for the canonical directions of **B** produces (exercise 6.5):

$$\begin{aligned} B_X(M_S) &= B_0 + [(D - 3E)/g\beta]\,(M_S + ½) \\ B_Y(M_S) &= B_0 + [(D + 3E)/g\beta]\,(M_S + ½) \\ B_Z(M_S) &= B_0 - (2D/g\beta)\,(M_S + ½) \\ M_S &= -S, -S+1, \ldots, S-1 \end{aligned}$$

We set $B_0 = h\nu/g\beta$. For $S = 1$ for example, the two transitions are characterised by:

$$B_X = B_0 - (D - 3E)/2g\beta,\ B_Y = B_0 - (D + 3E)/2g\beta,\ B_Z = B_0 + D/g\beta$$

$$B_X = B_0 + (D - 3E)/2g\beta,\ B_Y = B_0 + (D + 3E)/2g\beta,\ B_Z = B_0 - D/g\beta$$

In figure 6.9, the density of the resonance lines and the spectrum are shown. The spectral features can be used to determine g and the absolute values of D and E. Like in the case of axial symmetry, a low-intensity quasi-isotropic line forms at half-field [Gordy, 1980].

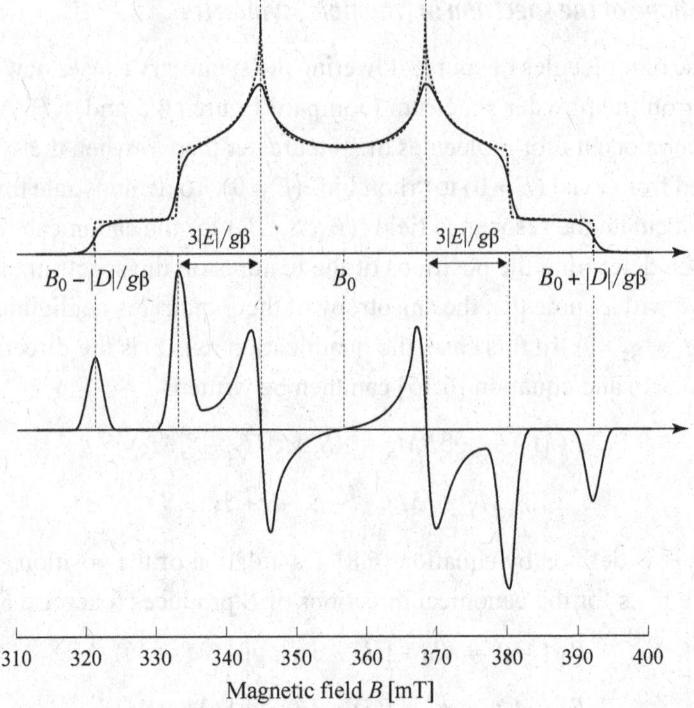

Figure 6.9 – Density (dashed line) and shape (continuous line) of the spectrum given by a centre of spin $S = 1$ with rhombic symmetry in the high-field situation. The spectrum is calculated for: $\nu = 10$ GHz, $g = 2.00$, $|D| = 0.033$ cm^{-1} (1000 MHz), $|E/D| = 0.1$.

6.6 – EPR spectrum for complexes of half-integer spin in the low-field situation. Kramers doublet

The low-field situation is achieved at X-band when $|D|$ exceeds a few cm^{-1}. It is therefore not observed with organic molecules in a triplet state, but it is frequent with transition ion complexes. For complexes with half-integer spin, we will see that calculation of the spectrum comes down to that for a centre of spin ½. It is therefore possible to use the results obtained in chapters 4 and 5. We will deal with the case of complexes with axial symmetry and generalise the results for any symmetry.

6.6.1 – Case of a complex with axial symmetry

Consider a complex characterised by the Hamiltonian [6.9], in which the principal term is the zero-field splitting term and the Zeeman term is a perturbation.

The zero-field splitting term is represented by a diagonal matrix in the basis $\{|S, M_S\rangle_z\}$ made of the eigenvectors shared by $(\hat{\mathbf{S}}^2, \hat{S}_z)$, and its eigenvalues are given by:

$$E_0(M_S) = D[M_S^2 - S(S+1)/3] \qquad [6.31]$$

As the eigenvectors $(|S, M_S\rangle_z, |S, -M_S\rangle_z)$ are associated with the same eigenvalue, there exist $(S + ½)$ *doubly degenerate* levels. The eigenvalues and eigenvectors for the Hamiltonian $\hat{H}_S$ are obtained to first order in perturbation theory by constructing the matrix representing $\hat{H}_{Zeeman}$ in each $\{|S, M_S\rangle_z, |S, -M_S\rangle_z\}$ sub-space of dimension 2, known as a "Kramers doublet". By noting $\theta = (z, \mathbf{B})$ and placing x in the $(z, \mathbf{B})$ plane (figure 6.3), the Zeeman term is written:

$$\hat{H}_{Zeeman} = \beta B\,(g_\perp \sin\theta\, \hat{S}_x + g_{//} \cos\theta\, \hat{S}_z) \qquad [6.32]$$

◇ ***Shape and intensity of the spectrum produced by the*** $\{|S, ½\rangle_z, |S, -½\rangle_z\}$ ***doublet***

▹ With the help of the matrix elements from appendix 7, we build the $(\tilde{\mathbf{S}}_\mathbf{x}, \tilde{\mathbf{S}}_\mathbf{y}, \tilde{\mathbf{S}}_\mathbf{z})$ matrices representing the $(\hat{S}_x, \hat{S}_y, \hat{S}_z)$ operators in the $\{|S, ½\rangle_z, |S, -½\rangle_z\}$ sub-space:

$$\tilde{\mathbf{S}}_\mathbf{x} = \begin{bmatrix} 0 & \frac{S+\frac{1}{2}}{2} \\ \frac{S+\frac{1}{2}}{2} & 0 \end{bmatrix}, \quad \tilde{\mathbf{S}}_\mathbf{y} = \begin{bmatrix} 0 & -i\frac{S+\frac{1}{2}}{2} \\ i\frac{S+\frac{1}{2}}{2} & 0 \end{bmatrix}, \quad \tilde{\mathbf{S}}_\mathbf{z} = \begin{bmatrix} \frac{1}{2} & 0 \\ 0 & -\frac{1}{2} \end{bmatrix}$$

If we compare these matrices to the ones representing the operators $(\hat{S}'_x, \hat{S}'_x, \hat{S}'_z)$ associated with a spin $S' = ½$ in the $\{|½, ½\rangle, |½, -½\rangle\}$ basis made of the eigenvectors shared by $(\hat{\mathbf{S}}'^2, \hat{S}'_z)$ (section 3.2.5), we observe that:

$$\tilde{\mathbf{S}}_\mathbf{x} = (S + ½)\, \tilde{\mathbf{S}}'_\mathbf{x}$$
$$\tilde{\mathbf{S}}_\mathbf{y} = (S + ½)\, \tilde{\mathbf{S}}'_\mathbf{y}$$
$$\tilde{\mathbf{S}}_\mathbf{z} = \tilde{\mathbf{S}}'_\mathbf{z}$$

We deduce that the matrix representing an operator with the form $(a_x\hat{S}_x + a_y\hat{S}_y + a_z\hat{S}_z)$ in the $\{|S, ½\rangle_z, |S, -½\rangle_z\}$ sub-space is identical to that representing the equivalent operator $(a_x^{\text{eff}}\hat{S}'_x + a_y^{\text{eff}}\hat{S}'_y + a_z^{\text{eff}}\hat{S}'_z)$ in the $\{|½, ½\rangle, |½, -½\rangle\}$ basis, with

$$a_x^{\text{eff}} = (S + ½)\, a_x, \quad a_y^{\text{eff}} = (S + ½)\, a_y, \quad a_z^{\text{eff}} = a_z$$

This equivalence applies in particular to the operators $\hat{H}_{Zeeman}$ (equation [6.32]) and $\hat{H}_1$ (equation [6.17]) which determine *the position* and the *intensity* of the resonance lines. As a result, the spectrum produced by the $\{|S, ½\rangle_z, |S, -½\rangle_z\}$

doublet is identical to that produced by a centre of spin ½ characterised by the "effective g values":

$$g_{//}^{\text{eff}} = g_{//};\ g_{\perp}^{\text{eff}} = (S + ½)\, g_{\perp} \qquad [6.33]$$

The position and the intensity of the resonance lines are determined by (equations [4.5] and [5.14]):

$$(g')^2 = (g_{//}^{\text{eff}})^2 \cos^2\theta + (g_{\perp}^{\text{eff}})^2 \sin^2\theta \qquad [6.34]$$

$$g_P(\theta) = (g_{\perp}^{\text{eff}})^2\,[1 + (g_{//}^{\text{eff}}/g'(\theta))^2\,]/8g'(\theta) \qquad [6.35]$$

For $g_{\perp} \approx 2$, we have $g_{\perp}^{\text{eff}} \approx 4$ if $S = 3/2$ and $g_{\perp}^{\text{eff}} \approx 6$ if $S = 5/2$. Measuring $g_{\perp}^{\text{eff}}$ on the spectrum therefore directly gives the value of S. It is important to note that the range of the field over which the spectrum for the doublet extends, as determined by $(g_{\perp}^{\text{eff}}, g_{//})$, is much broader than that of a spectrum for a centre of spin ½, determined by the numbers $(g_{\perp}, g_{//})$ which are both close to 2 (section 4.2.2).

When a hyperfine interaction between the unpaired electrons and a nucleus of spin I is described by the operator

$$\hat{H}_{hyperfine} = (A_{\perp}\,\hat{S}_x\hat{I}_x + A_{\perp}\,\hat{S}_y\hat{I}_y + A_{//}\,\hat{S}_z\hat{I}_z)$$

the hyperfine structure of the spectrum for the doublet is determined by (equation [4.11]):

$$(A')^2 = (A_{\perp}^{\text{eff}})^2\,(g_{\perp}^{\text{eff}}/g')^2 \sin^2\theta + (A_{//}^{\text{eff}})^2\,(g_{//}^{\text{eff}}/g')^2 \cos^2\theta \qquad [6.36]$$

with:

$$A_{//}^{\text{eff}} = A_{//};\ A_{\perp}^{\text{eff}} = (S + ½)\, A_{\perp} \qquad [6.37]$$

▹ We will now consider the *total intensity* $I_{1/2}$ of the spectrum for the doublet. If the paramagnetic centres are at thermal equilibrium at the temperature T (unsaturated regime), from equation [5.17] we deduce:

$$I_{1/2}(T) = \frac{\pi\beta}{2\hbar}(g_P)_{av}\, B_1 N_{1/2}(T) \tanh\left(\frac{h\nu}{2k_B T}\right) \qquad [6.38]$$

- $(g_P)_{av}$ represents the mean of $g_P(\theta)$ for all values of θ (equation [5.18]). Numerical analysis demonstrates that its value can be determined with the help of equation [5.19], with an error of less than 1.5 % as long as the effective g values ($g_x^{\text{eff}} \le g_y^{\text{eff}} < g_z^{\text{eff}}$) verify the inequality $0.2 < g_x^{\text{eff}}/g_y^{\text{eff}} < g_z^{\text{eff}}/g_y^{\text{eff}} < 8$ [Aasa and Vänngård, 1975].
- $N_{1/2}(T)$ is the *population of the doublet*, which is linked to the total number of centres N by:

$$N_{\frac{1}{2}}(T) = N \frac{\exp[-E_0(\frac{1}{2})/k_B T]}{\sum_{M_S=\frac{1}{2}}^{S} \exp[-E_0(M_S)/k_B T]} \quad [6.39]$$

where $E_0(M_S)$ is given by equation [6.31]. This expression is valid for $k_B T \gg g\beta B$. The temperature-dependence of $I_{½}(T)$ can be used to determine the sign and absolute value of the D parameter. Using equations [6.38] and [6.39], it is then possible to calculate N by comparing $I_{½}$ to the intensity of the spectrum produced by a reference sample (exercise 6.6).

◇ Other doublets

The matrices representing $(\hat{S}_x, \hat{S}_y, \hat{S}_z)$ in the $\{|S, M_S\rangle_z, |S, -M_S\rangle_z\}$ sub-spaces, where $M_S \neq \pm$ ½, are linked to those representing the operators $(\hat{S}'_x, \hat{S}'_x, \hat{S}'_z)$ associated with a spin ½ by:

$$\tilde{\mathbf{S}}_\mathbf{x} = 0, \; \tilde{\mathbf{S}}_\mathbf{y} = 0, \; \tilde{\mathbf{S}}_\mathbf{z} = 2|M_S|\tilde{\mathbf{S}}'_\mathbf{z}$$

For these doublets, equations [6.33] and [6.37] become:

$$g_{//}^{\text{eff}} = 2\,|M_S|\, g_{//}; \; g_\perp^{\text{eff}} = 0$$

$$A_{//}^{\text{eff}} = 2\,|M_S|\, A_{//}; \; A_\perp^{\text{eff}} = 0 \quad [6.40]$$

The intensity factor g_P is null for all values of θ (equation [6.35]): the transitions are forbidden.

◇ "Interdoublet" transitions

Transitions between adjacent doublets which verify $\Delta M_S = \pm 1$ are allowed. For example, consider the transitions between the $\{|S, ½\rangle_z, |S, -½\rangle_z\}$ and $\{|S, ³⁄₂\rangle_z, |S, -³⁄₂\rangle_z\}$ doublets which are separated by $2|D|$ in the absence of magnetic field (figure 6.10). Since βB is much less than $2|D|$ in the low-field situation, the energy of these transitions is close to $2|D|$ and the resonance condition can only be satisfied if $2|D| \approx h\nu$, where ν is the frequency of the radiation.

With an X-band spectrometer for which $h\nu \approx 0.3$ cm^{-1}, resonance is only possible if $|D| \approx 0.15$ cm^{-1}. This type of transition produces low-field lines which are often broad and difficult to observe, and information can only be extracted by numerical simulation of the spectrum. We will see the reason for this broadening in the section devoted to integer spin complexes.

To illustrate the results obtained in this section, figure 6.10 presents the energy level diagram for a complex of spin $S = 5/2$ with axial symmetry, and figure 6.11 shows the X-band spectrum due to the $\{|5/2, 1/2\rangle, |5/2, -1/2\rangle\}$ doublets of the 4 haemes in haemoglobin, where the Fe^{3+} ion is in a strong spin situation with $g_{//} = g_{\perp} = 2.00$ (section 4.2.2) and $D = 9$ cm^{-1} [Scholes *et al.*, 1971].

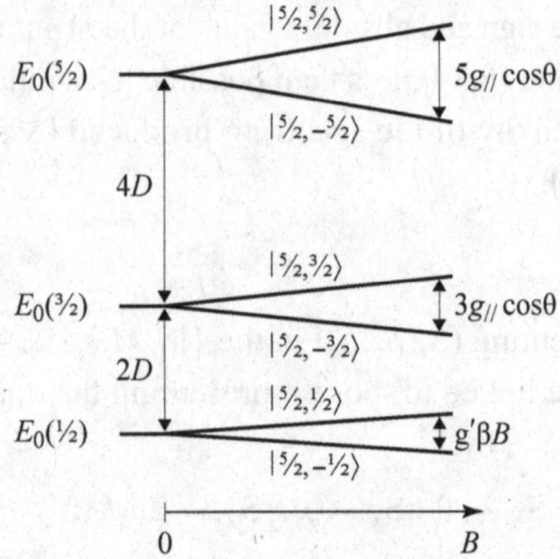

Figure 6.10 – Energy levels for a centre of spin $S = 5/2$ with axial symmetry in the low-field situation. The diagram is drawn for $D > 0$. Only the ground doublet produces a spectrum (example in figure 6.11). g' is given by equation [6.34], with $g_{//}^{\text{eff}} = g_{//}$; $g_{\perp}^{\text{eff}} = 3g_{\perp}$. θ is the angle between the symmetry axis and the field **B**.

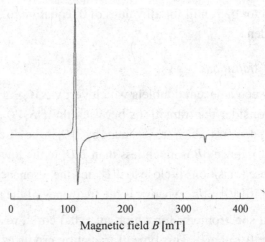

Figure 6.11 – X-band spectrum for haemoglobin. In this form of the protein, each of the four haemes contains an Fe^{3+} ion coordinated by four pyrrole nitrogen atoms and two axial ligands: a histidine nitrogen and a water molecule. Experimental conditions: temperature 10 K. Microwave frequency: 9.416 GHz, power 0.8 mW. Modulation frequency: 100 kHz, peak-to-peak amplitude 0.2 mT.

6.6.2 – Generalisation to a complex of any geometry

When the complex is not axially symmetric, (S + ½) doubly-degenerate energy levels still appear in the absence of a magnetic field, and their separation is determined by D and E. The Kramers doublets are *mixtures* of the kets $\{|S, M_S\rangle\}$, eigenvectors shared by ($\hat{\mathbf{S}}^2, \hat{S}_z$), and all possible "intra-doublet" transitions are allowed when the complex is placed in a magnetic field. Using the method described in the previous section, it can be shown that each doublet produces a spectrum similar to that given by a centre of spin ½ characterised by three effective g values: its shape is identical to that shown in figure 4.7, but its remarkable features are identified by $B_i = h\nu/\beta g_i^{\text{eff}}$, $i = x, y, z$. The effective g values are linked to the symmetry properties of the complex just as the principal values of the $\tilde{\mathbf{g}}$ matrix (section 4.2.1). Although the three numbers (g_x, g_y, g_z) are close to 2 (section 4.2.2), the effective g values are very different, causing the spectrum to spread over a much broader field range than the spectrum produced by centres of spin ½. When the $\tilde{\mathbf{g}}$ and $\tilde{\mathbf{D}}$ matrices for the complex have the same principal axes, the effective g values for each doublet can be expressed as a function of the principal values (g_x, g_y, g_z) and of the E/D *ratio*. This ratio is such that $0 \leq E/D \leq ⅓$ if the principal axes are named such that the principal values of the $\tilde{\mathbf{D}}$ matrix verify $|D_z| \geq |D_y| \geq |D_x|$(section 6.2.2). If we also set $g_x = g_y = g_z = 2$, we can calculate charts which give $(g_x^{\text{eff}}, g_y^{\text{eff}}, g_z^{\text{eff}})$ as a function of E/D for each Kramers doublet. From the numbers $(g_x^{\text{eff}}, g_y^{\text{eff}}, g_z^{\text{eff}})$ measured on the spectrum, and using these charts, we can deduce the value of S and estimate the E/D ratio [Hagen, 1992]. As an example, figure 6.12 represents the chart calculated for $S = 3/2$.

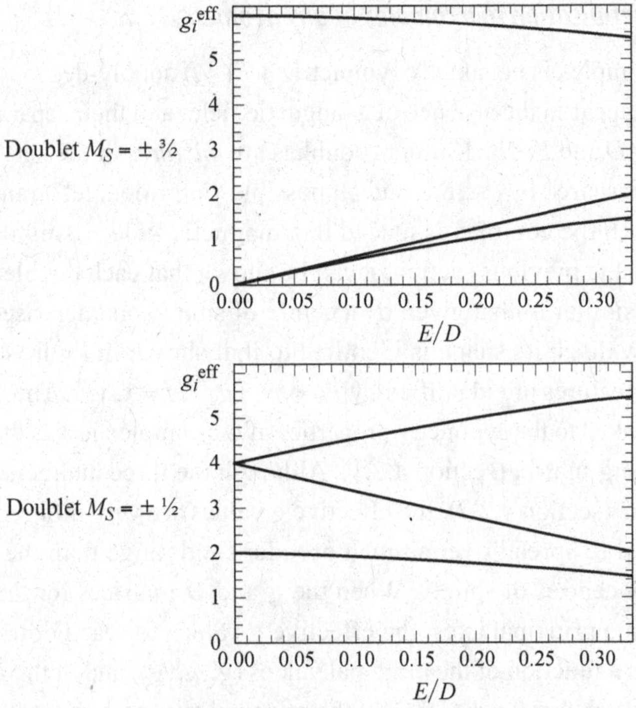

Figure 6.12 – Variation of the effective *g* values for the two Kramers doublets in a complex of spin $S = 3/2$ as a function of E/D. The doublets are identified by the values of M_S in the $E = 0$ limit (equation 6.31). Their separation is $\Delta = 2|D|\,[1+3(E/D)^2]^{1/2}$. The ground doublet is $M_S = \pm 1/2$ for $D > 0$.

The separation between doublets depends on the (D, E) parameters. When several doublets are populated at the temperature of the experiment, their spectra, weighted by their respective Boltzmann factors, are added together.

By studying the temperature-dependence of these spectra, (D, E) and the number of paramagnetic centres present in the sample can be determined by comparing the spectrum intensity to the intensity produced by a reference sample [Lanciano *et al.*, 2007].

6.7 – EPR spectrum for integer-spin complexes in the low-field situation

We will see that this situation is not favourable to extract information from the EPR spectrum. Consider a centre with axial symmetry for which the energy levels in the absence of magnetic field are given by equation [6.31].

▷ The ket $|S,0\rangle_z$ corresponds to the non-degenerate level $E_0(0)$. The $\hat{H}_{Zeeman}$ expression (equation [6.32]) shows that to first order in perturbation theory, the ${}_z\langle S,0|\hat{H}_{Zeeman}|S,0\rangle_z$ offset is null. Thus, according to the first order approximation, the interaction with the magnetic field modifies neither the energy nor the state of this level.

▷ The other levels are *doubly degenerate* in the absence of a magnetic field, and a Kramers doublet $\{|S,M_S\rangle_z, |S,-M_S\rangle_z\}$ is associated with each level. The application of a magnetic field removes the degeneracy, but equations [6.40] show that the $|S,M_S\rangle_z \leftrightarrow |S,-M_S\rangle_z$ transitions are *forbidden*.

▷ The $|S,0\rangle \leftrightarrow |S,1\rangle$ and $|S,0\rangle \leftrightarrow |S,-1\rangle$ transitions are allowed. In the low-field limit, the resonance condition is only satisfied if the $h\nu$ quantum for the spectrometer is close to $|D|$. In this situation, we show in complement 2 that the position of the resonance lines produced by the molecules with a given orientation relative to **B** is very dependent on the value of D. If this value varies slightly from one molecule to another in the sample, the lines produced by the different molecules become dispersed to form a very broad *inhomogeneous* line (complement 4 to chapter 5), and the spectrum resulting from the superposition of these inhomogeneous lines has no notable features. In this case, numerical simulation must be used to extract quantitative information from the spectrum.

These results can be generalised to complexes with any geometry. Transitions take place at low-field when the zero-field splitting is close to $h\nu$. These transitions give broad, unstructured spectra which are often difficult to observe. They can sometimes be more easily detected when the magnetic component of the radiation is *parallel* to **B**. To perform this type of experiment, rectangular "*dual mode*" cavities are used, in which $\mathbf{b}_1(t)$ is perpendicular to **B** for a certain resonance frequency (TE_{102} mode) and parallel to **B** for a slightly different frequency (TE_{012} mode).

6.8 – Points to consider in applications

The EPR spectrum for centres with a spin greater than ½ is very dependent on the zero-field splitting parameters and the frequency of the spectrometer. In the most general case, the parameters of the spin Hamiltonian can only be determined by using numerical simulation. But the spectrum can often be simplified by selecting the frequency so as to be in the high-field or low-field situation. This "multi-frequency" spectroscopy has become possible since the advent of EPR spectrometers for which the $h\nu$ quantum reaches tens of cm^{-1}.

6.8.1 – Organic molecules in a triplet state

▷ Numerous organic molecules can be prepared in a triplet state, either by photolysis and trapping in a solid matrix at low temperatures, or by continuously exposing the sample to ultraviolet radiation so as to create sufficient numbers of molecules in an excited state. The ground state of biradicals, in which two radical entities are coupled by exchange interaction, is generally a triplet state. Exchange coupling also confers specific magnetic properties to these molecules, which are described in chapter 7.

▷ In the case of organic molecules in a triplet state, the D and E parameters are small enough for the *high-field situation* to be practically achieved at Q-band and sometimes even at X-band. When the sample is a single crystal, the hyperfine structure is often resolved on the spectrum and its analysis provides the same information as in the case of radicals. Powder spectra present characteristic features centred at $g \approx 2$, which can be used to determine the principal values of the $\tilde{\mathbf{g}}$ matrix and the absolute value of the zero-field splitting parameters (figures 6.5, 6.8 and 6.9). The latter are mainly determined by dipolar magnetic interactions between the two unpaired electrons, and for a long time they were difficult to interpret due to the lack of an appropriate molecular model. Thanks to DFT-type calculation methods, they can now be exploited, and a detailed description of the electron distribution can be deduced [Sinnecker and Neese, 2006].

6.8.2 – Transition ion complexes in the high-field situation

It is generally necessary to use a high-field spectrometer to achieve this situation. The powder spectra are relatively simple to interpret, and by analysing their features, the principal values of the $\tilde{\mathbf{g}}$ matrix and the absolute value of the zero-field splitting parameters can be determined. Their *sign* can be deduced from the temperature-dependence of the intensity of the various features. Interpretation of the values of the zero-field splitting parameters once again requires the use of DFT-type methods [Duboc *et al.*, 2007; Neese, 2006].

6.8.3 – Transition ion complexes in the low-field situation

As the low-field situation is achieved at X-band when D exceeds a few cm^{-1}, it is often observed for transition ion complexes. The EPR properties are very different depending on whether the spin of the complex is integer or half-integer:

- ▹ When S is half-integer, the zero-field splitting terms produce $(S + ½)$ Kramers doublets and each doublet produces an EPR spectrum similar to that produced by a centre of spin ½, characterised by three "effective g" parameters, for which the values depend on S, (g_x, g_y, g_z), and the $|E/D|$ ratio. As the effective g are very different from $g_e = 2.0023$, the spectrum spreads over a broad field range and it is not always possible to observe all of its features. In favourable cases, comparison of the effective g values to the simplified charts calculated with $g_x = g_y = g_z = 2$ is sufficient to unambiguously determine the value of S, but not always that of the $|E/D|$ ratio [Lanciano *et al.*, 2007]. By studying the temperature-dependence of the intensity of the signals produced by the different doublets, the D and E parameters can be determined. To see the advantage of performing this type of measurement, we need simply consider iron complexes with axial symmetry of spin $S = 5/2$, such as the iron-porphyrins and haemes: while the $|5/2, -1/2\rangle \leftrightarrow |5/2, 1/2\rangle$ transition always produces a spectrum characterised by $g_{//}^{eff} = 2$ and $g_{\perp}^{eff} = 6$ (figure 6.11), the D parameter provides information on the *nature* of the axial ligands. We can also deduce the number of paramagnetic centres present in the sample from the intensity of the spectrum produced by the Kramers doublets [Bruschi *et al.*, 1992].
- ▹ Integer-spin complexes often produce a broad unstructured spectrum at low-field, which is more readily detected when the sample is placed in a parallel-mode cavity. By using an appropriate model to simulate this spectrum, in principle, it is possible to determine the spin S of the centre and the values of D and E. However, when these parameters are not too large, they can simply be measured by recording the spectrum at high-field.

6.8.4 – Spin-lattice relaxation for centres of spin greater than ½

- ▹ In the high-field situation, the spectrum is the result of transitions between $2S + 1$ energy levels. The paramagnetic centres reach thermal equilibrium thanks to all the relaxation transitions between these levels, and each transition is characterised by a rate constant. Relaxation is therefore faster than in the case of centres of spin ½, but it is generally multiphasic and it is not possible to define a spin-lattice relaxation time T_1. Nevertheless, at a given temperature, all the rate constants are determined by the same relaxation process, and the temperature-dependence of the relaxation is the same as for centres of spin ½ (section 5.4.1). In solid samples, the direct process is the most efficient at low temperature, and two-phonon processes dominate at higher temperatures [Roger *et al.*, 1980; Roger *et al.*, 1982]. In samples that are liquid at room temperature, the combination of rotational Brownian motion of the molecules and the anisotropy of the $\tilde{\mathbf{D}}$ matrix creates a very effective relaxation process which generally leads to broadening and disappearance of the spectrum. The case of Mn^{2+} ions in aqueous solution is unusual. These ions form octahedral $Mn(H_2O)_6{}^{2+}$ complexes with a spin $S = 5/2$. The $\tilde{\mathbf{g}}$ and $\tilde{\mathbf{A}}$ matrices are isotropic (sections 4.2.1 and 4.4.2) and the $\tilde{\mathbf{D}}$ matrix is null (section 6.2.1), such that the relaxation processes based on the anisotropy of these matrices do not occur. At room temperature an EPR spectrum with a well-resolved hyperfine structure is observed (figure 2.7), but the relaxation is relatively rapid since the spectrum is not saturated. The relaxation is controlled by time-dependent zero-field splitting terms created by geometric fluctuations in the complex [Carrington and McLachlan, 1979].

▹ In complexes of half-integer spin in the low-field situation, each Kramers doublet behaves like a centre of spin ½ and the relaxation processes are identical to those for such centres. However, the proximity of the excited doublets favours particularly efficient Orbach processes which entail a characteristic temperature-dependence of T_1 (section 5.4.1). This property can be used to determine the value of the zero-field splitting [Scholes *et al.*, 1971].

Complement 1 – Intensity of the resonance line at high temperature in the high-field limit

Here, we demonstrate some of the results used to calculate the intensity of the resonance lines in section 6.4.2.

◇ *Expression for the intensity factor*

As the energy of the transition is a linear function of B (equation [6.15]), the expression for the intensity factor $g_P(M_S)$ established in section 5.2.1 can be used. Here, this expression is written:

$$g_P(M_S) = |V(M_S)|_{av}^2/g'(\beta B_1)^2; \quad V(M_S) = {}_{z'}\langle S, M_S|\hat{H}_1|S, M_S+1\rangle_{z'}$$

By replacing $\hat{H}_1$ by its general expression (equation [6.18]), we obtain:

$$|V(M_S)|^2 = (\beta^2 B_1^2/4)\,(c_{x'}^2 + c_{y'}^2)\,[S(S+1) - M_S\,(M_S+1)]$$

$$|V(M_S)|_{av}^2 = (\beta^2 B_1^2/4)\,(c_{x'}^2 + c_{y'}^2)_{av}\,[S(S+1) - M_S\,(M_S+1)]$$

The average is calculated over the angle α defining the direction of $\mathbf{B}_1$ in the plane perpendicular to $\mathbf{B}$ (figure 5.4). The intensity factor for the $|S, M_S\rangle \leftrightarrow |S, M_S+1\rangle$ transition is therefore:

$$g_P(M_S) = [(c_{x'}^2 + c_{y'}^2)_{av}/4g']\,[S(S+1) - M_S\,(M_S+1)] \qquad (1)$$

This equation does not involve the zero-field splitting terms and it is valid for any value of S. In particular, for $S = ½$, it is written:

$$g_P(M_S = -½) = (c_{x'}^2 + c_{y'}^2)_{av}/4g'$$

As a consequence, the quantity $(c_{x'}^2 + c_{y'}^2)_{av}/4g'$ is nothing but the intensity factor g_P given by equation [5.15]. By introducing this notation into equation (1), we obtain:

$$g_P(M_S) = g_P\,[S(S+1) - M_S\,(M_S+1)]$$

◇ *Expression for the population difference in the high-temperature limit*

As the paramagnetic centres are at *thermal equilibrium* at temperature T, the population of the $E(M_S)$ energy level is written:

$$n(M_S) = n(\theta,\varphi)\frac{\exp[-E(M_S)/k_B T]}{\sum\limits_{M_S=-S}^{S} \exp[-E(M_S)/k_B T]}$$

We assume that the temperature is high enough for the approximation

$$\exp[-E(M_S)/k_B T] \approx 1 - E(M_S)/k_B T$$

to be valid for any value of M_S. Since the energy levels cover a range of around $2Sh\nu$, this "high temperature" condition requires a higher temperature when the values of S and ν are large. A few kelvins is enough at X-band, but it is necessary to exceed several tens of kelvins when a frequency of 360 GHz is used. When this condition is satisfied, the difference $\Delta n(M_S) = n(M_S) - n(M_S + 1)$ is given by:

$$\Delta n(M_S) = n(\theta, \varphi) \frac{E(M_S+1) - E(M_S)}{k_B T \sum_{M_S=-S}^{S} \left(1 - \frac{E(M_S)}{k_B T}\right)}$$

At resonance, the numerator is equal to $h\nu$. In addition, as the trace of the Hamiltonian is *null*, the sum of the $E(M_S)$ energies is null, and the sum in the denominator is equal to $(2S + 1)$. In this "high temperature" limit, the population difference $\Delta n(M_S)$ is the same for all transitions and is given by:

$$\Delta n \approx \frac{n(\theta, \varphi) h\nu}{(2S+1) k_B T}$$

Complement 2 – Shape of the low-field spectrum for a centre of spin $S = 1$

A sample contains transition ion complexes of spin $S = 1$, with axial symmetry, for which the lowest energy levels are described by the Hamiltonian [6.9] where D is assumed to be *positive*. For molecules with their z axis parallel or perpendicular to the field **B**, the variation of these energy levels as a function of B is represented in figures 6.1a and 6.1b, respectively. We are interested in the transition which takes place at low-field between the levels labelled $E(1)$, $E(0)$ in figure 6.1a, and E_+, E_- in figure 6.1b, for which the energy ΔE is close to D at low-field. We will determine the position of the resonance line for these "canonical" directions, then we will examine how the lines are affected by the *distribution* of D. For simplicity, we assume that $g_{//} = g_{\perp} = g$.

◇ *Expression for the resonance field for the canonical directions of* **B**

▹ For molecules in which the z axis is *parallel* to **B**, the transition energy is deduced from equation [6.10] (figure 6.1a):

$$\Delta E = D + g\beta B$$

The resonance field is

$$B_{//} = (h\nu - D)/g\beta \qquad (1)$$

The transition is only possible if $h\nu \geq D$. The position of the line can be identified by the number $g_{//}^* = h\nu/\beta B_{//}$, which is equal to:

$$g_{//}^* = g/(1 - D/h\nu) \qquad (2)$$

This number depends on the frequency of the spectrometer and is larger when $h\nu$ is close to D. For example, for $g = 2.0$ and $D = 0.28\ \text{cm}^{-1}$, $g_{//}^* = 2.6$ when $h\nu = 1.2\ \text{cm}^{-1}$ (Q-band) and $g_{//}^* = 30$ when $h\nu = 0.30\ \text{cm}^{-1}$ (X-band).

▹ For molecules in which the z axis is *perpendicular* to **B**, the transition energy is deduced from equations [6.11]:

$$\Delta E = [D^2 + (2g\beta B)^2]^{1/2}$$

The resonance field is

$$B_{\perp} = [(h\nu)^2 - D^2]^{1/2}/2g\beta \qquad (3)$$

and the position of the line is identified by:

$$g_{\perp}^* = 2g/[1 - (D/h\nu)^2]^{1/2} \qquad (4)$$

For $h\nu = 0.30\ \text{cm}^{-1}$, we obtain $g_{\perp}^* \approx 11$.

◇ *Effect of distribution of the D parameter on the shape of the spectrum*

If the numbers ($g_{//}$*, $g_\perp$*) could be measured from the spectral features, it would be possible to deduce the g and D parameters from equations [2] and [4]. Unfortunately, these features are generally masked as the line is quite broad due to a mechanism specific to this type of transition. In figure 6.13a we have represented the variations of $B_{//}$ (equation 1) and $B_\perp$ (equation 3) as a function of D, switching the abscissa and ordinate axes. This graph shows that $B_{//}$ and $B_\perp$ are very dependent on D. As this parameter is very sensitive to the molecule's geometry, its value varies slightly from one molecule to another such that the lines produced by the molecules with the same orientation relative to **B** become dispersed and produce an *inhomogeneous* line (complement 3 to chapter 5).

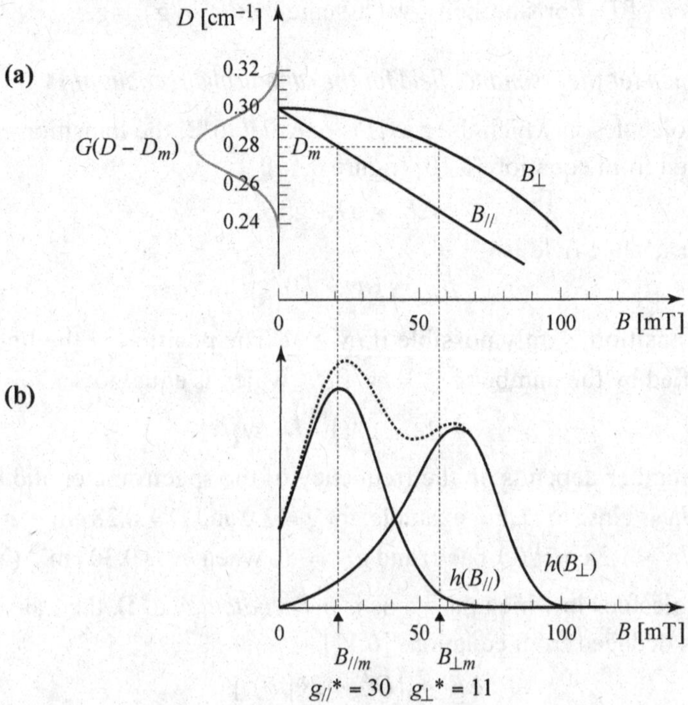

Figure 6.13 – **(a)** Variation of the resonance fields $B_{//}$ and $B_\perp$ as a function of D. **(b)** Densities of distributions of $B_{//}$ and $B_\perp$ due to the normal distribution of D described by $G(D - D_m)$. Their sum is represented by the dashed line.

To quantitatively describe this phenomenon, we assume that D is distributed with a *normal* density $G(D - D_m)$ characterised by its mean, D_m, and its standard deviation, σ_D (figure 6.13a).

▹ The *density* $h(B_{//})$ of the lines produced by the molecules for which the z axis is parallel to **B** is given by:

$$h(B_{//}) = G(D - D_m)/(dB_{//}/dD) \quad (5)$$

where $B_{//}$ and D are linked by equation (1). As this equation is *linear*, $dB_{//}/dD$ is constant and $h(B_{//})$ is a Gaussian centred at $B_{//m} = (h\nu - D_m)/g\beta$, with standard deviation $\sigma_{//} = \sigma_D/g\beta$, which is "truncated" when σ_D is large. This Gaussian is represented in figure 6.13b for $D_m = 0.28\ \mathrm{cm}^{-1}$ and $\sigma_D = 0.014\ \mathrm{cm}^{-1}$. We observe that a distribution of just 10 % of D leads to very extensive broadening of the inhomogeneous line, with a full width at half maximum of around 30 mT.

▹ The density $h(B_\perp)$ of the lines produced by the molecules for which the z axis is perpendicular to **B** is defined by an equation similar to equation (5), in which $B_\perp$ and D are linked by equation (3). As this equation is non-linear, this density centred near $B_{\perp m} = [(h\nu)^2 - D_m^2]^{1/2}/2g\beta$ is not Gaussian and its full width at half maximum is around 35 mT (figure 6.13b).

The overlap between these two densities is already significant. As the densities corresponding to the intermediate orientations form between $h(B_{//})$ and $h(B_\perp)$, it is easy to understand why the spectrum resulting from the superposition of all the densities has no specific features linked to the g and D values. Only numerical simulations based on a detailed model of the distribution of D can extract information from this type of spectrum [Hendrich and Debrunner, 1989].

References

AASA R. & VÄNNGÅRD T. (1975) *Journal of Magnetic Resonance* **19**: 308-315.

ABRAGAM A. & BLEANEY B. (1986) *Electron Paramagnetic Resonance of transition ions*, Dover, New York.

BRUSCHI M. *et al.* (1992) *Biochemistry* **31**: 3281- 3288.

CARRINGTON A. & MCLACHLAN A.D. (1979) *Introduction to Magnetic Resonance with Applications to Chemistry and Chemical Physics*, Chapman & Hall, London.

CATALA L. (1999) Nitronyl-nitroxide and imino-nitroxide oligoradicals: synthesis and study of the magnetic properties in the isolated state and in the crystalline state, PhD, Louis Pasteur University, Strasbourg.

DUBOC C. *et al.* (2007) *Inorganic Chemistry* **46**: 4905-4916.

GORDY W. (1980) *Theory and Applications of Electron Spin Resonance*, John Wileys and Sons, New York.

HENDRICH M.P. & DEBRUNNER P.G. (1989) *Biophysical Journal* **56**: 489-506.
LANCIANO P. *et al.* (2007) *Journal of Physical Chemistry* B **111**: 13632-13637
NEESE F. (2006) *Journal of the American Chemical Society* **128**: 10213-10222.
ROGER G. *et al.* (1980) *Journal de Physique* **41**: 169-175.
ROGER G. *et al.*(1982) *Journal de Physique* **43**: 285-291.
SCHOLES C.P. *et al.*(1971) *Biochimica et Biophysica Acta* **244**: 206-210.

Exercises

6.1. The dipolar interaction between two electrons A and B with spin angular momenta $\mathbf{s}_A$ and $\mathbf{s}_B$ is described by (appendix 3):

$$H_{dip} = (\mu_0/4\pi)\,(g_e\beta)^2/r^3[\mathbf{s}_A\cdot\mathbf{s}_B - 3\,(\mathbf{u}\cdot\mathbf{s}_A)(\mathbf{u}\cdot\mathbf{s}_B)]$$

where **r** is the vector $\overrightarrow{AB}$, r its modulus and **u** the unit vector along **r**. The order of magnitude of H_{dip} is that of the prefactor. Calculate the prefactor in cm^{-1} for $r = 3$ Å.

6.2. Using the matrix elements in appendix 7, show that the matrix representing the $\hat{H}_{ZFS}$ operator (equations [6.1] and [6.3]) is not null when $S \neq ½$ (hint: examine the diagonal elements).

6.3. Consider the sum $A = \sum_{M_S=-S}^{S-1} [S(S+1) - M_S(M_S+1)]$

a) Show that it is unaltered if we extend the sum to $M_S = S$.

b) Using equation [3.11], show that $A = 2S(S+1)(2S+3)/3$.

6.4. Following the approach described in complement 2 to chapter 4, determine the density of the resonance lines for which the positions are given by equation [6.28]. To simplify, we can set $\Delta B = -(D/g\beta)(M_S + ½)$.

6.5. In the high-field situation, the energy levels for a paramagnetic centre with rhombic symmetry are given by equation [6.30], where $\hat{H}_{ZFS}$ is defined by equation [6.8]. We are interested in the resonance fields for the canonical directions of **B**.

a) Assume **B** parallel to Z, and set $x' = X, y' = Y, z' = Z$. Using appendix 7, calculate the matrix element involved in equation [6.30]. From this, deduce the energy levels $E(M_S)$, the energies $\Delta E(M_S)$ for the transitions $|S, M_S\rangle_{z'} \leftrightarrow |S, M_S + 1\rangle_{z'}$ and the resonance field for the $2S$ transitions.

b) Perform the same calculation for **B** parallel to X, setting $x' = Y, y' = Z, z' = X$.

c) Perform the same calculation for **B** parallel to Y, setting $x' = Z, y' = X, z' = Y$.

6.6. Consider a frozen solution of N complexes of spin $S = 3/2$, with axial symmetry, characterised by $g_{//} \approx g_\perp \approx 2.0$. We assume that the low-field situation is achieved and we are interested in the $\{|3/2, -1/2\rangle, |3/2, 1/2\rangle\}$ doublet.

a) What are the effective g values for this doublet? Using equation [5.19], calculate the mean intensity factor $(g_P)_{av}$.

b) Determine the zero-field energy levels. From these levels, deduce the expression for the $N_{½}(T)$ population of the doublet. Represent this expression graphically as a function of T for $|D| = 5\ cm^{-1}$, distinguishing between the $D > 0$ and $D < 0$ cases.

c) Hereafter, we assume $D = 5\ cm^{-1}$. We record the spectrum for the doublet at X-band, at $T = 10$ K, in *non-saturating* conditions. Verify that the low-field situation is effectively achieved. The intensity $I_{½}$ determined by two successive integrations of the spectrum is equal to 3 times the intensity I_0 of the spectrum produced by a reference sample recorded in the same conditions (temperature, power), non-saturating for this sample. The reference sample contains N_{ref} molecules of spin ½ characterised by $g_x = g_y = g_z = 2.15$, for which the signal intensity follows Curie's law. From this information, deduce the N/N_{ref} ratio.

6.7. A frozen aqueous solution of Mn^{2+} ions (configuration $3d^5$) produces an EPR spectrum with 6 equidistant hyperfine lines centred at $g = 2.00$, visible at all temperatures (figure 2.7). In contrast, the Mn^{2+} ion placed in substitution in $K_4Fe(CN)_6$ produces a powder spectrum characterised by ($g_x = 0.63$, $g_y = 2.18$, $g_z = 2.62$) and ($|A_x| = 312$ MHz, $|A_y| = 138$ MHz, $|A_z| = 252$ MHz) which can only be observed at very low temperature ($T < 20$ K) [Carrington and McLachlan, 1979]. What is the reason for these differences, given that H_2O and CN^- are weak and strong ligands, respectively, in the spectrochemical series (see complement 1 to chapter 4)?

Answers to exercises

6.1. H_{dip} [cm^{-1}] = $1.7/r^3$[Å^3], or 0.063 cm^{-1} for r = 3 Å.

6.2. The diagonal elements of the matrix representing $\hat{H}_{ZFS}$ are written:

$$\begin{aligned}\langle M_S|\hat{H}_{ZFS}|M_S\rangle &= D_X\langle M_S|\hat{S}_x^2|M_S\rangle + D_Y\langle M_S|\hat{S}_y^2|M_S\rangle + D_Z\langle M_S|\hat{S}_z^2|M_S\rangle \\ &= (D_Z/2)\,[3M_S^2 - S(S+1)]\end{aligned}$$

The data from appendix 7 and equation [6.3] were used. We can readily verify that the quantity in square brackets vanishes only if S = ½.

6.3. a) As the supplementary term is null, the sum is not modified.

b) By developing and using

$$\sum_{M_S=-S}^{S} M_S = 0 \text{ and } \sum_{M_S=-S}^{S} (M_S)^2 = \frac{S(S+1)(2S+1)}{3} \text{ (equation [3.11])},$$

we obtain the forecast result.

6.4. The resonance field is written

$$B = B_0 + \Delta B\,(3\cos^2\theta - 1) \qquad [1]$$

The sign of ΔB depends on that of D and on the value of M_S. We will assume, for example, $\Delta B > 0$. When θ increases from 0 to 90°, B decreases from $B_{//} = B_0 + 2\Delta B$ to $B_\perp = B_0 - \Delta B$. The density $D(B)$ of resonance lines is therefore linked to the probability density $p(\theta)$ of θ by:

$$D(B)\,\mathrm{d}B = -p(\theta)\,\mathrm{d}\theta$$

Thus $D(B) = -p(\theta)/(\mathrm{d}B/\mathrm{d}\theta)$. By differentiating [1] to calculate $\mathrm{d}B/\mathrm{d}\theta$ and replacing $p(\theta)$ by $\sin\theta$ (complement 3 to chapter 4), we obtain $D(B) = 1/(6\Delta B\cos\theta)$. From [1] we deduce $\cos\theta = [(B - B_\perp)/3\Delta B]^{½}$, which leads to

$$D(B) = ½\,[(B_{//} - B_\perp)(B - B_\perp)]^{-½}$$

$\Delta B < 0$ can be dealt with in the same way, and produces the same expression.

6.5. a) $\hat{H}_{ZFS} = D\,[\hat{S}_{z'}^2 - S(S+1)/3] + E(\hat{S}_{x'}^2 - \hat{S}_{y'}^2)$

$E(M_S) = g\beta BM_S + D\,[M_S^2 - S(S+1)/3]$

$\Delta E(M_S) = g\beta B + 2D(M_S + ½)$

$B_Z = B_0 - (2D/g\beta)(M_S + ½)$

b) $\hat{H}_{ZFS} = D\,[\hat{S}_{y'}^2 - S(S+1)/3] + E(\hat{S}_{z'}^2 - \hat{S}_{x'}^2)$

$E(M_S) = g\beta BM_S + [(3E - D)/2]\,[M_S^2 - S(S+1)/3]$

$\Delta E(M_S) = g\beta B + (3E - D)\,(M_S + ½)$

$B_X = B_0 + [(D - 3E)/g\beta]\,(M_S + ½)$

c) $\hat{H}_{ZFS} = D\,[\hat{S}_{x'}^2 - S(S+1)/3] + E(\hat{S}_{y'}^2 - \hat{S}_{z'}^2)$

$E(M_S) = g\beta BM_S - [(3E + D)/2]\,[M_S^2 - S(S+1)/3]$

$\Delta E(M_S) = g\beta B - (3E + D)\,(M_S + ½)$

$B_Y = B_0 + [(D + 3E)/g\beta]\,(M_S + ½)$

6.6. a) $g_{//}^{\text{eff}} = g_{//} = 2.0$; $g_{\perp}^{\text{eff}} = 2g_{\perp} = 4.0$ (figure 6.12); $(g_P)_{av} = 0.855$.

b) The energy levels in the absence of magnetic field are given by equation [6.31]: $E_0(M_S = \pm ½) = -D$; $E_0(M_S = \pm ³⁄₂) = D$. Equation [6.39] produces:

$$N_{½}(T) = \frac{N}{1 + \exp(-2D/k_B T)}$$

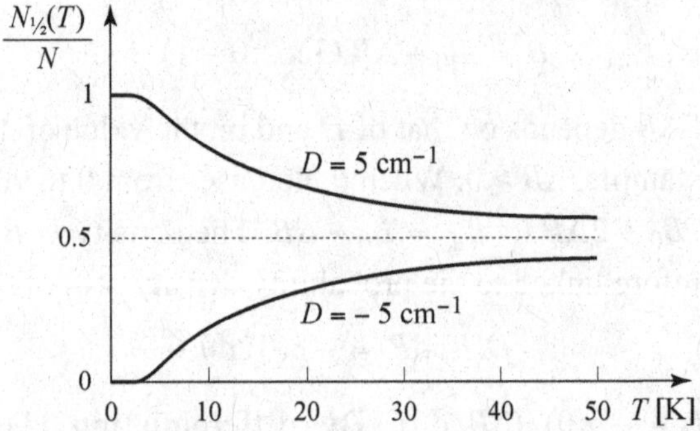

c) At X-band, $h\nu \approx 0.3\ \text{cm}^{-1} \ll D$: the low-field situation is achieved. The intensity of the spectrum for the doublet is given by equation [6.38]:

$$I_{½}(T) = (\pi\beta/2\hbar)\,(g_P)_{av}\,B_1\,N_{½}(T)\tanh(h\nu/2k_BT)$$

In the same conditions, the intensity of the reference sample is equal to (equation [5.17]):

$$I_{ref}(T) = (\pi\beta/2\hbar)\,(g_P)_{av}^{ref}\,B_1\,N_{ref}\tanh(h\nu/2k_BT)$$

with $(g_P)_{av}^{ref} = 2.15/4 = 0.537$. From these relations, we deduce

$$\frac{N}{N_{ref}} = \frac{(g_P)_{av}^{ref}\,I_{½}(T)[1 + \exp(-2D/k_BT)]}{(g_P)_{av}\,I_{ref}(T)}$$

The numerical application produces $N/N_{ref} = 2.3$.

6.7. In aqueous solution, Mn^{2+} ions form $Mn(H_2O)_6{}^{2+}$ complexes with octahedral symmetry where they are in a strong spin situation, $S = 5/2$. For the $3d^5$ configuration, the ground term is characterised by $L = 0$, which has several consequences. Firstly, the three principal values of the $\tilde{\mathbf{g}}$ matrix are practically equal to g_e (section 4.2.2) and the $\tilde{\mathbf{A}}$ matrix is isotropic with a principal value A_s (section 4.4.1 and appendix 2). In addition, the spin-lattice relaxation is not very effective, which explains why the signal is observable over a very broad temperature range.

The strong anisotropy of the $\tilde{\mathbf{g}}$ and $\tilde{\mathbf{A}}$ matrices of the $Mn(CN)_6{}^{4-}$ complex indicates that the symmetry is not cubic and that the orbital angular momentum of the electrons contributes significantly to the magnetic properties. The absence of effect of the zero-field splitting terms on the spectrum indicates that the Mn^{2+} ion is in a weak spin situation, $S = 1/2$, in line with the position of the CN^- ion in the spectrochemical series. The strong contribution of the orbital angular momentum explains the very effective spin-lattice relaxation of this complex, even at low temperatures.

Chapter 7

Effects of dipolar and exchange interactions on the EPR spectrum. Biradicals and polynuclear complexes

7.1 – Introduction

The magnetic interactions that we have considered up to now were internal to the paramagnetic centres and these centres were therefore *independent* from each other. But when the centres are close enough, the EPR spectrum can be modified by two interactions involving their unpaired electrons:

- ▹ *Dipolar* interaction between their *magnetic moments*. This interaction is proportional to $1/r^3$ and is generally weaker than the interaction with the magnetic field.
- ▹ *Exchange* interaction, which is an effect of intercentre *electrostatic* interactions due to overlap between orbitals occupied by the unpaired electrons. This interaction can be up to several hundred cm^{-1}, but decreases very rapidly (generally exponentially) when the intercentre distance increases.

The effects of these interactions on the spectrum depend on their intensity and on the number of centres involved. The following situations can be distinguished:

1. When the molecules contain a single paramagnetic centre and are diluted in the sample, the interaction is too weak to alter the spectrum. It nevertheless plays an important role, since it ensures the internal equilibrium of the paramagnetic centres, based on which the "populations" of their energy levels are defined; and it determines the "homogeneous" width of the resonance lines (section 5.2.1 and complement 3 to chapter 5).

P. Bertrand, *Electron Paramagnetic Resonance Spectroscopy*,
https://doi.org/10.1007/978-3-030-39663-3_7

2. When two paramagnetic centres are present in the same molecule, they are often close enough for their spectra to be altered due to dipolar and exchange interaction:
   - If these interactions are weaker than the interaction with the magnetic field, they cause *splitting* of the resonance lines. The intensity of the spectra remains the same, but their shape is very sensitive to the relative arrangement of the centres. Quantitative studies of these shapes can provide detailed structural information, as we will see in section 7.3.
   - In biradicals and polynuclear transition ion complexes, the strong intercentre exchange interaction profoundly modifies the magnetic properties, to the point that the EPR spectrum may disappear at low temperature. The shape and temperature-dependence of the spectrum provides information on the magnitude of this interaction and on the nature of the centres. This case will be studied in section 7.4.
3. When the paramagnetic centres in a solid are coupled by exchange interaction, *collective phenomena* appear such as ferromagnetism and antiferromagnetism. The spectra produced by this type of material (ferromagnetic and antiferromagnetic resonance spectroscopies) are interpreted using models which are very different to those used up to now. These models are not described here, but these spectroscopies and their applications are presented in [Heinrich, 2005] and in chapter 12 of the second volume.

7.2 – Origin and description of intercentre interactions

In this section, we provide information on the origin of the dipolar and exchange interactions, and we show how their effects can be described using spin operators.

7.2.1 – The true nature of exchange interaction

The specificity of exchange interaction is due to the fact that its very *existence* is linked to the *mathematical form* of the wave function describing the state of the electrons in atoms and molecules, and the fact that this form is imposed by the *indistinguishable* nature of the electrons.

In classical physics, a well-defined trajectory is associated with any particle in movement and it is implicitly assumed that two particles can always be distinguished, even when they are identical. In quantum physics, the notion of trajectory does not exist, but at any time the probability that a particle is at a particular point in space can be defined by a probability density. This probability density is equal to the square of the modulus of the wave function describing the particle's state. If the wave functions for two identical particles overlap at any given time, these particles become indistinguishable at later times. Because of this fundamental indeterminacy, identical particles must be considered *indistinguishable* at the atomic scale. This constraint is mathematically expressed as follows: if the wave function $\Psi(a_1, a_2, \ldots, a_n, t)$ describes the state of an ensemble of n identical particles, where $\{a_i\}$ represents the set of orbital (position) and spin variables defining the state of the particle i, the quantity $|\Psi(a_1, a_2, \ldots, a_n, t)|^2$ must remain unchanged when the sets of variables for any two particles i and j are *permuted*:

$$|\Psi(a_1, a_2, \ldots, a_i, \ldots, a_j, \ldots, a_n, t)|^2 = |\Psi(a_1, a_2, \ldots, a_j, \ldots, a_i, \ldots, a_n, t)|^2$$

This leads to

$$\Psi(a_1, a_2, \ldots, a_i, \ldots, a_j, \ldots, a_n, t) = \varepsilon\, \Psi(a_1, a_2, \ldots, a_j, \ldots, a_i, \ldots, a_n, t)$$

where ε is a complex number with a modulus of 1. Further analysis of the properties of the "permutation group" shows that ε can only take the values +1 or −1 [Ayant and Belorizky, 2000; Cohen-Tannoudji, 2015]. In the first case, the function is said to be symmetric, and in the second it is antisymmetric. The symmetric or antisymmetric nature of the wave function describing a system of identical particles has extensive consequences on its physical properties, and all of the experiments performed up to now demonstrate that the wave functions for electrons and other matter particles of spin ½ (protons, neutrons) are always *antisymmetric*. This remarkable property was demonstrated by W. Pauli on the basis of very general hypotheses, and one of its consequences is the existence of exchange interaction, as demonstrated by the following example.

Consider two centres, A and B, which each have an unpaired electron. If the centres are sufficiently distant to avoid *overlap* of the orbitals φ_A and φ_B occupied by the unpaired electrons, electrons 1 and 2 will be *distinguishable* (figure 7.1a). The intercentre interactions are assumed to be negligible and the orbital $\varphi_A(\mathbf{r}_1)$ describing the state of electron 1 is assumed to verify Schrödinger's equation:

$$[T(1) + V_A(\mathbf{r}_1)]\, \varphi_A(\mathbf{r}_1) = E_A\, \varphi_A(\mathbf{r}_1) \qquad [7.1]$$

where $T(1)$ is the differential operator $(-\hbar^2/2m_e)(\partial^2/\partial x_1{}^2 + \partial^2/\partial y_1{}^2 + \partial^2/\partial z_1{}^2)$ associated with its kinetic energy and $V_A(\mathbf{r}_1)$ is the potential energy resulting from the electrostatic attraction of the "core" made up of the nuclei and the other electrons in centre A. Similarly, the orbital $\varphi_B(\mathbf{r}_2)$ which describes the state of electron 2 verifies the partial differential equation:

$$[T(2) + V_B(\mathbf{r}_2)]\,\varphi_B(\mathbf{r}_2) = E_B\,\varphi_B(\mathbf{r}_2) \qquad [7.2]$$

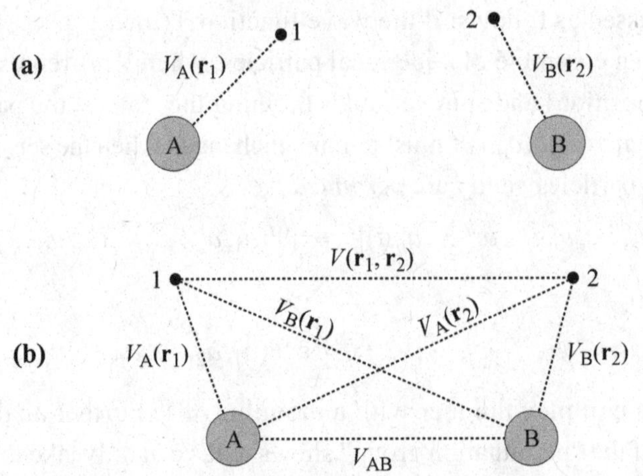

Figure 7.1 – Electrostatic interactions to which unpaired electrons from two centres are subjected. **(a)** The electrons are distinguishable. **(b)** The electrons are indistinguishable.

Hereafter, we will assume that the functions $\varphi_A(\mathbf{r}_1)$ et $\varphi_B(\mathbf{r}_2)$ are *real*. Now, consider the system (A, B) formed by the *combination of the two centres*. The orbital state of this system is described by a wave function $\Phi_0(\mathbf{r}_1, \mathbf{r}_2)$ which verifies the equation:

$$H_0\,\Phi_0(\mathbf{r}_1, \mathbf{r}_2) = E_0\,\Phi_0(\mathbf{r}_1, \mathbf{r}_2)$$

with:

$$H_0 = T(1) + V_A(\mathbf{r}_1) + T(2) + V_B(\mathbf{r}_2)$$

Using equations [7.1] and [7.2], it is easy to verify that the function $\Phi_0(\mathbf{r}_1, \mathbf{r}_2) = \varphi_A(\mathbf{r}_1)\varphi_B(\mathbf{r}_2)$ satisfies this equation, with:

$$E_0 = E_A + E_B$$

We now introduce the interactions between the two centres (figure 7.1b). The function $\Phi(\mathbf{r}_1, \mathbf{r}_2)$ describing the state of the two electrons verifies the following equation:

$$H\,\Phi(\mathbf{r}_1,\mathbf{r}_2) = E\,\Phi(\mathbf{r}_1,\mathbf{r}_2) \qquad [7.3]$$

$$H = V_{AB} + h(1) + h(2) + V(\mathbf{r}_1,\mathbf{r}_2)$$

$$h(i) = T(i) + V_A(\mathbf{r}_i) + V_B(\mathbf{r}_i), \quad i = 1,2$$

The quantity V_{AB} represents the interaction between the two "cores". It is independent of the coordinates of electrons 1 and 2, and is therefore a simple additive constant. $V(\mathbf{r}_1,\mathbf{r}_2)$ is the repulsion interaction between the two electrons. As the electrons are now *indistinguishable*, the wave function for the (A, B) system must be antisymmetric with respect to the permutation of their orbital and spin variables. Since the interactions described by H do not involve the spin variables, the wave function is written in the form of a product

$$\Psi(1,2) = \Phi(\mathbf{r}_1,\mathbf{r}_2)\,\chi(1,2)$$

where $\chi(1,2)$ describes the *spin state* of the electrons. The function $\Psi(1,2)$ must be antisymmetric, but each of its factors may be symmetric or antisymmetric. It can be shown that the two lowest-energy functions verifying equation [7.3] are a *symmetric* function $\Phi_S(\mathbf{r}_1,\mathbf{r}_2)$ and an *antisymmetric* function $\Phi_{AS}(\mathbf{r}_1,\mathbf{r}_2)$. When the overlap between φ_A and φ_B is small, these functions and their energy can be written [Kahn, 1993]:

$$\Phi_S(\mathbf{r}_1,\mathbf{r}_2) = \frac{1}{\sqrt{2}}[\varphi_A(\mathbf{r}_1)\varphi_B(\mathbf{r}_2) + \varphi_B(\mathbf{r}_1)\varphi_A(\mathbf{r}_2)]$$

$$E_S = E_0 + V_{AB} + K + \frac{J}{2}$$

$$\Phi_{AS}(\mathbf{r}_1,\mathbf{r}_2) = \frac{1}{\sqrt{2}}[\varphi_A(\mathbf{r}_1)\varphi_B(\mathbf{r}_2) - \varphi_B(\mathbf{r}_1)\varphi_A(\mathbf{r}_2)] \qquad [7.4]$$

$$E_{AS} = E_0 + V_{AB} + K - \frac{J}{2}$$

▹ The "Coulomb" contribution K which appears in E_S and E_{AS} is given by:

$$K = \int \varphi_A{}^2(\mathbf{r})\,V_B(\mathbf{r})\,d^3\mathbf{r} + \int \varphi_B{}^2(\mathbf{r})\,V_A(\mathbf{r})\,d^3\mathbf{r}$$
$$+ \int \varphi_A{}^2(\mathbf{r}_1)\,\varphi_B{}^2(\mathbf{r}_2)\,V(\mathbf{r}_1,\mathbf{r}_2)\,d^3\mathbf{r}_1\,d^3\mathbf{r}_2$$

In these expressions, $d^3\mathbf{r}$, $d^3\mathbf{r}_1$ and $d^3\mathbf{r}_2$, respectively, represent the products $(dx\,dy\,dz)$, $(dx_1\,dy_1\,dz_1)$ and $(dx_2\,dy_2\,dz_2)$, and the integration is performed over the whole space. The first two terms are the energy due to attraction between an electron with a probability density $\varphi_A{}^2(\mathbf{r})$ and the core B, and the energy due to attraction between an electron with a probability density

$\varphi_B^2(\mathbf{r})$ and the core A, the third is the energy due to the repulsion between the two electrons.

- The "exchange" contribution J, which appears with a different sign in E_S and E_{AS}, is written:

$$J = J_1 + J_2$$

$$J_1 = 4\left(\int \varphi_A(\mathbf{r})\varphi_B(\mathbf{r})\, d^3\mathbf{r}\right)\left(\int \varphi_A(\mathbf{r}_1)\, h(1)\, \varphi_B(\mathbf{r}_1)\, d^3\mathbf{r}_1\right)$$

$$J_2 = 2\int \varphi_A(\mathbf{r}_1)\varphi_B(\mathbf{r}_1)\, V(\mathbf{r}_1,\mathbf{r}_2)\, \varphi_A(\mathbf{r}_2)\, \varphi_B(\mathbf{r}_2)\, d^3\mathbf{r}_1\, d^3\mathbf{r}_2 \qquad [7.5]$$

To understand the meaning of this contribution, assume that φ_A and φ_B do not *overlap*. As a result, the product $\varphi_A(\mathbf{r})\varphi_B(\mathbf{r})$ is null for any value of $\mathbf{r}$, producing $J_1 = J_2 = 0$. According to equations [7.4], the states represented by $\Phi_S(\mathbf{r}_1, \mathbf{r}_2)$ and $\Phi_{AS}(\mathbf{r}_1, \mathbf{r}_2)$ then have the same energy ($E_0 + V_{AB} + K$).

The states represented by the functions

$$\varphi_A(\mathbf{r}_1)\varphi_B(\mathbf{r}_2) = \frac{1}{\sqrt{2}}[\Phi_S(\mathbf{r}_1,\mathbf{r}_2) + \Phi_{AS}(\mathbf{r}_1,\mathbf{r}_2)]$$

and

$$\varphi_B(\mathbf{r}_1)\varphi_A(\mathbf{r}_2) = \frac{1}{\sqrt{2}}[\Phi_S(\mathbf{r}_1,\mathbf{r}_2) - \Phi_{AS}(\mathbf{r}_1,\mathbf{r}_2)]$$

in which the electrons are *distinguishable*, also have this energy: when the orbitals do not overlap, the distinguishability (or not) of the electrons has no effect on the energy levels. Conversely, when φ_A and φ_B *overlap* the electrons are indistinguishable, the energies of the states $\Phi_S(\mathbf{r}_1, \mathbf{r}_2)$ and $\Phi_{AS}(\mathbf{r}_1, \mathbf{r}_2)$ are different and the difference is equal to J.

It can be shown that $J_1 \leq 0$ and $J_2 \geq 0$ (exercise 7.1). It can be observed that J_1 cancels out when *the overlap integral* between φ_A and φ_B is null, i.e., when the orbitals are *orthogonal*. In addition, the function $\Phi_{AS}(\mathbf{r}_1, \mathbf{r}_2)$ cancels out for $\mathbf{r}_1 = \mathbf{r}_2$, i.e., when the two electrons occupy the same position (equation [7.4]). Therefore, in this state, the probability that the electrons are close is weak, which minimises their electrostatic repulsion. In contrast, the two electrons can occupy the same position in the state described by $\Phi_S(\mathbf{r}_1, \mathbf{r}_2)$, which explains why the term J_2 tends to destabilise this state compared to that described by $\Phi_{AS}(\mathbf{r}_1, \mathbf{r}_2)$. Since the antisymmetric property requires $\Phi_S(\mathbf{r}_1, \mathbf{r}_2)$ and $\Phi_{AS}(\mathbf{r}_1, \mathbf{r}_2)$ to be associated with antisymmetric and symmetric (relative to the spin variables) spin functions, respectively, we will now examine how such functions can be constructed. Electron 1 can occupy two spin states described by the functions $\{\alpha(1), \beta(1)\}$ which verify:

$$\hat{\mathbf{s}}_1^{\,2}\,\alpha(1) = \tfrac{3}{4}\,\alpha(1);\quad \hat{\mathbf{s}}_1^{\,2}\,\beta(1) = \tfrac{3}{4}\,\beta(1)$$

$$\hat{s}_{1z}\,\alpha(1) = \tfrac{1}{2}\,\alpha(1);\quad \hat{s}_{1z}\,\beta(1) = -\tfrac{1}{2}\,\beta(1)$$

where $\mathbf{s}_1$ is the spin angular momentum of electron 1 and z is any direction in the Euclidean space (in the space of spin states $\mathcal{E}_{1/2}$, $\alpha(1)$ and $\beta(1)$ are represented by the kets $|\tfrac{1}{2},\tfrac{1}{2}\rangle$ and $|\tfrac{1}{2},-\tfrac{1}{2}\rangle$). The functions $\{\alpha(2),\beta(2)\}$ describing the possible spin states for electron 2 are defined in the same way. From these four functions, we can contruct

- ▹ an *antisymmetric* function: $\frac{1}{\sqrt{2}}\,[\alpha(1)\beta(2) - \beta(1)\alpha(2)]$
- ▹ three *symmetric* functions: $\frac{1}{\sqrt{2}}\,[\alpha(1)\beta(2) + \beta(1)\alpha(2)];\ \alpha(1)\alpha(2);\ \beta(1)\beta(2)$

These can be recognised as the eigenfunctions shared by the operators $(\hat{\mathbf{S}}^2, \hat{S}_z)$ where $\mathbf{S} = \mathbf{s}_1 + \mathbf{s}_2$ is the *total spin* of the (A, B) system, which can also be written (appendix 4):

$$\chi(S = 0, M_S = 0)$$

$$\chi(S = 1, M_S = 0);\ \chi(S = 1, M_S = 1);\ \chi(S = 1, M_S = -1) \qquad [7.6]$$

The functions $\Psi(1,2)$ associated with the energy levels E_S and E_{AS} are obtained by multiplying $\Phi_S(\mathbf{r}_1,\mathbf{r}_2)$ by $\chi(S = 0)$, and $\Phi_{AS}(\mathbf{r}_1,\mathbf{r}_2)$ by one of the three functions $\chi(S = 1)$. The results are schematically represented in figure 7.2a, where we assumed $J > 0$. Because of the overlap between φ_A and φ_B, the *electrostatic* interactions between A and B cause splitting between the states characterised by different values of S. This splitting is (incorrectly) said to be due to the "exchange interaction" between A and B. The same phenomenon exists in *atoms*, where J_1 does not exist and where the positive term J_2 preferentially stabilises the state with maximum total spin. This is the source of Hund's rule, described in appendix 1.

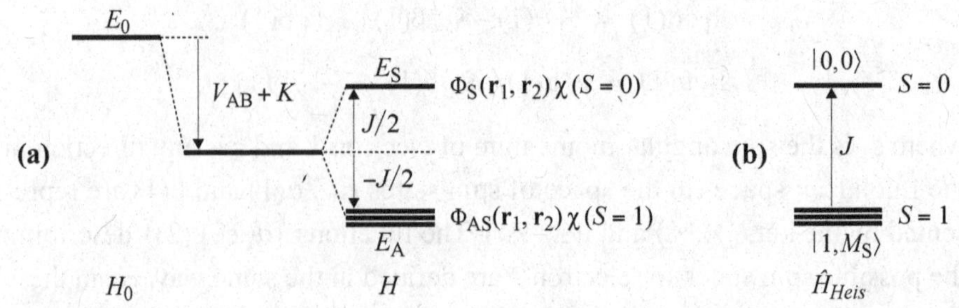

Figure 7.2 – **(a)** Effect of electrostatic interactions between two centres, each with an unpaired electron, when the orbitals occupied by these electrons overlap (we assume $J > 0$). **(b)** Phenomenological description by the Heisenberg operator.

7.2.2 – Phenomenological description of exchange interaction

A splitting J between the spin state $S = 0$ and the three $S = 1$ spin states can be produced using the "Heisenberg" operator $\hat{H}_{Heis} = -J\,\mathbf{s}_1 \cdot \mathbf{s}_2$.

Indeed, the matrix representing this operator in the "coupled basis" defined by equations [7.6] is diagonal, and its eigenvalues are given by $-J/2\ [S(S+1) - s_1(s_1+1) - s_2(s_2+1)]$ (exercise 7.2). The splitting between the state $S = 0$ and the three $S = 1$ states is equal to J (figure 7.2b). Although the $\hat{H}_{Heis}$ operator *reproduces* the splitting due to exchange interaction, it provides no information on its origins.

We can generalise the foregoing in the case where the centres A and B have *several unpaired electrons* [Kahn, 1993]. Due to the *overlap* between the orbitals occupied by these electrons, the intercentre electrostatic interactions cause the energy levels of the (A, B) system to split. When this interaction is not too strong and does not involve the spin variables, this splitting can be reproduced by the operator:

$$\hat{H}_{Heis} = -J\,\mathbf{S}_A \cdot \mathbf{S}_B \qquad [7.7]$$

where S_A and S_B are the spins for the two centres. The energy levels are given by $E(S) = -(J/2)\,S(S+1)$ up to an additive constant, where S associated with $\mathbf{S} = \mathbf{S}_A + \mathbf{S}_B$ varies from $S_{min} = |S_A - S_B|$ to $S_{max} = S_A + S_B$ in unit steps (appendix 4). The exchange parameter, J, includes a positive component J_2 and a negative component J_1. The first tends to stabilise the state characterised by S_{max} and is sometimes qualified as "ferromagnetic" since J is positive in metallic iron. By analogy, the negative component, which tends to stabilise the state characterised by S_{min}, is qualified as "antiferromagnetic". As the exchange parameter is determined by the orbital overlap, it is very sensitive to structural details and its value can vary over a very broad range. Thus, negative values of J of several hundred cm^{-1} have been measured for some polynuclear complexes, whereas in other complexes J is much smaller and has a variable sign [Owen and Harris, 1972]. In organic biradicals, J is often around 0.1 to 1 cm^{-1}, but can sometimes be up to several tens of cm^{-1} [Catala *et al.*, 2001; Wautelet *et al.*, 2003]. Values as low as 10^{-3} cm^{-1} have been measured for paramagnetic centres separated by 20 Å [More et *al.*, 2005]. It is important to note that the prefactor of the Heisenberg operator is sometimes denoted J or $-2J$ in the literature. It is therefore essential to check the convention used when comparing numerical values.

7.2.3 – "Anisotropic components" of exchange interaction

The splitting of the energy levels caused by exchange interaction can be described by a Heisenberg operator if we consider that the electrostatic interactions do not affect the spin states of the electrons, i.e., if we ignore the spin-orbit interaction. When this interaction is taken into account, the effects of exchange interaction are described by the following operator [Bencini and Gatteschi, 1990]:

$$\hat{H}_{exch} = -J\,\mathbf{S}_A \cdot \mathbf{S}_B + \hat{H}_{exch}^{anis} \qquad [7.8]$$
$$\hat{H}_{exch}^{anis} = \mathbf{S}_A \cdot \tilde{\mathbf{J}}_{\mathbf{exch}}^{\mathbf{anis}} \cdot \mathbf{S}_B$$

where $\tilde{\mathbf{J}}_{\mathbf{exch}}^{\mathbf{anis}}$ is a matrix with *null-trace*, in which the elements are much smaller than J. The second term in equation [7.8] is said to represent the *anisotropic* components of the exchange interaction. This nomenclature is misleading. Indeed, as the exchange interaction is due to orbital overlap, the parameter J in the Heisenberg operator is strongly dependent on the relative orientation of the centres. The qualifiers "isotropic" and "anisotropic" used here indicate that, when two centres belong to the same molecule, the effect of the $\hat{H}_{Heis}$ term on the spectrum is independent of the molecule's orientation relative to **B**, whereas that of the $\hat{H}_{exch}^{anis}$ term depends on it: in the *isotropic regime*, the effects of $\hat{H}_{exch}^{anis}$ on the spectrum average out to zero due to movement, whereas those of $\hat{H}_{Heis}$ persist.

The $\tilde{\mathbf{J}}_{\mathbf{exch}}^{\mathbf{anis}}$ matrix is sometimes written in the form

$$\tilde{\mathbf{J}}_{\mathbf{exch}}^{\mathbf{anis}} = \tilde{\mathbf{J}}_{\mathbf{s}} + \tilde{\mathbf{J}}_{\mathbf{as}}$$

where $\tilde{\mathbf{J}}_{\mathbf{s}}$ is *symmetric* and $\tilde{\mathbf{J}}_{\mathbf{as}}$ is *antisymmetric*. The $\tilde{\mathbf{J}}_{\mathbf{as}}$ matrix is null when the (A, B) system has a centre of symmetry [Bencini and Gatteschi, 1990]. The $\mathbf{S}_A \cdot \tilde{\mathbf{J}}_{\mathbf{as}} \cdot \mathbf{S}_B$ operator is called the "Dzyaloshinskii-Moriya interaction" by solid-state physicists, which makes things unnecessarily more complicated.

7.2.4 – Magnetic dipolar interaction

Consider two paramagnetic centres, radicals or mononuclear transition ion complexes, characterised by the spins S_A and S_B and the matrices $\tilde{\mathbf{g}}_{\mathbf{A}}$ and $\tilde{\mathbf{g}}_{\mathbf{B}}$. When their distance exceeds the extension of the orbitals occupied by their unpaired electrons (point dipole approximation), the effects of the intercentre dipolar interactions can be described by the following operator:

$$\hat{H}_{dip} = (\mu_0/4\pi)\beta^2 \, [(\tilde{\mathbf{g}}_{\mathbf{A}} \cdot \mathbf{S}_A)\cdot(\tilde{\mathbf{g}}_{\mathbf{B}} \cdot \mathbf{S}_B) - 3(\mathbf{r}\cdot\tilde{\mathbf{g}}_{\mathbf{A}}\cdot\mathbf{S}_A)(\mathbf{r}\cdot\tilde{\mathbf{g}}_{\mathbf{B}}\cdot\mathbf{S}_B)/r^2]/r^3 \quad [7.9]$$

$\mu_0 = 4\pi \times 10^{-7}$ H m^{-1} is the vacuum permeability, $\mathbf{r}$ is the vector $\overrightarrow{AB}$ and r its length. For the typical values $g_A = g_B = 2$, H_{dip} is of the same order of magnitude as $(\mu_0/\pi)\beta^2/r^3$ which is $1.7/r^3[\text{Å}^3]$ in cm^{-1}, i.e., 1.4×10^{-2} cm^{-1} (400 MHz) for $r = 5$ Å and 1.7×10^{-3} cm^{1} (50 MHz) for $r = 10$ Å. These energies are comparable to those of the hyperfine and superhyperfine interactions, respectively, in a Cu^{2+} complex (figure 4.11).

7.3 – Effects of weak intercentre interactions on the spectrum

The spin Hamiltonian describing all of the interactions involving the unpaired electrons from the two centres A and B can be written:

$$\hat{H} = \hat{H}_A + \hat{H}_B + \hat{H}_{exch} + \hat{H}_{dip} \quad [7.10]$$

The $\hat{H}_A$ and $\hat{H}_B$ operators describe the "local" interactions occurring within each centre: each includes a Zeeman term and may include terms for hyperfine interactions, as well as zero-field splitting terms when the spin exceeds ½. The $\hat{H}_{exch}$ and $\hat{H}_{dip}$ operators describe the intercentre interactions (equations [7.8] and [7.9]). In this section, we assume that these interactions are much weaker than those with the field **B**. This situation can occur in crystals incorporating transition ions or rare earth ions, and it is frequent in biological macromolecules containing several paramagnetic centres separated by 10 to 20 Å. To occur at X-band, the J parameter must be less than 0.03 cm^{-1}. In these conditions, the anisotropic components of the exchange interaction can be neglected. To show how the shape of the spectrum depends on the relative arrangement of the centres, we consider the simple case of centres with spin ½ in the absence of hyperfine coupling.

In this case, the local terms are reduced to the Zeeman terms, and the Hamiltonian [7.10] becomes:

$$\hat{H} = \beta\, \mathbf{B}\cdot\tilde{\mathbf{g}}_\mathbf{A}\cdot\mathbf{S}_\mathrm{A} + \beta\, \mathbf{B}\cdot\tilde{\mathbf{g}}_\mathbf{B}\cdot\mathbf{S}_\mathrm{B} + \hat{H}_{Heis} + \hat{H}_{dip} \qquad [7.11]$$

We will see that intercentre interactions cause the resonance lines to split without modifying the total intensity of the two spectra. These effects are similar to those produced by the hyperfine coupling (section 5.2.4) and zero-field splitting terms in the high-field limit (section 6.4.2), but here they affect the spectra for both centres. We will first examine the effects of the two interactions separately.

7.3.1 – Effects of the dipolar interactions

We first deal with the case where the $\tilde{\mathbf{g}}_\mathbf{A}$ and $\tilde{\mathbf{g}}_\mathbf{B}$ matrices are isotropic, then with the case where their principal axes are parallel.

◇ *The* $\tilde{\mathbf{g}}$ *matrices for the two centres are isotropic*

The Hamiltonian [7.11] becomes:

$$\hat{H} = g_\mathrm{A}\beta\mathbf{B}\cdot\mathbf{S}_\mathrm{A} + g_\mathrm{B}\beta\mathbf{B}\cdot\mathbf{S}_\mathrm{B} + \hat{H}_{dip} \qquad [7.12]$$

$$\hat{H}_{dip} = (\mu_0/4\pi)\, g_\mathrm{A} g_\mathrm{B}\beta^2\, [\mathbf{S}_\mathrm{A}\cdot\mathbf{S}_\mathrm{B} - 3(\mathbf{S}_\mathrm{A}\cdot\mathbf{r})(\mathbf{S}_\mathrm{B}\cdot\mathbf{r})/r^2]/r^3 \qquad [7.13]$$

The principal term of Hamiltonian [7.12] is the Zeeman term, which can be written

$$\hat{H}_{Zeeman} = g_\mathrm{A}\beta B\, \hat{S}_{\mathrm{A}Z} + g_\mathrm{B}\beta B\, \hat{S}_{\mathrm{B}Z} \qquad [7.14]$$

where Z is the direction of $\mathbf{B}$. If $\{|M_\mathrm{A}\rangle, M_\mathrm{A} = \pm ½\}$ and $\{|M_\mathrm{B}\rangle, M_\mathrm{B} = \pm ½\}$ are the respective eigenvectors of $(\hat{S}_{\mathrm{A}Z}, \hat{\mathbf{S}}_\mathrm{A}^2)$ and $(\hat{S}_{\mathrm{B}Z}, \hat{\mathbf{S}}_\mathrm{B}^2)$, the kets of the product basis $\{|M_\mathrm{A}\rangle|M_\mathrm{B}\rangle\}$ verify:

$$\hat{H}_{Zeeman}|M_\mathrm{A}\rangle|M_\mathrm{B}\rangle = (g_\mathrm{A}\, M_\mathrm{A} + g_\mathrm{B}\, M_\mathrm{B})\beta B\, |M_\mathrm{A}\rangle|M_\mathrm{B}\rangle$$

They are therefore the eigenvectors of $\hat{H}_{Zeeman}$ associated with the eigenvalues:

$$E_0(M_\mathrm{A}, M_\mathrm{B}) = (g_\mathrm{A}\, M_\mathrm{A} + g_\mathrm{B}\, M_\mathrm{B})\beta B$$

The four corresponding energy levels are represented in figure 7.3, assuming $g_\mathrm{A} > g_\mathrm{B}$. They produce the expected lines when A and B are independent (exercise 7.3). First order perturbation theory can be used to determine the effect of $\hat{H}_{dip}$ on these levels in two situations:

1. When the dipolar term is smaller than the difference $(g_A - g_B)\beta B$, the expression which applies to non-degenerate energy levels is used:

$$E(M_A, M_B) = E_0(M_A, M_B) + \langle M_A|\langle M_B|\hat{H}_{dip}|M_A\rangle|M_B\rangle \quad [7.15]$$

Only the $\hat{S}_{AZ}\hat{S}_{BZ}$ terms in $\hat{H}_{dip}$ contribute to the matrix element. According to equation [7.13], these terms can be written $D(1 - 3\cos^2\gamma)\,\hat{S}_{AZ}\hat{S}_{BZ}$, where we have set

$$D = (\mu_0/4\pi)g_A g_B \beta^2/r^3 \quad [7.16]$$

and γ is the angle between the Z axis (direction of **B**) and the intercentre vector **r**. Equation [7.15] becomes:

$$E(M_A, M_B) = E_0(M_A, M_B) + D(1 - 3\cos^2\gamma)\, M_A M_B$$

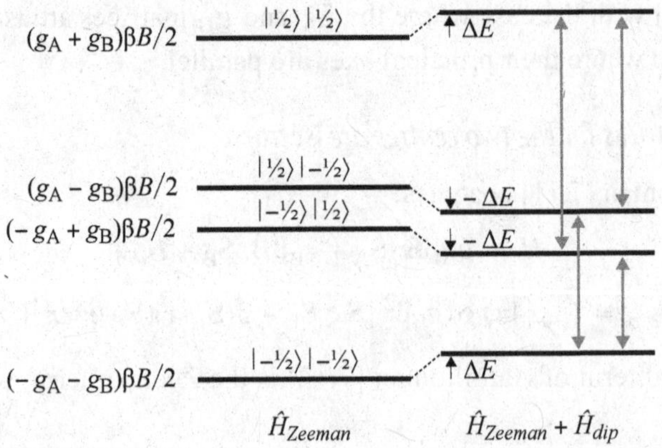

Figure 7.3 – Effect of dipolar interactions on the energy levels for two centres of spin ½ with isotropic $\tilde{\mathbf{g}}$ matrices. The $\Delta E = D(1-3\cos^2\gamma)/4$ shift is assumed to be positive. The four allowed transitions are indicated.

The resulting energy levels are illustrated in figure 7.3. To the first order, the eigenvectors are the kets $\{|M_A\rangle|M_B\rangle\}$. The four transitions have the same probability as in the absence of interaction, and their energies are $g_A\beta B \pm D(1 - 3\cos^2\gamma)/2$ and $g_B\beta B \pm D(1 - 3\cos^2\gamma)/2$. From these expressions, we can deduce the resonance fields:

$$B = B_{0A} \pm \Delta B_A/2;\; B_{0A} = h\nu/g_A\beta;\; \Delta B_A = (1 - 3\cos^2\gamma)(\mu_0/4\pi)g_B\beta/r^3$$

$$B = B_{0B} \pm \Delta B_B/2;\; B_{0B} = h\nu/g_B\beta;\; \Delta B_B = (1 - 3\cos^2\gamma)(\mu_0/4\pi)g_A\beta/r^3 \quad [7.17]$$

Two patterns of two lines with the same intensity are obtained, centred at B_{0A} and B_{0B}, and split by ΔB_A and ΔB_B, respectively (figure 7.4a). The splitting varies with γ and cancels out at the "magic angle" $\gamma_m = 54.74°$ (figure 7.4b).

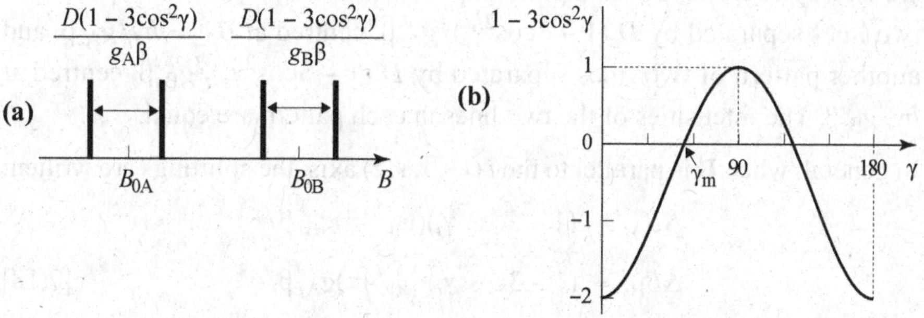

Figure 7.4 – **(a)** Splitting of resonance lines created by the dipolar interactions. The D parameter is defined by equation [7.16], γ is the angle between **B** and **r**. **(b)** Variation of $(1 - 3\cos^2\gamma)$ and definition of the magic angle γ_m.

In equations [7.17], the angular dependence of the resonance fields is identical to that of a centre with axial symmetry of spin $S = 1$ in the high-field situation (equation [6.28]). The density of the lines and the shape of the powder spectra are therefore the same as those shown in figure 6.5. As mentioned above, the intensity of each spectrum is equal to that of the spectrum in the absence of any interaction.

2. The foregoing calculation is not valid when the dipolar term is not small relative to the $(g_A - g_B)\beta B$ difference. This situation occurs, in particular, when $g_A = g_B$. If the calculation is repeated for $g_A = g_B = g$, equations [7.17] become $B = B_0 \pm \Delta B/2$, with $B_0 = h\nu/g\beta$ and $\Delta B = \frac{3}{2}(1 - 3\cos^2\gamma)(\mu_0/4\pi)g\beta/r^3$ (exercise 7.4). The splitting is multiplied by $\frac{3}{2}$ relative to the $g_A \neq g_B$ case.

◇ *The principal axes of the $\tilde{\mathbf{g}}_A$ and $\tilde{\mathbf{g}}_B$ matrices are parallel*

We will note (g_{Ax}, g_{Ay}, g_{Az}) and (g_{Bx}, g_{By}, g_{Bz}) the principal values of the two matrices, and $(\gamma_x, \gamma_y, \gamma_z)$ the angles between the intercentre vector **r** and the principal axes $\{x, y, z\}$. We will show that the splitting that appears for the "canonical directions" of **B** can be described using the expressions established in the previous section. Consider for example molecules which have their principal axis z parallel to **B**. For these molecules, the Zeeman term is written:

$$\hat{H}_{Zeeman} = g_{Az}\beta B\,\hat{S}_{Az} + g_{Bz}\beta B\,\hat{S}_{Bz}$$

This term takes the same form as equation [7.14]. Similarly, the $\hat{H}_{dip}$ term involved in perturbation theory, which is written $D_z(1 - 3\cos^2\gamma_z)\, \hat{S}_{Az}\, \hat{S}_{Bz}$ with $D_z = (\mu_0/4\pi)g_{Az}g_{Bz}\beta^2/r^3$, takes the same form as when the $\tilde{\mathbf{g}}$ matrices are isotropic. It is deduced that the dipolar interactions produce a pattern of two lines separated by $D_z(1 - 3\cos^2\gamma_z)/g_{Az}\beta$, centred at $B_{Az} = h\nu/g_{Az}\beta$, and another pattern of two lines separated by $D_z(1 - 3\cos^2\gamma_z)/g_{Bz}\beta$, centred at $h\nu/g_{Bz}\beta$. The intensities of the two lines in each pattern are equal.

In general, when **B** is parallel to the i ($i = x, y, z$) axis, the splittings are written:

$$\Delta B_{Ai} = (1 - 3\cos^2\gamma_i)(\mu_0/4\pi)g_{Bi}\,\beta/r^3$$

$$\Delta B_{Bi} = (1 - 3\cos^2\gamma_i)(\mu_0/4\pi)g_{Ai}\,\beta/r^3 \qquad [7.18]$$

These expressions are valid if the quantities $(g_{Ai} - g_{Bi})\beta B$ are greater than the dipolar term. When two principal values are equal, the splitting is multiplied by 3/2.

The splittings created by dipolar interaction in the three canonical directions are *not independent*. Indeed, the γ_i angles are linked by

$$\cos^2\gamma_x + \cos^2\gamma_y + \cos^2\gamma_z = 1 \qquad [7.19]$$

and as the anisotropy of the $\tilde{\mathbf{g}}_\mathbf{A}$ and $\tilde{\mathbf{g}}_\mathbf{B}$ matrices is generally weak, equations [7.18] show that the splittings approximately verify the relations

$$\Delta B_{Ax} + \Delta B_{Ay} + \Delta B_{Az} = 0; \; \Delta B_{Bx} + \Delta B_{By} + \Delta B_{Bz} = 0$$

On the EPR spectrum, only the *absolute value* of the splitting can be measured. These relations imply that the absolute value of the largest splitting is approximately equal to the sum of the absolute values of the two others; this is a "signature" of dipolar interactions.

To illustrate these results, consider two centres of spin $S_A = S_B = 1/2$, characterised by ($g_{Ax} = 1.95, g_{Ay} = 2.00, g_{Az} = 2.05$) and ($g_{Bx} = 1.75, g_{By} = 1.80, g_{Bz} = 1.85$). In the absence of dipolar interaction, the EPR spectrum is the sum of the spectra produced by the two centres (figure 7.5a). We have numerically calculated the interaction spectrum from the Hamiltonian [7.11] (without the exchange term) for different *directions* of the vector **r** relative to $\{x, y, z\}$ (figures 7.5b to 7.5e), with $r = 9$ Å. For this distance, the calculations give $(\mu_0/4\pi)g_B\beta/r^3 = 2.3$ mT for $g_B = 1.8$ and $(\mu_0/4\pi)g_A\beta/r^3 = 2.5$ mT for $g_A = 2.0$. We will examine to what extent the features of these spectra are described by the expressions given by the first order approximation of perturbation theory (equations [7.18]).

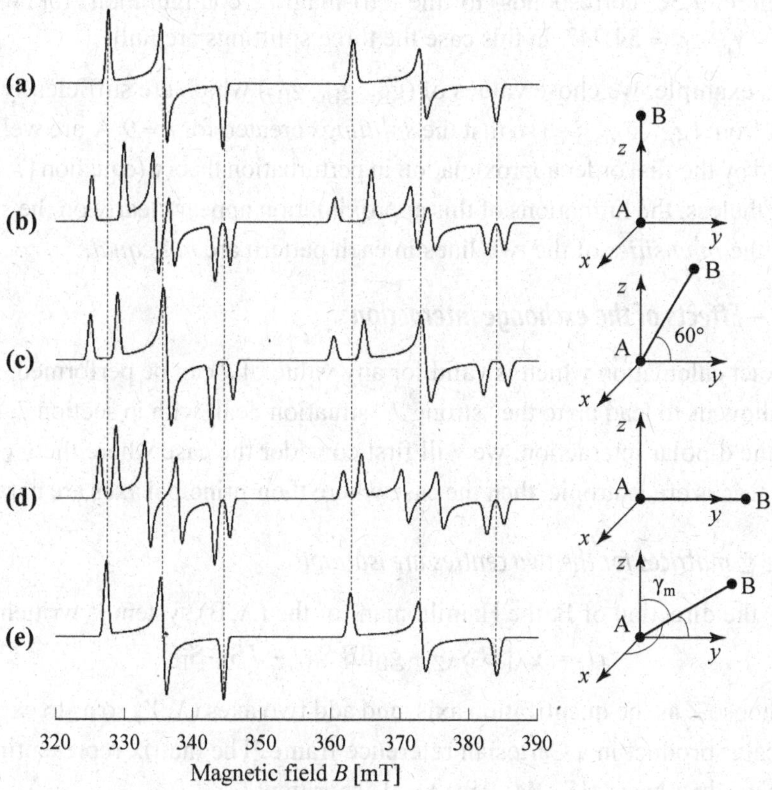

Figure 7.5 – Effect of dipolar interactions on X-band spectra (ν = 9.400 GHz) for two centres A and B characterised by g_{Ax} = 1.95, g_{Ay} = 2.00, g_{Az} = 2.05; g_{Bx} = 1.75, g_{By} = 1.80, g_{Bz} = 1.85. The intercentre distance is r = 9 Å and their relative arrangement is shown on the right.

- ▹ For **r** parallel to z ($\gamma_z = 0, \gamma_x = \gamma_y = 90°$), on the spectrum the following values can be measured: $|\Delta B_{Az}|$ = 4.6 mT, $|\Delta B_{Ax}| = |\Delta B_{Ay}|$ = 2.3 mT, and $|\Delta B_{Bz}|$ = 5 mT, $|\Delta B_{Bx}| = |\Delta B_{By}|$ = 2.5 mT (figure 7.5b), in agreement with equations [7.18].
- ▹ When **r** turns in the (y,z) plane (figures 7.5c and 7.5d), the angles γ_y and γ_z vary, but γ_x remains constant: ΔB_{Ax} and ΔB_{Bx} remain almost unchanged. For γ_y = 60°, which is close to the magic angle (figure 7.4b), the ΔB_{Ay} and ΔB_{By} splittings are too weak to be resolved (figure 7.5c). γ_z is equal to 30°, and $|\Delta B_{Az}|$ and $|\Delta B_{Bz}|$ are observed to be slightly larger than 2.3 mT and 2.5 mT, respectively, in line with figure 7.4b. In figure 7.5d, the splittings are reversed relative to those shown in figure 7.5b.

▹ Figure 7.5e corresponds to the “tri-magic” configuration for which $\gamma_x = \gamma_y = \gamma_z = 54.74°$; in this case the three splittings are null.

In this example, we chose values of (g_{Bx}, g_{By}, g_{Bz}) which are sufficiently different from (g_{Ax}, g_{Ay}, g_{Az}) so that the *splittings* created for $r = 9$ Å are well described by the first order approximation in perturbation theory (equation [7.18]). Nevertheless, the limitations of this approximation appear clearly on the spectrum: the *intensities* of the two lines in each pattern are *not equal*.

7.3.2 – Effects of the exchange interaction

An exact calculation which is valid for any value of J can be performed. This will allow us to lead up to the “strong J” situation dealt with in section 7.4. As with the dipolar interaction, we will first consider the case where the $\tilde{\mathbf{g}}_{\mathbf{A}}$ and $\tilde{\mathbf{g}}_{\mathbf{B}}$ matrices are isotropic, then the case where their principal axes are parallel.

◇ *The $\tilde{\mathbf{g}}$ matrices for the two centres are isotropic*

If Z is the direction of **B**, the Hamiltonian for the (A, B) system is written:

$$\hat{H} = g_A \beta B \hat{S}_{AZ} + g_B \beta B \hat{S}_{BZ} - J \mathbf{S}_A \cdot \mathbf{S}_B \qquad [7.20]$$

We choose Z as the quantization axis, and add two axes (X, Y) so as to express the scalar product in a Cartesian reference frame. The matrix representing $\hat{H}$ in the product basis $\{|S_A, M_A\rangle|S_B, M_B\rangle\}$ is written:

	$\|½\rangle\|½\rangle$	$\|½\rangle\|-½\rangle$	$\|-½\rangle\|½\rangle$	$\|-½\rangle\|-½\rangle$
$\langle ½\|\langle ½\|$	$(g_A+g_B)\beta B/2 - J/4$	0	0	0
$\langle ½\|\langle -½\|$	0	$\Delta g \beta B/2 + J/4$	$-J/2$	0
$\langle -½\|\langle ½\|$	0	$-J/2$	$-\Delta g \beta B/2 + J/4$	0
$\langle -½\|\langle -½\|$	0	0	0	$\beta B(-g_A - g_B)/2 - J/4$

We set $\Delta g = g_A - g_B > 0$. Two eigenvalues can be obtained directly along with their associated eigenvectors:

$$E(½, ½) = \beta B(g_A + g_B)/2 - J/4; \quad |½\rangle|½\rangle$$

$$E(-½, -½) = -\beta B(g_A + g_B)/2 - J/4; \quad |-½\rangle|-½\rangle \qquad [7.21]$$

By resolving a second degree equation, the two other eigenvalues can be determined:

$$E_+ = J/4 + ½\,[(\Delta g \beta B)^2 + J^2]^{½}$$

$$E_- = J/4 - ½\,[(\Delta g \beta B)^2 + J^2]^{½} \qquad [7.22]$$

The eigenvectors are linear combinations of the kets $|½\rangle|-½\rangle$ and $|-½\rangle|½\rangle$ which are obtained by applying the method described in section 3.2.4:

$$E_+: |u\rangle = c_{1+}|½\rangle|-½\rangle + c_{2+}|-½\rangle|½\rangle;\ c_{2+}/c_{1+} = (\Delta g\beta B - [(\Delta g\beta B)^2 + J^2]^{½})/J$$
$$E_-: |v\rangle = c_{1-}|½\rangle|-½\rangle + c_{2-}|-½\rangle|½\rangle;\ c_{2-}/c_{1-} = (\Delta g\beta B + [(\Delta g\beta B)^2 + J^2]^{½})/J \quad [7.23]$$

The probabilities of transitions between the four energy levels are proportional to the square of the modulus of the matrix elements of $\hat{H}_1 = g_A\beta B_1\hat{S}_{AX} + g_B\beta B_1\hat{S}_{BX}$, where X is the direction of $\mathbf{B}_1$. The four allowed transitions are characterised by:

$$\Delta E_1 = \beta B(g_A + g_B)/2 - J/2 - ½\,[(\Delta g\beta B)^2 + J^2]^{½};\ |\langle ½|\langle ½|\hat{H}_1|u\rangle|^2 = (c_{2+}\,g_A + c_{1+}\,g_B)^2(\beta B_1)^2/4$$

$$\Delta E_2 = \beta B(g_A + g_B)/2 - J/2 + ½\,[(\Delta g\beta B)^2 + J^2]^{½};\ |\langle ½|\langle ½|\hat{H}_1|v\rangle|^2 = (c_{2-}\,g_A + c_{1-}\,g_B)^2(\beta B_1)^2/4$$

$$\Delta E_3 = \beta B(g_A + g_B)/2 + J/2 + ½\,[(\Delta g\beta B)^2 + J^2]^{½};\ |\langle -½|\langle -½|\hat{H}_1|u\rangle|^2 = (c_{1+}\,g_A + c_{2+}\,g_B)^2(\beta B_1)^2/4$$

$$\Delta E_4 = \beta B(g_A + g_B)/2 + J/2 - ½\,[(\Delta g\beta B)^2 + J^2]^{½};\ |\langle -½|\langle -½|\hat{H}_1|v\rangle|^2 = (c_{1-}\,g_A + c_{2-}\,g_B)^2(\beta B_1)^2/4$$

The sum of the squares of the matrix elements is $(g_A{}^2 + g_B{}^2)(\beta B_1)^2/2$, which is the value obtained in the absence of interaction (exercise 7.5). Equations [7.23] suggest two limit situations:

1. $J^2 \ll (\Delta g\beta B)^2$. We then have $c_{1+} = c_{2-} = 1$, $c_{2+} = c_{1-} = 0$.
- The transitions of energy ΔE_2 and ΔE_3 produce two lines of intensity proportional to $(g_A\beta B_1)^2/4$, centred at $B_{0A} = h\nu/g_A\beta$ and split by $J/g_A\beta$.
- The transitions of energy ΔE_1 and ΔE_4 produce two lines with an intensity proportional to $(g_B\beta B_1)^2/4$, centred at $B_{0B} = h\nu/g_B\beta$ and split by $J/g_B\beta$ (figure 7.6a).

In this situation, the exchange interaction has the same effect as the dipolar interaction (figure 7.4a), but the splitting produced does not have an *angular dependence*.

2. $J^2 \gg (\Delta g\beta B)^2$. For $J > 0$, we have $c_{2+}/c_{1+} = -1$, $c_{2-}/c_{1-} = 1$.
- The transitions of energy ΔE_2 and ΔE_4 produce two *central lines* at $B_0 = h\nu/((g_A + g_B)\beta/2)$ with an intensity proportional to $(g_A + g_B)^2(\beta B_1)^2/8$.
- The transitions of energy ΔE_1 and ΔE_3 produce two *lateral lines* at $B_0 \pm J/[(g_A + g_B)\beta/2]$ with an intensity proportional to $(g_A - g_B)^2(\beta B_1)^2/8$.

When J increases starting from situation **1**, the "internal" lines get closer together and their positions tend towards B_0. In contrast, the "external" lines move further away to produce lateral lines of much weaker intensity than the central lines (figure 7.6b). When $J < 0$ the same results are produced. When $g_A = g_B = g$, the intensity of the lateral lines is null for all values of J and only two lines persist at $B_0 = h\nu/g\beta$: *the exchange interaction has no effect on the spectrum.*

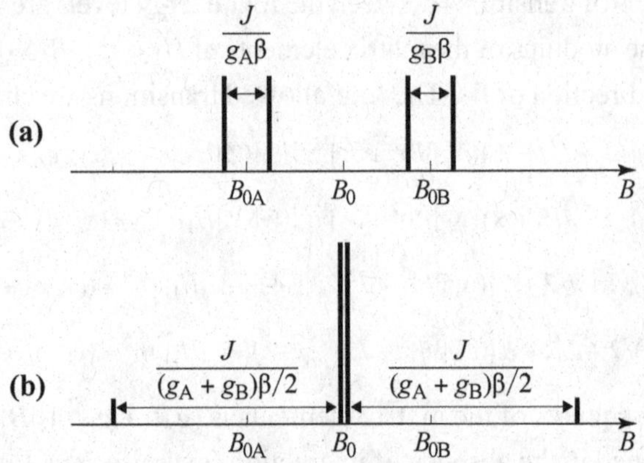

Figure 7.6 – Splitting of the resonance lines created by exchange interaction in two limit cases: **(a)** $J^2 \ll [(g_A - g_B)\beta B]^2$; **(b)** $J^2 \gg [(g_A - g_B)\beta B]^2$.

◇ *The principal axes of the $\tilde{\mathbf{g}}_\mathbf{A}$ and $\tilde{\mathbf{g}}_\mathbf{B}$ matrices are parallel*

When **B** is parallel to the principal axis i ($i = x, y, z$), the Hamiltonian is written:

$$\hat{H} = g_{Ai}\,\beta B\,\hat{S}_{Ai} + g_{Bi}\,\beta B\,\hat{S}_{Bi} - J\,\mathbf{S}_A{\cdot}\mathbf{S}_B \qquad [7.24]$$

It has the same form as Hamiltonian [7.20], and the expressions established in the previous section can be used. These results will now be applied to the spectra calculated numerically from Hamiltonian [7.11] (without the dipolar term) in two specific situations:

1. The two paramagnetic centres have the following principal values:

$$g_{Ax} = g_{Bx} = 1.95,\ g_{Ay} = g_{By} = 2.00,\ g_{Az} = 2.05;\ g_{Bz} = 2.08$$

Their powder spectra at X-band are represented in figure 7.7a. The structures of the two spectra are identical at B_x and B_y, and at low-field, peaks are observed at $B_{Az} = h\nu/g_{Az}\beta$ and $B_{Bz} = h\nu/g_{Bz}\beta$. The interaction spectra calculated for increasing values of $|J|$ are represented in figures 7.7b to 7.7d.

When **B** is parallel to x or to y, the position of the lines is determined by Hamiltonian [7.24] where the values of g_{Ai} and g_{Bi} are *equal*: the exchange interaction causes no splitting at B_x and B_y.

In contrast, splitting does appear at B_{Az} and B_{Bz}, which vary with $|J|$ in line with figure 7.6: the pattern progressively shifts from two sets of 2 lines centred at B_{Az} and B_{Bz} with splitting $J/g_{Az}\beta$ and $J/g_{Bz}\beta$ (figure 7.7b) to a central line at $h\nu/[(g_{Az} + g_{Bz})\beta/2]$ and two low-intensity lateral lines separated by $2J/[(g_{Az} + g_{Bz})\beta/2]$ (figures 7.7c and 7.7d).

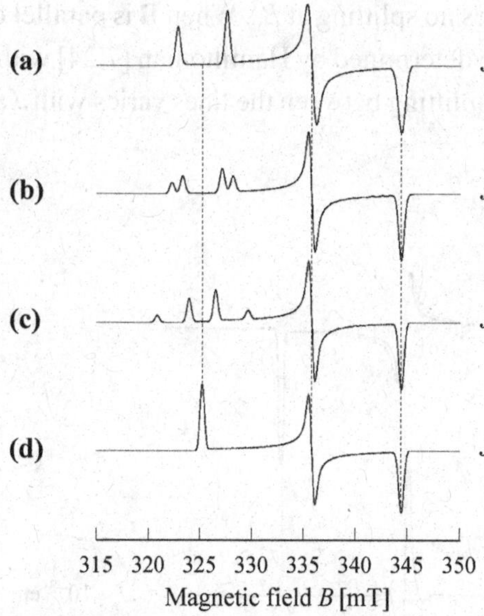

Figure 7.7 – Effect of the exchange interaction on the X-band spectrum (ν = 9.400 GHz) for two centres characterised by $g_{Ax} = g_{Bx} = 1.95$; $g_{Ay} = g_{By} = 2.00$; $g_{Az} = 2.05$; $g_{Bz} = 2.08$ and increasing values of J.

2. We will now consider two centres where the $\tilde{\mathbf{g}}$ matrices have the same principal values (1.95; 2.00; 2.05). In the absence of interaction, their powder spectra are *identical* (figure 7.8a). The effect of exchange interaction considerably depends on the relative orientation of the axes that correspond to the same principal value:
   - When the three axes are parallel in pairs:

$$\begin{aligned} g_{Ax} &= 1.95, \; g_{Ay} = 2.00, \; g_{Az} = 2.05 \\ g_{Bx} &= 1.95, \; g_{By} = 2.00, \; g_{Bz} = 2.05 \end{aligned}$$

the splitting is null for the three canonical directions of **B**. The calculated spectrum is practically superposable on the spectrum in the absence of interaction, even for a value of J of up to 5×10^{-2} cm^{-1} (figure 7.8a).

- We will now assume that the principal axes of $\tilde{\mathbf{g}}_B$ are deduced from those of $\tilde{\mathbf{g}}_A$ by a 90° rotation around the y axis:

$$g_{Ax} = 1.95;\ g_{Ay} = 2.00;\ g_{Az} = 2.05$$

$$g_{Bx} = 2.05;\ g_{By} = 2.00;\ g_{Bz} = 1.95$$

the spectra from figures 7.8b and 7.8c are obtained. Since $g_{Ay} = g_{By}$, the exchange interaction causes no splitting at B_y. When **B** is parallel to x or z, the positions of the lines are determined by Hamiltonian [7.24] with $g_{Ax} = 1.95$ and $g_{Bx} = 2.05$, and the splitting between the lines varies with J as predicted in figure 7.6.

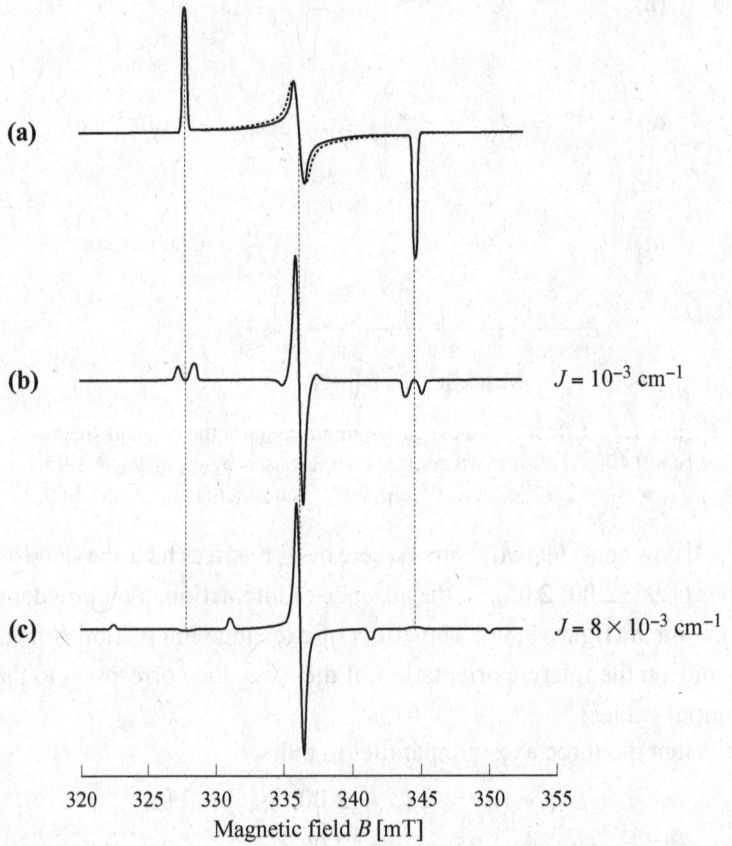

Figure 7.8 – Effect of the exchange interaction on the X-band spectrum (ν = 9.400 GHz) for two centres with the same principal values (1.95; 2.00; 2.05). **(a)** continuous lines: spectrum in the absence of interaction, dashed lines: $J = 5 \times 10^{-2}$ cm^{-1} with parallel axes corresponding to the same principal values. **(b)** and **(c)** after 90° rotation about y, for two values of J.

7.3.3 – General case

We first review the results obtained when the axes of the $\tilde{g}$ matrices are *parallel*:

◇ *Dipolar interaction alone*

When **B** is parallel to the principal axis i ($i = x, y, z$), the resonance lines split *symmetrically* (figure 7.4a). The splitting is determined by the intercentre distance r and the angle γ_i between the principal axis i and the intercentre vector (equation [7.18]) and it is not very dependent on the anisotropy of the principal values of the $\tilde{g}_A$ and $\tilde{g}_B$ matrices.

◇ *Exchange interaction alone*

When **B** is parallel to the principal axis i, the pattern created by the exchange interaction is determined by the $(g_{Ai} - g_{Bi})\beta B$ difference and by the J parameter.

1. When the first term dominates, the pattern is *symmetric*, like that produced by dipolar interaction (figure 7.6a), and the splitting is equal to $J/g_{Ai}\beta$ and $J/g_{Bi}\beta$.
2. When J dominates, the pattern is *dissymetric*: two lines appear at the "mean position" defined by $(g_{Ai} + g_{Bi})/2$, and two lateral lines of lower intensity located at $\pm J/[(g_A + g_B)\beta/2]$ from this mean position (figure 7.6b). Exchange interaction has no effect when $g_{Ai} = g_{Bi}$.

In general, the two interactions occur simultaneously. When the $\hat{H}_{dip}$ and $\hat{H}_{Heis}$ terms are both small relative to $|g_{Ai} - g_{Bi}|\beta B$, the calculation from section 7.3.1 can be repeated by replacing $\hat{H}_{dip}$ by $\hat{H}_{dip} + \hat{H}_{Heis}$ and expressions [7.18] become:

$$\Delta B_{Ai} = (1 - 3\cos^2\gamma_i)(\mu_0/4\pi)g_{Bi}\,\beta/r^3 - J/g_{Ai}\beta$$

$$\Delta B_{Bi} = (1 - 3\cos^2\gamma_i)(\mu_0/4\pi)g_{Ai}\,\beta/r^3 - J/g_{Bi}\beta$$

When the $\tilde{g}_A$ and $\tilde{g}_B$ matrices are weakly anisotropic and $|J|$ is larger than the dipolar contribution, these equations and equation [7.19] imply that the mean splitting observed for the three canonical directions is approximately equal to $|J/g_A\beta|$ for the spectrum of A and to $|J/g_B\beta|$ for the spectrum of B, and $|J|$ can be directly determined from the spectra.

All of these results were obtained by assuming the principal axes of the $\tilde{g}_A$ and $\tilde{g}_B$ matrices to be *parallel*. In general, these axes are different and the resonance fields depend on their relative orientation which is defined by a set of three Euler angles.

Equations [7.18] can still be used to analyse the effects of dipolar interactions, by replacing $\{\gamma_i\}$ by $\{\gamma_{Ai}\}$, the angle between the intercentre vector and the principal axes of $\tilde{\mathbf{g}}_\mathbf{A}$ if we are interested in the splitting of the A lines, or by $\{\gamma_{Bj}\}$, the angle between the intercentre vector and the principal axes of $\tilde{\mathbf{g}}_\mathbf{B}$ when studying the splitting of the B lines.

For a given direction of **B**, the results from figure 7.6 can still be used to analyse the effect of the exchange interaction, on the condition that $(g_{Ai}-g_{Bi})\beta B$ is replaced by $(g'_A-g'_B)\beta B$, where g'_A and g'_B identify the position of the resonance lines for the centres A and B in the absence of interaction *for this direction of* **B**.

In this section, we have voluntarily ignored the effects of hyperfine coupling on the interaction spectrum. A detailed analysis of these effects can be found in [Smith and Pilbrow, 1974].

7.4 – Effects of strong exchange interaction on the spectrum Biradicals and polynuclear complexes

7.4.1 – Introduction

The exchange interactions between paramagnetic centres are frequently much stronger than their interactions with the field **B**. For this to occur at X-band, $|J|$ must only exceed a few cm^{-1}, which is the case in "biradicals" composed of two coupled radical entities and in numerous polynuclear transition ion complexes. "Pairs" of paramagnetic centres can also form in specific conditions. For example, irradiating molecular solids at low temperature triggers a sequence of reactions which can result in broken chemical bonds and the formation of strongly interacting pairs of radicals [Gordy, 1980; Atherton, 1993; Bencini and Gatteschi, 1990].

In this "strong exchange" situation, the magnetic properties of the (A, B) system are completely different from those of the two centres considered separately, and they are highly influenced by the *sign* of the *J* parameter. Consider, for example, the case of two centres of spin ½ dealt with in section 7.3.2. When $|J|$ is large relative to the Zeeman terms, equations [7.21] to [7.23] show that when **B** is parallel to the principal axis *i*, the energy levels and eigenvectors are given by:

$$3J/4: \quad (|\tfrac{1}{2}\rangle|-\tfrac{1}{2}\rangle - |-\tfrac{1}{2}\rangle|\tfrac{1}{2}\rangle)/\sqrt{2}$$

$$-J/4 + (g_{Ai} + g_{Bi})\beta B/2: \quad |½\rangle|½\rangle$$
$$-J/4: \quad (|½\rangle|-½\rangle + |-½\rangle|½\rangle)/\sqrt{2}$$
$$-J/4 - (g_{Ai} + g_{Bi})\beta B/2: \quad |-½\rangle|-½\rangle$$

We once again find the kets $\{|0,0\rangle, |1,1\rangle, |1,0\rangle, |1,-1\rangle\}$, eigenvectors shared by $(\hat{\mathbf{S}}^2, \hat{S}_i)$, where $\mathbf{S} = \mathbf{S}_A + \mathbf{S}_B$ is the *total spin* for the (A, B) system (appendix 4).

▹ If J is negative, the ground state is described by the ket $|0,0\rangle$ from which no transition is allowed: no EPR spectrum is produced.

▹ If J is positive, the "triplet state" transitions $|1,-1\rangle \leftrightarrow |1,0\rangle$ and $|1,0\rangle \leftrightarrow |1,1\rangle$ produce a line at $B = h\nu/[(g_{Ai} + g_{Bi})\beta/2]$.

Both situations are observed in copper-nitroxide complexes where the oxygen from a nitroxide radical serves as a ligand for a Cu^{2+} ion. When the oxygen is in the equatorial position, the overlap between the $d_{x^2-y^2}$ orbital for the Cu^{2+} ion and the π^* orbital for the radical creates a strong antiferromagnetic interaction (component J_1 in equations [7.5]); the J parameter is negative and no EPR spectrum is observed. But when the oxygen is in the axial position, these orbitals are almost orthogonal and the J_1 component is very weak. In this case, the positive J_2 component dominates and the ground state is characterised by $S = 1$ [Caneschi *et al.*, 1988].

We will see how to deal with this problem for any spin and any $\tilde{\mathbf{g}}$ matrices.

7.4.2 – Construction of equivalent Hamiltonians for a pair of paramagnetic centres

The spin Hamiltonian describing the interactions to which the unpaired electrons for two centres of spin S_A and S_B are subjected is written (equation [7.10]):

$$\hat{H} = \hat{H}_A + \hat{H}_B + \hat{H}_{exch} + \hat{H}_{dip} \qquad [7.25]$$

$$\hat{H}_A = \beta\,\mathbf{B}\cdot\tilde{\mathbf{g}}_A\cdot\mathbf{S}_A + \mathbf{S}_A\cdot\tilde{\mathbf{a}}_A\cdot\mathbf{I}_A + \mathbf{S}_A\cdot\tilde{\mathbf{D}}_A\cdot\mathbf{S}_A \qquad [7.26]$$
$$\hat{H}_B = \beta\,\mathbf{B}\cdot\tilde{\mathbf{g}}_B\cdot\mathbf{S}_B + \mathbf{S}_B\cdot\tilde{\mathbf{a}}_B\cdot\mathbf{I}_B + \mathbf{S}_B\cdot\tilde{\mathbf{D}}_B\cdot\mathbf{S}_B$$

For simplicity, we have assumed that the electrons from each centre interact with a single nucleus, but the treatment below is entirely general. The exchange and dipolar terms are given by equations [7.8] and [7.9], respectively. In the "strong exchange" situation that we are interested in, the intercentre distance is small and the relevance of the "point dipole" approximation (section 7.2.4) is unclear. However, it can be shown that expression [7.9] remains valid if we consider r as an *effective distance* [Owen and Harris, 1972].

To determine the eigenvalues and eigenvectors for Hamiltonian [7.25], the matrix representing it in a basis of dimension $(2S_A + 1)(2S_B + 1)$ must be diagonalised. Because of this dimension and of the numerous terms included in $\hat{H}$, the problem is much more difficult that those dealt with in chapters 4 and 6, but we will see that it is considerably simplified in the frequent situation where the Heisenberg operator dominates *all the terms* in the Hamiltonian.

◇ *Equivalent Hamiltonians*

Hamiltonian [7.25] can be written in the form

$$\begin{aligned} \hat{H} &= -J\,\mathbf{S}_A \cdot \mathbf{S}_B + \hat{H}_p \\ \hat{H}_p &= \hat{H}_A + \hat{H}_B + \hat{H}^{anis}_{exch} + \hat{H}_{dip} \end{aligned} \qquad [7.27]$$

where $\hat{H}_p$ is a perturbation. The eigenvalues and eigenvectors of the Heisenberg operator are obtained by introducing the total spin angular momentum $\mathbf{S} = \mathbf{S}_A + \mathbf{S}_B$ (exercise 7.2 and appendix 4). For each possible value of S

$$S = |S_A - S_B|, |S_A - S_B| + 1, \ldots, S_A + S_B \qquad [7.28]$$

there is a corresponding energy level $E(S)$ and a set of $(2S + 1)$ eigenvectors defined by:

$$\begin{aligned} &E(S) = -J/2\,[S(S+1) - S_A(S_A+1) - S_B(S_B+1)]; \\ &\{|S, M_S\rangle, M_S = -S, -S+1, \ldots S\} \end{aligned} \qquad [7.29]$$

The kets $\{|S, M_S\rangle\}$, eigenvectors shared by $(\hat{\mathbf{S}}^2, \hat{S}_z)$ where z is any direction in the Euclidean space, make up the basis for a sub-space $\mathcal{E}_S$ of dimension $(2S + 1)$. As each energy level $E(S)$ is $(2S + 1)$-fold degenerate, the matrix representing the $\hat{H}_p$ perturbation (equation [7.27]) must be constructed in the basis $\{|S, M_S\rangle\}$ for each $\mathcal{E}_S$ sub-space. To do so, $\hat{H}_p$ can be replaced by an "equivalent Hamiltonian", $\hat{H}_S$, for which the operators are expressed as a function of the components of $\mathbf{S}$.

- ▷ First, consider the Zeeman and hyperfine terms in $\hat{H}_A$ and $\hat{H}_B$ (equation [7.26]). In the $\mathcal{E}_S$ sub-space, $\mathbf{S}_A$ and $\mathbf{S}_B$ can be replaced by $K_A\mathbf{S}$ and $K_B\mathbf{S}$, respectively, where K_A and K_B are "spin coupling coefficients" (appendix 4):

$$K_A = \frac{S(S+1) + S_A(S_A+1) - S_B(S_B+1)}{2S(S+1)} \qquad [7.30]$$

K_B is obtained by permuting the A and B indices ($K_A + K_B = 1$). Within $\mathcal{E}_S$, we therefore have the *equivalence*:

$$\beta\, \mathbf{B}\cdot\tilde{\mathbf{g}}_A\cdot\mathbf{S}_A + \mathbf{S}_A\cdot\tilde{\mathbf{a}}_A\cdot\mathbf{I}_A + \beta\, \mathbf{B}\cdot\tilde{\mathbf{g}}_B\cdot\mathbf{S}_B + \mathbf{S}_B\cdot\tilde{\mathbf{a}}_B\cdot\mathbf{I}_B$$
$$\equiv \beta\, \mathbf{B}\cdot\tilde{\mathbf{g}}\cdot\mathbf{S} + \mathbf{S}\cdot\tilde{\mathbf{A}}_A\cdot\mathbf{I}_A + \mathbf{S}\cdot\tilde{\mathbf{A}}_B\cdot\mathbf{I}_B$$

$$\tilde{\mathbf{g}} = K_A\,\tilde{\mathbf{g}}_A + K_B\,\tilde{\mathbf{g}}_B \qquad [7.31]$$

$$\tilde{\mathbf{A}}_A = K_A\,\tilde{\mathbf{a}}_A;\ \tilde{\mathbf{A}}_B = K_B\,\tilde{\mathbf{a}}_B \qquad [7.32]$$

▹ Similarly, within $\mathcal{E}_S$ the zero-field splitting terms are replaced by an equivalent operator:

$$\mathbf{S}_A\cdot\tilde{\mathbf{D}}_A\cdot\mathbf{S}_A + \mathbf{S}_B\cdot\tilde{\mathbf{D}}_B\cdot\mathbf{S}_B \equiv \mathbf{S}\cdot\tilde{\mathbf{D}}_{ZFS}\cdot\mathbf{S}$$

with:

$$\tilde{\mathbf{D}}_{ZFS} = d_A\,\tilde{\mathbf{D}}_A + d_B\tilde{\mathbf{D}}_B \qquad [7.33]$$

The coefficients d_A and d_B are expressed as a function of (S_A, S_B, S) [Bencini and Gatteschi, 1990].

▹ Dealing with the interaction terms is less straightforward. The matrix representing $(\hat{H}_{exch}^{anis} + \hat{H}_{dip})$ in the basis $\{|S, M_S\rangle\}$ can be written as the sum of a symmetric matrix and an antisymmetric matrix and an equivalent operator of the form $\mathbf{S}\cdot\tilde{\mathbf{D}}_{Int}\cdot\mathbf{S}$ can only be defined for the symmetric part [Bencini and Gatteschi, 1990]. The antisymmetric contribution is therefore often neglected and the zero-field splitting terms and the interaction terms are combined in a $\mathbf{S}\cdot\tilde{\mathbf{D}}\cdot\mathbf{S}$ term where

$$\tilde{\mathbf{D}} = \tilde{\mathbf{D}}_{ZFS} + \tilde{\mathbf{D}}_{Int} \qquad [7.34]$$

Finally, within the $\mathcal{E}_S$ sub-space, $\hat{H}_p$ is replaced by the equivalent Hamiltonian:

$$\hat{H}_S = \beta\, \mathbf{B}\cdot\tilde{\mathbf{g}}\cdot\mathbf{S} + \mathbf{S}\cdot\tilde{\mathbf{A}}_A\cdot\mathbf{I}_A + \mathbf{S}\cdot\tilde{\mathbf{A}}_B\cdot\mathbf{I}_B + \mathbf{S}\cdot\tilde{\mathbf{D}}\cdot\mathbf{S} \qquad [7.35]$$

This Hamiltonian is identical to that of a paramagnetic centre of spin S, characterised by the $\tilde{\mathbf{g}}$ and $\tilde{\mathbf{D}}$ matrices, which interacts with two nuclei of spin I_A and I_B. However, the meaning of the parameters included in [7.35] is not the same as that of those defined in chapters 4 and 6: the $\tilde{\mathbf{g}}$, $\tilde{\mathbf{A}}_A$, $\tilde{\mathbf{A}}_B$ and $\tilde{\mathbf{D}}$ matrices depend on the local parameters for the two centres and on the spin coupling coefficients (and therefore on S_A, S_B, S), and the $\tilde{\mathbf{D}}$ matrix also depends on the parameters describing the intercentre interactions.

◇ *The EPR spectrum*

The eigenvalues (E_{Sk}, $k = 1, 2, \ldots\ 2S + 1$) and the eigenvectors for the matrix representing $\hat{H}_S$ in the basis $\{|S, M_S\rangle\}$ can be calculated for each value of S

using the methods described for a single centre in chapters 4 and 6. The energy levels for the Hamiltonian $\hat{H}$ (equation [7.27]) are given to first order in perturbation theory by:

$$E(S,k) = E(S) + E_{Sk}$$

where $E(S)$ is defined by equation [7.29] (figure 7.9). The energies of the *transitions between the* $2S + 1$ *levels* depend only on the eigenvalues $\{E_{Sk}\}$. The transition probabilities are determined by the operator

$$\hat{H}_1 = \beta\, \mathbf{B}_1 \cdot \tilde{\mathbf{g}}_\mathbf{A} \cdot \mathbf{S}_\mathrm{A} + \beta\, \mathbf{B}_1 \cdot \tilde{\mathbf{g}}_\mathbf{B} \cdot \mathbf{S}_\mathrm{B}$$

This operator can be replaced within the sub-space $\mathcal{E}_S$ by the equivalent operator

$$\hat{H}_{1S} = \beta\, \mathbf{B}_1 \cdot \tilde{\mathbf{g}} \cdot \mathbf{S}$$

which is the same as that used in chapters 5 and 6 to calculate the intensity of the lines for a single centre. The *shape* and *intensity* of the EPR spectrum produced by the $E(S)$ level are therefore identical to those of the spectrum produced by a paramagnetic centre of spin S characterised by the Hamiltonian $\hat{H}_S$ (equation [7.35]). The same spectral profiles as those in the figures in chapters 2, 4 and 6 are found. For example, the half-field line for $S = 1$ is observed [Fournel *et al.*, 1998]. When S is half-integer greater than ½ and the principal term in the Hamiltonian [7.35] is the zero-field splitting term, Kramers' doublets appear which can be characterised by effective parameters [Crouse *et al.*, 1995].

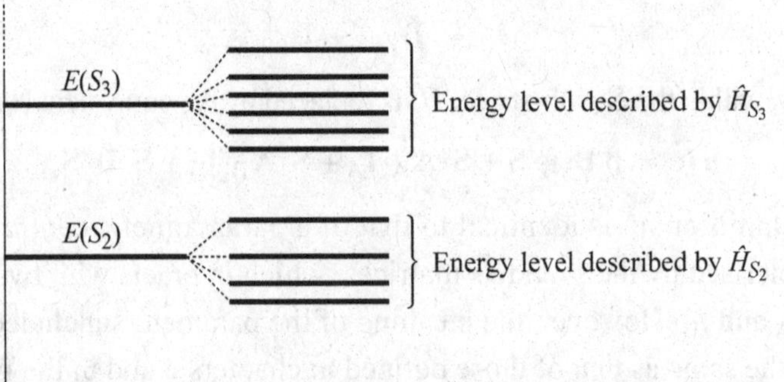

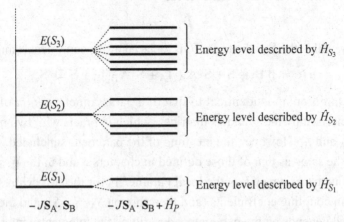

Figure 7.9 – Energy levels for a system with two centres A and B coupled by a strong exchange interaction. $E(S)$ is given by equation [7.29]. The equivalent Hamiltonians ($\hat{H}_{S_1}, \hat{H}_{S_2}$ ...) describe the splitting produced by $\hat{H}_P$ (equation [7.27]).

At thermal equilibrium, the EPR spectrum for the (A, B) pair is the result of the *superposition* of the spectra produced by all the $E(S)$ energy levels, each weighted by its Boltzmann factor. As this factor depends on J (equation [7.29]), by studying the *temperature-dependence* of the spectrum it is possible to determine the exchange parameter.

7.4.3 – Equivalent Hamiltonians and EPR spectra for a few typical pairs

To illustrate these results, we will detail the expression for the equivalent Hamiltonians for a few pairs of paramagnetic centres and we will present some examples of EPR spectra.

◇ *Organic biradical*

A strong exchange interaction between two radicals produces two states characterised by $S = 0$ and $S = 1$ split by $|J|$; the ground state is generally the triplet state. Its spectrum is determined by the Hamiltonian [7.35], where $\tilde{\mathbf{g}}$, $\tilde{\mathbf{A}}_A$, $\tilde{\mathbf{A}}_B$ and $\tilde{\mathbf{D}}$ are defined by equations [7.31], [7.32], [7.34] with $K_A = K_B = ½$ (equation [7.30]).

In solid medium, the combined effects of the zero-field splitting terms and the hyperfine terms often produce complex spectral shapes; their analysis is facilitated when the sample is a single crystal [Gordy, 1980]. When the spectrum can be observed in solution at room temperature, it is considerably simplified in the *isotropic regime* where the effects of the anisotropic components of all the magnetic interactions disappear (section 4.5.2).

▹ For bis-nitroxide derivatives, for example, the equivalent Hamiltonian is written:

$$\hat{H}_{S=1} = g\beta \mathbf{B}\cdot\mathbf{S} + A_A \mathbf{S}\cdot\mathbf{I}_A + A_B \mathbf{S}\cdot\mathbf{I}_B \qquad [7.36]$$

$I_A = I_B = 1$ are the spins of the ^{14}N nuclei, and the g, A_A and A_B parameters are given by:

$$g = (g_{A\,iso} + g_{B\,iso})/2; \; A_A = a_{A\,iso}/2; \; A_B = a_{B\,iso}/2$$

The positions of the lines is determined by applying the method described in section 2.4. The hyperfine terms produce a pattern of 3 × 3 = 9 hyperfine lines. When the two nuclei are equivalent, $A_A = A_B$ and the (1 : 2 : 3 : 2 : 1) pattern from figures 2.11b and 2.12 is produced. However, the spectrum is different to that for a single *radical* with two equivalent ^{14}N nuclei as the hyperfine constant is *two-fold smaller* and its intensity is increased by a factor of 8/3 (see note following equation [6.22]).

▹ For bis (nitronyl-nitroxide) derivatives with 4 equivalent ^{14}N nuclei [Catala *et al.*, 2001], the Hamiltonian is written:

$$\hat{H}_{S=1} = g\beta \mathbf{B}\cdot\mathbf{S} + A\,(\mathbf{S}\cdot\mathbf{I}_{A1} + \mathbf{S}\cdot\mathbf{I}_{A2} + \mathbf{S}\cdot\mathbf{I}_{B1} + \mathbf{S}\cdot\mathbf{I}_{B2})$$

with $A = a_{iso}/2$. From Pascal's triangle (complement 2 to chapter 2), we predict a pattern of 9 lines separated by $A/g\beta$, with relative intensities (1 : 4 : 10 : 16 : 19 : 16 : 10 : 4 : 1). The X-band spectrum for a liquid solution containing this type of biradical is shown in figure 7.10. Simulation of this spectrum gives $g = g_{A\,iso} = g_{B\,iso} = 2.0067$ and $A = 10.4$ MHz, which is effectively half of the hyperfine constant $a_{iso} \approx 20$ MHz measured for a nitronyl-nitroxide radical (figure 2.12a and exercise 2.6) (exercise 7.6).

The spectrum produced by a frozen solution of these molecules is represented in figure 6.8. From the temperature-dependence of its intensity, it was deduced that $J = 16$ cm^{-1} [Catala *et al.*, 2001].

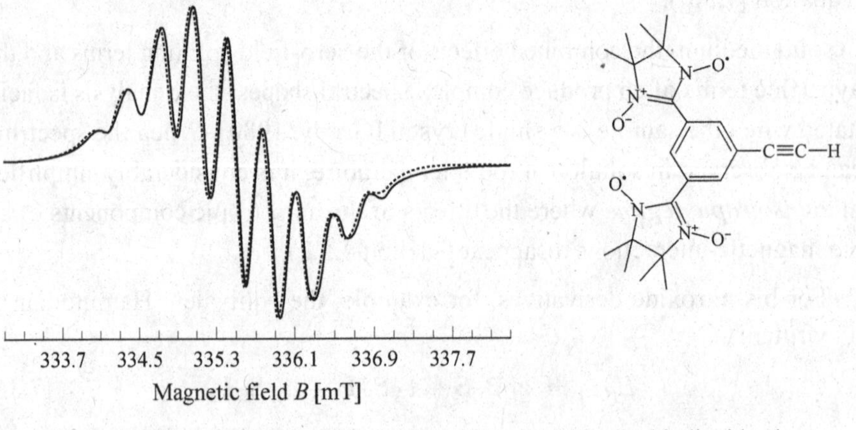

Figure 7.10 – Spectrum for a bis (nitronyl-nitroxide) biradical in liquid solution at room temperature. Microwaves: frequency 9.42145 GHz, power 2 mW. Modulation: frequency 100 kHz, peak-to-peak amplitude 0.01 mT. The parameters used in the simulation are given in the text.

◇ *The (Fe^{3+}, Fe^{2+}) pair*

Some proteins contain dinuclear (Fe^{3+}, Fe^{2+}) complexes. In these complexes, the iron ions in a high-spin situation ($S_A = 5/2$, $S_B = 2$) are bridged by sulfur or oxygen ligands. The orbital overlaps lead to strong *antiferromagnetic* exchange interactions and the $|J|$ parameter varies between 10 and around 300 cm^{-1} depending on the complex [Bertrand *et al.*, 1991]. As only 2 % of iron nuclei are paramagnetic (table 2.2), the hyperfine interactions can be ignored and the Hamiltonian is written:

$$\hat{H} = -J\,\mathbf{S}_\mathrm{A}\cdot\mathbf{S}_\mathrm{B} + \hat{H}_p$$

$$\hat{H}_p = \beta\,\mathbf{B}\cdot\tilde{\mathbf{g}}_\mathbf{A}\cdot\mathbf{S}_\mathrm{A} + \mathbf{S}_\mathrm{A}\cdot\tilde{\mathbf{D}}_\mathrm{A}\cdot\mathbf{S}_\mathrm{A} + \beta\,\mathbf{B}\cdot\tilde{\mathbf{g}}_\mathbf{B}\cdot\mathbf{S}_\mathrm{B} + \mathbf{S}_\mathrm{B}\cdot\tilde{\mathbf{D}}_\mathrm{B}\cdot\mathbf{S}_\mathrm{B} + \hat{H}_{exch}^{anis} + \hat{H}_{dip}$$

The Heisenberg operator produces a ground state of spin $S = ½$ and excited states characterised by $S = 3/2, 5/2, 7/2, 9/2$. At cryogenic temperatures, only the ground state is populated and produces an EPR spectrum. At temperatures where the excited states start to become populated, their spectra are broadened due to spin-lattice relaxation and are therefore not observed.

Within the $\mathcal{E}_{1/2}$ sub-space, $\hat{H}_p$ can be replaced by the equivalent Hamiltonian:

$$\hat{H}_{S=1/2} = \beta\,\mathbf{B}\cdot\tilde{\mathbf{g}}\cdot\mathbf{S}$$

where $\tilde{\mathbf{g}}$ is given by equation [7.31] with $K_\mathrm{A} = 7/3$, $K_\mathrm{B} = -4/3$ (equation [7.30]). We have omitted the $\mathbf{S}\cdot\tilde{\mathbf{D}}\cdot\mathbf{S}$ term from this equation as it has no effect on the spectrum for $S = ½$ (section 6.2.1). The shape and intensity of the spectrum produced by a frozen solution of molecules are identical to those for a single centre of spin $S = ½$, characterised by the $\tilde{\mathbf{g}}$ matrix. If we assume that the $\tilde{\mathbf{g}}_\mathbf{A}$ matrix for the ferric site is *isotropic* with $g_\mathrm{A} = g_e = 2.0023$ (section 4.2.2), the principal values of the $\tilde{\mathbf{g}}$ matrix can be written:

$$g_x = 7/3\,g_e - 4/3\,g_{\mathrm{B}x},\quad g_y = 7/3\,g_e - 4/3\,g_{\mathrm{B}y},\quad g_z = 7/3\,g_e - 4/3\,g_{\mathrm{B}z}$$

where $\{x,y,z\}$ are the principal axes of the $\tilde{\mathbf{g}}_\mathbf{B}$ matrix. If we set $g_{\mathrm{B}i} = g_e + \Delta g_i$ with $\Delta g_i \geq 0$ $(i = x,y,z)$ for the Fe^{2+} ion (section 4.2.2), we obtain $g_i = g_e - 1.33\,\Delta g_i$. This model predicts three principal values which are less than g_e. This is effectively what is observed with several (Fe^{3+}, Fe^{2+}) complexe [Bertrand *et al.*, 1991], but not with 2Fe – 2S centres for which one principal value is greater than g_e (figure 7.11) [Guigliarelli and Bertrand, 1999]. This specificity is linked to the coordination of the Fe^{3+} ion by four sulfur atoms, which produces a g_A value greater than g_e [Schneider *et al.*, 1968; Sweeney and Coffman, 1972]. Long before the crystal structure of molecules containing 2Fe – 2S centres was known, the EPR spectrum was attributed to a strongly coupled (Fe^{3+}, Fe^{2+}) pair in which each iron ion is coordinated to 4 sulfur atoms [Gibson *et al.*, 1966].

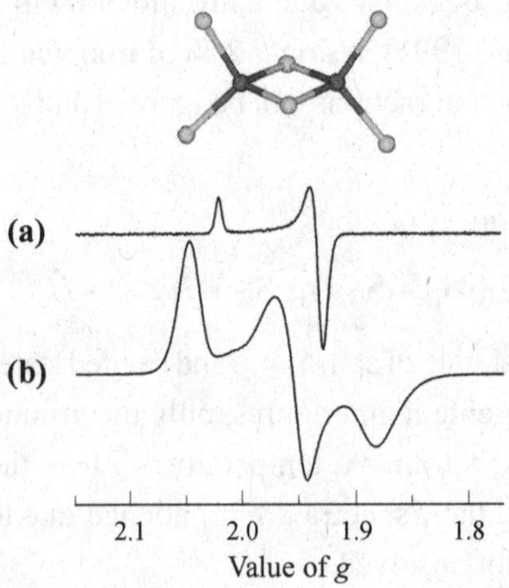

Figure 7.11 – Structure of a 2Fe – 2S centre and X-band spectra for two proteins containing this centre in the reduced state (Fe^{3+}, Fe^{2+}). Spectra are plotted on a *g-scale* obtained by replacing B with the dimensionless quantity $h\nu/\beta B$. **(a)** Protein from the adrenal glands: the centre is characterised by $g_\perp = 1.938$, $g_{//} = 2.023$. **(b)** Spinach protein: the centre is characterised by $g_x = 1.884$, $g_y = 1.957$, $g_z = 2.050$. Experimental conditions: temperature 15 K, frequency 9.4123 GHz, power 0.4 mW. Modulation: frequency 100 kHz, peak-to-peak amplitude **(a)** 0.8 mT **(b)** 2 mT.

Oxidation of a (Fe^{3+}, Fe^{2+}) complex produces an (Fe^{3+}, Fe^{3+}) pair. The strong antiferromagnetic interaction between these two ferric ions of spin $S_A = S_B = 5/2$ produces a ground state with spin $S = 0$, as a consequence no EPR spectrum is observed.

◇ *The (Mn^{3+}, Mn^{4+}) pair*

Polynuclear complexes of manganese have been extensively studied since it was discovered that a tetranuclear manganese complex plays an essential role in photosynthesis as performed by plants. The simplest complexes are dinuclear with the Mn^{3+} ($S_A = 2$) and Mn^{4+} ($S_B = 3/2$) ions, linked by one or two μ–oxo bridges, coupled by a strong antiferromagnetic exchange interaction. The Hamiltonian for this pair is written:

$$\hat{H} = -J\,\mathbf{S}_A\cdot\mathbf{S}_B + \hat{H}_p$$

$$\begin{aligned}\hat{H}_p = {} & \beta\,\mathbf{B}\cdot\tilde{\mathbf{g}}_A\cdot\mathbf{S}_A + \mathbf{S}_A\cdot\tilde{\mathbf{a}}_A\cdot\mathbf{I}_A + \mathbf{S}_A\cdot\tilde{\mathbf{D}}_A\cdot\mathbf{S}_A + \beta\,\mathbf{B}\cdot\tilde{\mathbf{g}}_B\cdot\mathbf{S}_B \\ & + \mathbf{S}_B\cdot\tilde{\mathbf{a}}_B\cdot\mathbf{I}_B + \mathbf{S}_B\cdot\tilde{\mathbf{D}}_B\cdot\mathbf{S}_B + \hat{H}_{exch}^{anis} + \hat{H}_{dip}\end{aligned}$$

$I_A = I_B = 5/2$ are the spins of the ^{55}Mn nuclei (table 2.2). The Heisenberg operator which dominates this Hamiltonian ($|J|$ is around 300 cm^{-1}) leads to a ground state of spin $S = 1/2$ and to excited states characterised by $S = 3/2, 5/2, 7/2$. Like in the case of (Fe^{3+}, Fe^{2+}) complexes, only the ground state contributes to the EPR spectrum. Within the $\mathcal{E}_{1/2}$ sub-space, $\hat{H}_p$ can be replaced by the following equivalent Hamiltonian:

$$\hat{H}_{S=½} = \beta\,\mathbf{B}\cdot\tilde{\mathbf{g}}\cdot\mathbf{S} + \mathbf{S}\cdot\tilde{\mathbf{A}}_A\cdot\mathbf{I}_A + \mathbf{S}\cdot\tilde{\mathbf{A}}_B\cdot\mathbf{I}_B \qquad [7.37]$$

where $\tilde{\mathbf{g}}$, $\tilde{\mathbf{A}}_A$ and $\tilde{\mathbf{A}}_B$ are given by equations [7.31] and [7.32] with $K_A = 2$, $K_B = -1$ (equation [7.30]). The $\tilde{\mathbf{g}}$ and $\tilde{\mathbf{a}}$ matrices for the manganese ions are weakly anisotropic and their principal values are not very dependent on their valence state. We can therefore predict the general shape of the spectrum by setting:

$$g_A = g_B = g;\; a_A = a_B = a$$

Hamiltonian [7.37] therefore becomes:

$$\hat{H}_{S=½} = g\beta\mathbf{B}\cdot\mathbf{S} + A_A\,\mathbf{S}\cdot\mathbf{I}_A + A_B\,\mathbf{S}\cdot\mathbf{I}_B$$

with $A_A = 2a$, $A_B = -a$. The energies for the 36 allowed transitions are given to the first order by (section 2.4.2):

$$\begin{aligned}\Delta E(M_{IA}, M_{IB}) &= g\beta B + M_{IA}A_A + M_{IB}A_B \\ &= g\beta B + (2M_{IA} - M_{IB})a\end{aligned}$$

From this, the expression for the resonance fields can be deduced:

$$B(M_{IA}, M_{IB}) = h\nu/g\beta - (2M_{IA} - M_{IB})a/g\beta \qquad [7.38]$$

where ν is the frequency of the spectrometer. When M_{IA} and M_{IB} vary between $-5/2$ and $5/2$, the quantity $(2M_{IA} - M_{IB})$ takes the values $-15/2$, $-15/2+1$, ..., $15/2$, but some of them appear several times (figure 7.12a). We therefore expect to observe a pattern of 16 equidistant lines split by $a/g\beta$, the intensity of which varies in accordance with figure 7.12b.

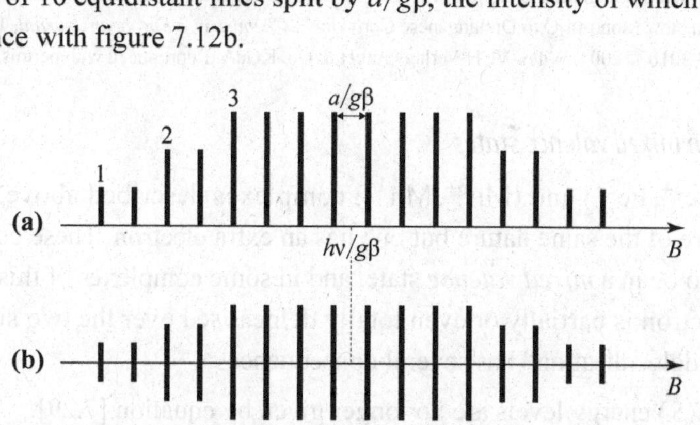

Figure 7.12 – **(a)** Result of stacking of the 36 lines defined by equation [7.38]. **(b)** Schematic representation of the spectrum produced by a (Mn^{3+}, Mn^{2+}) pair with the same isotropic $\tilde{\mathbf{g}}$ and $\tilde{\mathbf{A}}$ matrices.

The X-band spectrum for a $Mn^{3+}(\mu - oxo)_2Mn^{4+}$ complex recorded at 100 K is represented in figure 7.13 [Hureau *et al.*, 2004]. The expected pattern is observed, with minor differences due to the weak anisotropy of the $\tilde{\mathbf{A}}_A$ and $\tilde{\mathbf{A}}_B$ matrices and to the effects of second order perturbation theory (exercise 7.7). The very specific shape of this spectrum is linked to the $K_A = 2$, $K_B = -1$ values of the spin coupling coefficients and it is therefore typical of a (Mn^{3+}, Mn^{4+}) pair.

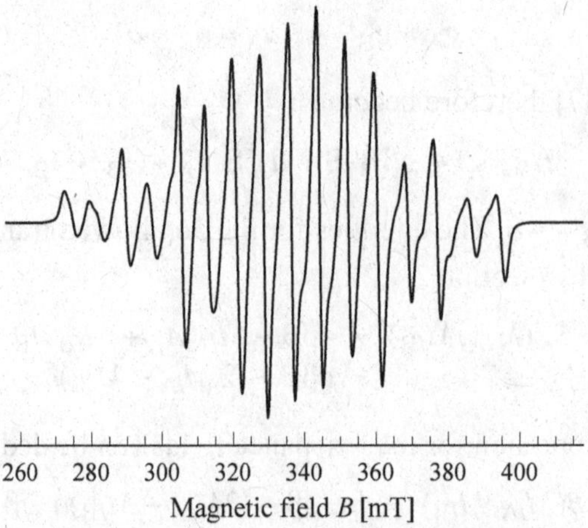

Figure 7.13 – X-band spectrum for the $[(L_4{}^2)Mn^{3+}(\mu\text{-}O)_2Mn^{4+}(L_4{}^2)]^{3+}$ complex at 100 K. Microwaves: frequency 9.3850 GHz, power 2 mW. Modulation: frequency 100 kHz, peak-to-peak amplitude 0.5 mT.
[Hureau C. *et al.*: "Synthesis, Structure, and Characterisation of a New Phenolato-Bridged Manganese Complex $[Mn_2(mL)_2]^{2+}$: Chemical and Electrochemical Access to a New Mono-m-Oxo Dimanganese Core Unit", *Chemistry: a European Journal*, **10**, 1998–2010 © 2004, Wiley-VCH Verlag GmbH & Co. KGaA. Reproduced with permission]

◇ *Note on mixed valence states*

In the (Fe^{3+}, Fe^{2+}) and (Mn^{3+}, Mn^{4+}) complexes described above, the two cations are of the same nature but one has an extra electron. These complexes are said to be in a *mixed valence* state, and in some complexes of this type the extra electron is partially or even totally delocalised over the two sites. This "valence delocalisation" has several consequences:

- ▹ The $E(S)$ energy levels are no longer given by equation [7.29].
- ▹ Expressions [7.31], [7.32] and [7.33] are modified.

Details on these points can be found in [Blondin and Girerd, 1990; Guigliarelli and Bertrand, 1999].

◇ *The Mn^{2+} (μ–OAc)$_2$ Mn^{2+} complex*

This complex differs from the previous complex on several points. As the spins of the Mn^{2+} ions are $S_A = S_B = 5/2$, the antiferromagnetic exchange interaction results in a ground state with $S = 0$ and in excited states characterised by $S = 1, 2, 3, 4, 5$. The J parameter is not very large in this complex (around 10 cm^{-1}), such that the excited states are rapidly populated when the temperature increases, and their contributions are added to produce the X-band and Q-band spectra represented in figure 7.14 [Blanchard *et al.*,2003].

To analyse these spectra, the spectrum produced by each $E(S)$ level can be calculated using a Hamiltonian similar to the one in equation [7.35] with an $\mathbf{S} \cdot \tilde{\mathbf{D}} \cdot \mathbf{S}$ term since S is greater than ½ in this case. As the $\tilde{\mathbf{g}}_A$ and $\tilde{\mathbf{a}}_A$ matrices for the Mn^{2+} ions are practically isotropic, the zero-field splitting terms are responsible for the features shown in figure 7.14. In this case, the hyperfine structure is masked by the large width of the inhomogeneous lines, typical of centres with integer values of S (complement 3 to chapter 6). Since the spectrum recorded at 4.3 K is unchanged at 6.8 K, it is produced by the first excited energy level of spin $S = 1$. This spectrum is very different from those predicted in the high-field and low-field limits (chapter 6), which shows that the Zeeman terms and zero-field splitting terms are of the same order of magnitude in this case. Further study and application of a regression procedure produced the simulations shown in figure 7.14 and allowed the zero-field splitting parameters for the $S = 1$ and $S = 2$ levels to be determined [Blanchard *et al.*, 2003]. The values of these parameters are difficult to analyse, since they contain contributions from the zero-field splitting matrices for the Mn^{2+} centres and from the dipolar and exchange interactions (equations [7.33] and [7.34]).

The method described in this section can be applied to complexes with more than two coupled centres, although the calculations rapidly become more difficult as the number of centres increases. To show how this generalisation is performed, in complement 1 we present the case of a *trinuclear* complex.

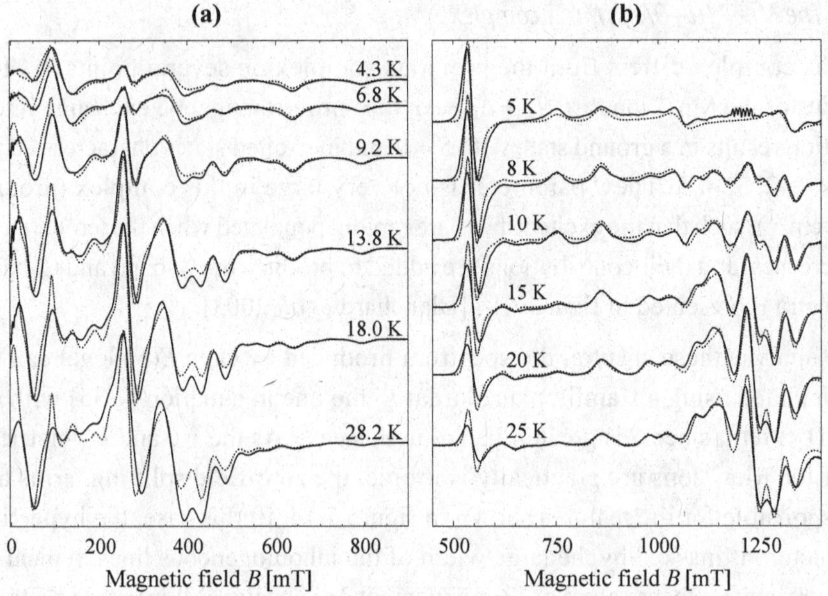

Figure 7.14 – **(a)** X-band (ν = 9.40 GHz) and **(b)** Q-band (ν = 34 GHz) spectra for the $[(Bpmp)Mn_2(\mu\text{-}OAc)_2](ClO_4)$ complex recorded at different temperatures. Power: X-band 0.5 or 2 mW, Q-band: 0.58 or 2.3 mW. Modulation: frequency 100 kHz, peak-to-peak amplitude 0.5 mT. Numerical simulations are represented by the dashed lines.
[Reproduced with permission from Blanchard S. *et al.*, *Inorganic Chemistry*, **42**, 4568–4578 © 2003, American Chemical Society]

7.5 – How intercentre interactions affect the intensity of the spectrum and the relaxation properties

7.5.1 – EPR spectrum intensity

The intensity of the spectrum produced by two interacting paramagnetic centres is simply expressed in the two limit situations studied in sections 7.3 and 7.4.

◇ Situation with weak intercentre interactions

Within this limit, the intensity of the spectrum for each centre is unmodified. If the sample contains *N* (A, B) pairs, the total intensity is equal to the sum of the intensities of the spectra produced by *N* centres A and *N* centres B (equations [5.17] and [5.19]). When the experimental spectrum is integrated, a large enough field range must be chosen to include any lateral lines due to exchange interaction (section 7.3.2).

◇ Strong exchange situation

In this situation, the EPR spectrum produced by N (A, B) pairs is the result of superposition of the components produced by the different levels $E(S)$ (figure 7.9), each one weighted by its Boltzmann factor. The intensity of the spectrum is therefore equal to the weighted sum of the intensities of these components. For example, in the case of the (Fe^{3+}, Fe^{2+}) and (Mn^{3+}, Mn^{4+}) complexes treated in section 7.4.3, only the component due to the $S = ½$ ground state is observed. Its intensity is obtained by multiplying the expression produced by equation [5.17], calculated for N paramagnetic centres with $(g_P)_{av}$ deduced from the principal values measured on the spectrum (equation [5.19]), by the Boltzmann factor for the $E(½)$ level which is written:

$$\frac{2\exp[-E(\frac{1}{2})/k_B T]}{\sum_S (2S+1)\exp[-E(S)/k_B T]}$$

where $E(S)$ is given by equation [7.29]. By studying the temperature-dependence of the intensity of the spectrum, the J parameter can be determined.

The same procedure can be applied when several energy levels contribute to the spectrum, but then we must calculate the intensity of the spectrum produced by a centre of spin greater than ½, which is not always simple (sections 6.4.2 and 6.6.1).

7.5.2 – Relaxation properties

◇ Situation with weak intercentre interactions

When intercentre interactions and relaxation processes are combined, complex phenomena emerge. In this section, we examine some effects which play a significant role in practical EPR spectroscopy. We will assume that T_{1A}, the spin-lattice relaxation time for centre A, is much shorter than that for centre B.

1. The spectral patterns created by the intercentre interactions *depend on the value of* T_{1A}. This phenomenon is very similar to chemical exchange in NMR, and it can be modelled in the same way [Caldeira *et al.*, 2000].
   - If T_{1A} is long enough to verify the relation

$$\hbar/g_A\beta T_{1A} \ll \Delta B_A \qquad [7.39],$$

where ΔB_A is the splitting created by the intercentre interactions on the spectrum for A (section 7.3), the splitting is observed on both spectra. This relation means that the broadening of the lines for the A spectrum which is linked to the relaxation time T_{1A} (equation [5.28]) is much less than ΔB_A.

- If $\hbar/g_A\beta T_{1A}$ and ΔB_A are of the same magnitude, the two lines in the pattern merge into a single broad central line.
- If T_{1A} is short enough to verify the relation

$$\hbar/g_A\beta T_{1A} \gg \Delta B_A \qquad [7.40]$$

the central line narrows to produce the line that would be observed in the absence of interaction.

At very low temperatures, T_{1A} is generally long enough for condition [7.39] to be verified, and the splitting is visible on the interaction spectra. When the temperature increases, T_{1A} decreases rapidly and the lines are seen to "merge". When the temperature is such that condition [7.40] is satisfied, the effects of intercentre interactions completely disappear from the spectra. This spectacular temperature-dependence is very useful when attempting to interpret the complex spectral shapes sometimes observed at low temperature [Guigliarelli *et al.*, 1995].

2. Intercentre interactions *accelerate relaxation* of the B centre, for which the spin-lattice and spin-spin relaxation times become:

$$\begin{aligned} 1/T_{1B} &= 1/T_{1B}{}^0 + k_{1dip} + k_{1exch} \\ 1/T_{2B} &= 1/T_{2B}{}^0 + k_{2dip} + k_{2exch} \end{aligned} \qquad [7.41]$$

where $T_{1B}{}^0$ and $T_{2B}{}^0$ are the relaxation times in the absence of interaction. The conditions in which these equations are valid and the expression for the (k_{1dip}, k_{2dip}) rate constants due to the dipolar coupling and the (k_{1exch}, k_{2exch}) rate constants due to exchange interaction, are detailed in [Guigliarelli *et al.*, 1995]. We will simply highlight the fact that these values depend on (T_{1A}, T_{2A}), that k_{1dip} and k_{2dip} are inversely proportional to the 6th power of the distance r and that k_{1exch} and k_{2exch} are proportional to J^2.

◇ *Strong exchange situation*

Within this limit, the (A, B) pair behaves like a single paramagnetic centre. Its ground state is characterised by its spin S and the spin Hamiltonian defined by equation [7.35]. For $S = ½$, the relaxation properties are identical to those described in section 5.4, but the (A, B) pair then has low-lying excited states and Orbach-type relaxation processes are often very efficient (section 5.4.3 and figure 5.11). If S is greater than ½, the observations from section 6.8.4 apply.

7.6 – Points to consider in applications

7.6.1 – How weak intercentre interactions affect the spectra

Within this limit, the centres retain their individual nature and the interactions cause the resonance lines to split without altering their intensity. This splitting is considerably influenced by the relative arrangement of the centres, and as a result a quantitative study of the interaction spectrum can provide structural information. This type of study is difficult due to the large number of parameters required to describe the "local" characteristics ($\tilde{\mathbf{g}}$, $\tilde{\mathbf{A}}$ matrices) and the relative arrangement of the two centres. The local characteristics can be determined from the spectrum "in the absence of interaction" produced either by preparing the (A, B) pairs in a state where only one centre is paramagnetic, or by recording the spectrum at a high enough temperature to satisfy condition [7.40]. The shape of the interaction spectrum depends on the *frequency* of the spectrometer. Indeed, increasing the frequency and thus the magnitude of the magnetic field increases the spreading of the spectrum due to the anisotropy of the $\tilde{\mathbf{g}}_\mathbf{A}$ and $\tilde{\mathbf{g}}_\mathbf{B}$ matrices, and simplifies the interpretation of the spectrum. It also increases terms such as $(g_{Ai} - g_{Bi})\beta B$, which modifies the splitting pattern (section 7.3.3). Comparison of spectra recorded at different frequencies is therefore very useful to interpret the interaction spectra, and their simulation using the same set of parameters constitutes a good test of uniqueness [More *et al.*,1996; Fournel *et al.*, 1998; More *et al.* 2005].

When the intercentre interactions are too weak for the splitting of the lines to be resolved on the spectrum, they can be quantitatively studied using Pulsed ELectron-electron DOuble Resonance (PELDOR) pulsed EPR techniques [Schweiger and Jeschke, 2001].

7.6.2 – How strong exchange interaction affects the spectrum

A strong exchange interaction between the paramagnetic centres can be considered to be a weak chemical bond: the centres behave like a single entity, and the $E(S)$ energy levels are described by the Heisenberg operator. It is therefore not surprising that the EPR spectrum is extensively modified by this type of interaction. The examples dealt with in section 7.4.3 show that analysis of the hyperfine structure provides direct information on the spin coupling coefficients, and consequently on the nature of the interacting centres. For a given value of S, the spin coupling coefficients indicate how the normalised spin density is distributed over the different centres (appendix 5). This explains the direct relationship between the spin coupling coefficients and the hyperfine matrix (equation [7.32]). The principal values of the $\tilde{\mathbf{g}}$ matrix also depend on the spin coupling coefficients (equation [7.31]), but their analysis is less straightforward than that of the hyperfine structure, as shown by the example of the (Fe^{3+}, Fe^{2+}) pair dealt with in section 7.4.3.

7.6.3 – Dynamic effects of intercentre interactions

In contrast to the "static" effects of intercentre interactions which appear on the spectra, the effects on relaxation times are qualified as "dynamic" (equations [7.41]). If the relaxation times in the absence of interaction $(T_{1B}{}^0, T_{2B}{}^0)$ and (T_{1A}, T_{2A}) are known, information on the intercentre distance and on the exchange parameter can be obtained by measuring (T_{1B}, T_{2B}) [Guigliarelli *et al.*, 1995].

Dynamic effects can also be observed in other conditions. For example, consider a solution of two paramagnetic molecules A and B for which only the EPR spectrum for the B molecules is observable at room temperature. Collisions between molecules create transient complexes:

$$\mathrm{A} + \mathrm{B} \overset{k}{\rightleftarrows} (\mathrm{A}, \mathrm{B})$$

and we assume that the ground state of the (A, B) complex is non-paramagnetic. If the exchange between the free (B) and complexed (A, B) forms of B is slow, the position of the resonance lines for B remains unchanged, but the lifetime for its spin states is shortened. The relaxation rates $1/T_{1B}$ and $1/T_{2B}$ increase, and if some conditions are verified, it is possible to write [Altenbach *et al.*, 2005]:

$$1/T_{1B} = 1/T_{1B}{}^0 + k\,[\mathrm{A}]$$
$$1/T_{2B} = 1/T_{2B}{}^0 + k\,[\mathrm{A}]$$

By studying how $1/T_{1B}$ or $1/T_{2B}$ are affected by the concentration [A], the rate constant k can be determined.

Complement 1 – Equivalent Hamiltonian for a trinuclear complex

The spin Hamiltonian describing the interactions to which the unpaired electrons in an (A, B, C) complex are subjected takes the form

$$\hat{H} = \hat{H}_A + \hat{H}_B + \hat{H}_C + \hat{H}_{exch} + \hat{H}_{dip} \qquad [1]$$

where $\hat{H}_A$, $\hat{H}_B$, $\hat{H}_C$ describe the *local* interactions (equation [7.26]) and $\hat{H}_{exch}$, $\hat{H}_{dip}$ account for the *intercentre* interactions. If the Heisenberg operator

$$\hat{H}_{Heis} = -J_{AB}\,\mathbf{S}_A\cdot\mathbf{S}_B - J_{BC}\,\mathbf{S}_B\cdot\mathbf{S}_C - J_{AC}\,\mathbf{S}_A\cdot\mathbf{S}_C \qquad [2]$$

in $\hat{H}_{exch}$ is much larger than the other terms in $\hat{H}$, the lowest energy levels can be described by an equivalent Hamiltonian similar to that of a single centre.

◇ *Construction of a coupled basis*

To determine the eigenvalues and eigenvectors of $\hat{H}_{Heis}$, a "coupled basis" is defined from the total spin $\mathbf{S} = \mathbf{S}_A + \mathbf{S}_B + \mathbf{S}_C$, starting, for example, by adding $\mathbf{S}_B$ and $\mathbf{S}_C$. We set

$$\mathbf{S}_{BC} = \mathbf{S}_B + \mathbf{S}_C$$

For each possible value of S_{BC}:

$$S_{BC} = |S_B - S_C|,\ |S_B - S_C| + 1,\ \ldots,\ S_B + S_C \qquad [3]$$

there is a corresponding set of $(2S_{BC} + 1)$ kets $\{|S_{BC}, M_{BC}\rangle\}$ which verify the following relations (appendix 3):

$$\hat{\mathbf{S}}_{BC}^2|S_{BC}, M_{BC}\rangle = S_{BC}(S_{BC}+1)\,|S_{BC}, M_{BC}\rangle$$

$$\hat{\mathbf{S}}_B^2|S_{BC}, M_{BC}\rangle = S_B(S_B+1)\,|S_{BC}, M_{BC}\rangle$$

$$\hat{\mathbf{S}}_C^2|S_{BC}, M_{BC}\rangle = S_C(S_C+1)\,|S_{BC}, M_{BC}\rangle$$

$\mathbf{S}$ is obtained by adding $\mathbf{S}_{BC}$ and $\mathbf{S}_A$. To each value of S_{BC} (equation [3]) the following values of S correspond:

$$S = |S_{BC} - S_A|,\ |S_{BC} - S_A| + 1,\ \ldots,\ S_{BC} + S_A \qquad [4]$$

and for each value of S, there exists a set of $(2S + 1)$ kets denoted $\{|S_{BC}, S, M\rangle\}$ which verify:

$$\hat{\mathbf{S}}^2|S_{BC}, S, M\rangle = S(S+1)\,|S_{BC}, S, M\rangle \qquad [5]$$

$$\hat{\mathbf{S}}_{BC}^2|S_{BC}, S, M\rangle = S_{BC}(S_{BC}+1)\,|S_{BC}, S, M\rangle \qquad [6]$$

$$\hat{\mathbf{S}}_A^2|S_{BC},S,M\rangle = S_A(S_A+1)\,|S_{BC},S,M\rangle \quad [7]$$

It can be shown that these kets also verify:

$$\hat{\mathbf{S}}_B^2|S_{BC},S,M\rangle = S_B(S_B+1)\,|S_{BC},S,M\rangle \quad [8]$$

$$\hat{\mathbf{S}}_C^2|S_{BC},S,M\rangle = S_C(S_C+1)\,|S_{BC},S,M\rangle \quad [9]$$

The $\{|S_{BC},S,M\rangle\}$ set of kets defined from equations [3] and [4] make up a "coupled basis" in the $\mathcal{E}$ space of dimension $(2S_A+1)(2S_B+1)(2S_C+1)$.

Two other coupled bases can be constructed from $\mathbf{S}_{AB} = \mathbf{S}_A + \mathbf{S}_B$ and from $\mathbf{S}_{AC} = \mathbf{S}_A + \mathbf{S}_C$. When the three exchange parameters are different, there is no advantage to using any particular basis when diagonalising $\hat{H}_{Heis}$, but this is no longer true when two exchange parameters are equal.

◇ *Case where two exchange parameters are equal*

Assume for example that $J_{AB} = J_{AC}$. We set

$$J_{AB} = J_{AC} = J; \; J_{BC} = J(1+\alpha)$$

In these conditions, the Heisenberg operator [2] becomes:

$$\hat{H}_{Heis} = -J\,\mathbf{S}_A\cdot(\mathbf{S}_B+\mathbf{S}_C) - J(1+\alpha)\mathbf{S}_B\cdot\mathbf{S}_C$$

The angular momentum $\mathbf{S}_{BC} = \mathbf{S}_B + \mathbf{S}_C$ is naturally introduced and we can write:

$$\begin{aligned}\hat{H}_{Heis} &= -J\,\mathbf{S}_A\cdot\mathbf{S}_{BC} - J\,\mathbf{S}_B\cdot\mathbf{S}_C - \alpha J\,\mathbf{S}_B\cdot\mathbf{S}_C \\ &= -(J/2)(\hat{\mathbf{S}}^2 - \hat{\mathbf{S}}_A^2 - \hat{\mathbf{S}}_B^2 - \hat{\mathbf{S}}_C^2) - \alpha(J/2)(\hat{\mathbf{S}}_{BC}^2 - \hat{\mathbf{S}}_B^2 - \hat{\mathbf{S}}_C^2)\end{aligned}$$

According to equations [5] to [9], this operator is represented by a diagonal matrix in the $\{|S_{BC},S,M\rangle\}$ basis and its eigenvalues are given by:

$$\begin{aligned}E(S_{BC},S) = &-(J/2)\,[S(S+1) - S_A(S_A+1) - S_B(S_B+1) - S_C(S_C+1)] \\ &-\alpha(J/2)[S_{BC}(S_{BC}+1) - S_B(S_B+1) - S_C(S_C+1)]\end{aligned} \quad [10]$$

The separation between the energy levels characterised by different values of S_{BC} is determined by the $\alpha J/2$ term and therefore by the *difference* between the exchange parameters. When this separation is large relative to the other terms in Hamiltonian [1], an equivalent Hamiltonian can be constructed within each $\{|S_{BC},S,M\rangle\}$ sub-space. In particular, $(\mathbf{S}_A, \mathbf{S}_B, \mathbf{S}_C)$ can be replaced by $(K_A\mathbf{S}, K_B\mathbf{S}, K_C\mathbf{S})$ in the Zeeman and hyperfine terms, where (K_A, K_B, K_C) are spin coupling coefficients which verify $K_A + K_B + K_C = 1$. These spin coupling coefficients are obtained as follows:

▹ For a given value of S_{BC}, $\mathbf{S}_B$ and $\mathbf{S}_C$ can be replaced by K_B^{BC} $\mathbf{S}_{BC}$ and K_C^{BC} $\mathbf{S}_{BC}$, respectively, with (equation [7.30]):

$$K_B^{BC} = \frac{S_{BC}(S_{BC}+1) + S_B(S_B+1) - S_C(S_C+1)}{2S_{BC}(S_{BC}+1)} \quad ; \quad K_B^{BC} + K_C^{BC} = 1$$

▹ For a given value of S, $\mathbf{S}_{BC}$ and $\mathbf{S}_A$ can be replaced by K_{BC}^S $\mathbf{S}$ and K_A^S $\mathbf{S}$ with

$$K_{BC}^S = \frac{S(S+1) + S_{BC}(S_{BC}+1) - S_A(S_A+1)}{2S(S+1)} \quad ; \quad K_{BC}^S + K_A^S = 1$$

From these relations, we deduce:

$$K_A = K_A^S, \; K_B = K_B^{BC} K_{BC}^S, \; K_C = K_C^{BC} K_{BC}^S$$

◇ Application to a Fe^{3+} trinuclear centre

A, B, C are Fe^{3+} centres with spins $S_A = S_B = S_C = 5/2$. We assume that the Heisenberg operator dominates the spin Hamiltonian with $J < 0$ (dominant antiferromagnetic interaction), and we seek to determine in what conditions the ground state is characterised by $S = 1/2$. The possible values of S_{BC} (equation [3]) are: 0, 1, 2, 3, 4, 5.

▹ Equation [4] shows that $S = 1/2$ can be obtained in two different ways with the following energies (equation [10]):

$$S_{BC} = 2: E(2, 1/2) = 51J/4 + 23\alpha J/4$$

$$S_{BC} = 3: E(3, 1/2) = 51J/4 + 11\alpha J/4$$

▹ Similarly, $S = 3/2$ can be obtained in 4 different ways:

$$S_{BC} = 1: E(1, 3/2) = 45J/4 + 31\alpha J/4$$

$$S_{BC} = 2: E(2, 3/2) = 45J/4 + 23\alpha J/4$$

$$S_{BC} = 3: E(3, 3/2) = 45J/4 + 11\alpha J/4$$

$$S_{BC} = 4: E(4, 3/2) = 45J/4 - 5\alpha J/4$$

These equations show that α must verify the inequality $0 < \alpha < 3/4$ in order for the ground state to be characterised by $S = 1/2$. The ground sub-space is therefore $\{|2, 1/2, M\rangle\}$, with $K_A = 1/3$, $K_B = -2/3$, $K_C = -2/3$, and the equivalent Hamiltonian is written:

$$\hat{H}_{S=1/2} = \beta\, \mathbf{B} \cdot \tilde{\mathbf{g}} \cdot \mathbf{S}$$

$$\tilde{\mathbf{g}} = 1/3\, \tilde{\mathbf{g}}_A - 2/3\, \tilde{\mathbf{g}}_B - 2/3\, \tilde{\mathbf{g}}_C$$

General methods have been developed to determine the expressions for the energy levels and the $\tilde{g}$ matrix for complexes with nuclearity greater than 3 [Belorizky and Fries, 1993].

References

ALTENBACH C. *et al.* (2005) *Biophysical Journal* **89**: 2103-2112.

ATHERTON N.M. (1993) *Principles of Electron Spin Resonance*, Ellis Horwood PTR Prentice Hall, New York.

AYANT Y. & BELORIZKY E. (2000) *Cours de mécanique quantique*, Dunod, Paris.

BELORIZKY E. & FRIES P.H. (1993) *Journal de Chimie Physique* **90**: 1077-1100.

BENCINI A. & GATTESCHI D. (1990) *EPR of Exchange Coupled Systems*, Springer-Verlag, Berlin.

BERTRAND P., GUIGLIARELLI B. & MORE C. (1991) *New Journal of Chemistry* **15**: 445-454.

BLANCHARD S. *et al.* (2003) *Inorganic Chemistry* **42**: 4568-4578.

BLONDIN G. & GIRERD J.-J. (1990) *Chemical Review* **90**: 1359-1376.

CALDEIRA J. *et al.* (2000) *Biochemistry* **39**: 2700-2707.

CANESCHI A. *et al.* (1988) *Inorganic Chemistry* **27**: 1031-1035.

CATALA L. *et al.* (2001) *Chemistry: A European Journal* **7**: 2466-2480.

CROUSE B.R. *et al.* (1995) *Journal of the American Chemical Society* **117**: 9612-9613.

FOURNEL A. *et al.* (1998) *Journal of Chemical Physics* **109**: 10905-10913.

GIBSON J.F. *et al.* (1966) *Proceeding of the National Academy of Sciences of the USA* **56**: 987-990.

GORDY W. (1980) *Theory and Applications of Electron Spin Resonance*, John Wileys and Sons, New York.

GUIGLIARELLI B. *et al.* (1995) *Biochemistry* **34**: 4781-4790.

GUIGLIARELLI B. & BERTRAND P. (1999) *Advances in Inorganic Chemistry* **47**: 421-497.

HEINRICH B. (2005) "Radio Frequency Techniques", in *Ultrathin Magnetic Structures* II, HEINRICH B. & BLAND J.A.C. eds, Springer Verlag, Berlin.

HUREAU C. *et al.* (2004) *Chemistry-A European Journal* **10**: 1998-2010.

KAHN O. (1993) *Molecular Magnetism*, Wiley-VCH Publishers, New York.

MORE C. *et al.* (1996) *Journal of Biological Chemistry* **1**: 152-161.

MORE C. *et al.* (2005) *Biochemistry* **44**: 11628-11635.

OWEN J. & HARRIS E.A. (1972) " Pair Spectra and Exchange Interactions" in *Electron Paramagnetic Resonance*, GESCHWIND S. ed. Plenum Press, New York.

SMITH T.D. & PILBROW J.R. (1974) *Coordination Chemistry Reviews* **13**: 173-278.

SCHNEIDER J. *et al.* (1968) *Journal of Physics and Chemistry of Solids* **29**: 451-462.
SCHWEIGER A. & JESCHKE G. (2001) *Principles of Pulse Electron Paramagnetic Resonance*, Oxford University Press, Oxford.
SWEENEY W.V. & COFFMAN R.E. (1972) *Biochimica et Biophysica Acta* **286**: 26-35.
WAUTELET P. *et al.* (2003) *Journal of Organic Chemistry* **68**: 8025-8036.

Exercises

7.1. What can be said about the *sign* of the various terms in the Coulomb contribution K and that of the exchange terms J_1 and J_2 (equations [7.5])?

7.2. Using the method described in section 1.4 of appendix 4, determine the eigenvalues and eigenvectors for the $\hat{H}_{Heis}$ operator defined by equation [7.7].

7.3. Verify that the energy levels and spin states in figure 7.3 which are due to $\hat{H}_{Zeeman}$ produce resonance lines with positions and intensities corresponding to two *independent* centres.

7.4. Return to the calculation in section 7.3.1 with $g_A = g_B = g$.

a) What effect does $\hat{H}_{dip}$ have on the $E_0(½, ½)$ and $E_0(-½, -½)$ energy levels to first order in perturbation theory?

b) Construct the matrix representing $\hat{H}_{dip}$ in the $\{|½\rangle|-½\rangle, |-½\rangle|½\rangle\}$ sub-space. From this matrix, deduce the effect of $\hat{H}_{dip}$ on the $E_0(½, -½) = E_0(-½, ½)$ energy level.

c) Determine the allowed transitions and the positions of the lines.

7.5. Consider the matrix elements determining the intensities of the four allowed transitions when exchange interaction occurs (section 7.3.2). Using the fact that the kets $|u\rangle$ et $|v\rangle$ are normalised and orthogonal, show that the sum of the intensities of the four transitions is proportional to $(g_A{}^2 + g_B{}^2)(\beta B_1)^2/2$, which is the value obtained in the absence of interaction (see exercise 7.3).

7.6. Determine the g and A parameters defined in section 7.4.3 for the bis(nitronyl-nitroxide) derivative, for which the spectrum is shown in figure 7.10. Rationalise the following series of observations:

- For a nitroxide radical, the interaction with the ^{14}N nucleus is characterised by $A_{iso} \approx 40$ MHz (figure 2.6a and exercise 2.4).
- For a nitronyl-nitroxide radical, $A_{iso} \approx 20$ MHz (figure 2.12a and exercise 2.6).
- For a bis(nitronyl-nitroxide) derivative, $A_{iso} \approx 10$ MHz.

7.7. Determine the g and a parameters defined in section 7.4.3 (figure 7.12) for the spectrum in figure 7.13. Compare their values to the numbers ($g_x = 2.001$, $g_y = 1.996$, $g_z = 1.984$), ($a_x{}^{3+} = 245$ MHz, $a_y{}^{3+} = 223$ MHz, $a_z{}^{3+} = 171$ MHz), ($a_x{}^{4+} = 207$ MHz, $a_y{}^{4+} = 228$ MHz, $a_z{}^{4+} = 216$ MHz) deduced from the numerical simulation of the spectrum [Hureau *et al.*, 2004].

Answers to exercises

7.1. As the potential energies V_A and V_B correspond to *attractions*, the first two terms of K are negative. The third is positive since $V(\mathbf{r}_1, \mathbf{r}_2)$ corresponds to *repulsion*.

The first factor of J_1 is the overlap integral between the orbitals φ_A and φ_B. Its sign is determined by that of the product $\varphi_A(\mathbf{r})\varphi_B(\mathbf{r})$ in the region of space where these orbitals overlap. The same product appears in the second factor, multiplied by $h(1)$ which represents an attractive interaction. The signs of the two factors are therefore different and J_1 is negative.

The integral defining J_2 is determined by the region of space where φ_A and φ_B overlap and where the electrons are close together. For the values of $(\mathbf{r}_1, \mathbf{r}_2)$ which determine J_2, the products $\varphi_A(\mathbf{r}_1)\varphi_B(\mathbf{r}_1)$ and $\varphi_A(\mathbf{r}_2)\varphi_B(\mathbf{r}_2)$ have the same sign. Since $V(\mathbf{r}_1, \mathbf{r}_2)$ corresponds to a *repulsion*, J_2 is positive.

7.2. The Heisenberg operator $\hat{H}_{Heis} = -J\,\mathbf{S}_A \cdot \mathbf{S}_B$ can be written $\hat{H}_{Heis} = -(J/2)(\hat{\mathbf{S}}^2 - \hat{\mathbf{S}}_A^2 - \hat{\mathbf{S}}_B^2)$, where $\mathbf{S} = \mathbf{S}_A + \mathbf{S}_B$. By following the procedure described in appendix 4, from any direction z in the Euclidean space a coupled basis $\{|S, M_S\rangle\}$ can be defined in which the $(\hat{\mathbf{S}}^2, \hat{S}_z, \hat{\mathbf{S}}_A^2, \hat{\mathbf{S}}_B^2)$ operators are represented by diagonal matrices. In this basis, $\hat{H}_{Heis}$ is represented by a diagonal matrix and its eigenvalues are given by:

$$E(M_S) = -(J/2)\,[S(S+1) - S_A(S_A+1) - S_B(S_B+1)]$$

7.3. For example, consider the transition $|½\rangle|½\rangle \leftrightarrow |-½\rangle|½\rangle$. Its energy is $g_A\beta B$ and its intensity is proportional to $(\langle ½|\langle ½|\hat{H}_1|-½\rangle|½\rangle)^2$, with

$$\hat{H}_1 = \hat{H}_{1A} + \hat{H}_{1B};\ \hat{H}_{1A} = g_A\beta\mathbf{S}_A \cdot \mathbf{B}_1;\ \hat{H}_{1B} = g_B\beta\mathbf{S}_B \cdot \mathbf{B}_1$$

As $\mathbf{B}_1$ is in the (X, Y) plane perpendicular to $\mathbf{B}$, these operators can be written:

$$\hat{H}_{1A} = g_A\beta B_1\,(\hat{S}_{AX}\cos\alpha + \hat{S}_{AY}\sin\alpha);\ \hat{H}_{1B} = g_B\beta B_1\,(\hat{S}_{BX}\cos\alpha + \hat{S}_{BY}\sin\alpha)$$

where $\alpha = (X, \mathbf{B}_1)$. Only $\hat{H}_{1A}$ contributes to the matrix element: a resonance line is produced at $B_{0A} = h\nu/g_A\beta$ with an intensity proportional to $(g_A\beta B_1)^2/4$. The transition $|½\rangle|-½\rangle \leftrightarrow |-½\rangle|-½\rangle$ produces the same line with the same intensity. The total intensity of this line is therefore proportional to $(g_A\beta B_1)^2/2$. The two other transitions produce a line at $B_{0B} = h\nu/g_B\beta$ with a total intensity proportional to $(g_B\beta B_1)^2/2$.

7.4. a) As the energy levels $E_0(½, ½)$ and $E_0(-½, -½)$ are non-degenerate, equation [7.15] can be used:

$$E(½, ½) = g\beta B + D(1 - 3\cos^2\gamma)/4;\ E(-½, -½) = -g\beta B + D(1 - 3\cos^2\gamma)/4$$

b) To construct this matrix, we note that:

- The diagonal elements due to the $D(1 - 3\cos^2\gamma)\hat{S}_{AZ}\hat{S}_{BZ}$ term are equal to $- D(1 - 3\cos^2\gamma)/4$.
- The off-diagonal elements are produced by the terms in $\hat{S}_{AX}\hat{S}_{BX}$, $\hat{S}_{AY}\hat{S}_{BY}$, $\hat{S}_{AX}\hat{S}_{BY}$, $\hat{S}_{AY}\hat{S}_{BX}$. If X is taken in the $(\mathbf{r}, \mathbf{B})$ plane, these terms can be written $D(\hat{S}_{AX}\hat{S}_{BX}(1 - 3\sin^2\gamma) + \hat{S}_{AY}\hat{S}_{BY})$. The following matrix is obtained:

	$\lvert ½\rangle\lvert -½\rangle$	$\lvert -½\rangle\lvert ½\rangle$
$\langle ½\rvert\langle -½\rvert$	$-D(1 - 3\cos^2\gamma)/4$	$-D(1 - 3\cos^2\gamma)/4$
$\langle -½\rvert\langle ½\rvert$	$-D(1 - 3\cos^2\gamma)/4$	$-D(1 - 3\cos^2\gamma)/4$

From this matrix, we deduce the eigenvalues and eigenvectors:

$$E_1 = 0;\ \frac{1}{\sqrt{2}}(\lvert ½\rangle\lvert -½\rangle - \lvert -½\rangle\lvert ½\rangle)$$

$$E_2 = -\frac{D}{2}(1 - 3\cos^2\gamma);\ \frac{1}{\sqrt{2}}(\lvert ½\rangle\lvert -½\rangle + \lvert -½\rangle\lvert ½\rangle)$$

c) Allowed transitions take place between:

- The energy levels E_2 and $E(½, ½)$: $\Delta E = g\beta B + ¾\, D(1 - 3\cos^2\gamma)$
- The energy levels E_2 and $E(-½, -½)$: $\Delta E = g\beta B - ¾\, D(1 - 3\cos^2\gamma)$

These transitions produce two lines of equal intensity centred at $B_0 = h\nu/g\beta$, separated by $^3/_2\ (D/g\beta)(1 - 3\cos^2\gamma)$.

7.5. As the kets $\lvert u\rangle$ and $\lvert v\rangle$ are normalised and orthogonal, we can set:

$$c_{1+} = \cos\varphi,\ c_{2+} = \sin\varphi,\ c_{1-} = \sin\varphi,\ c_{2-} = -\cos\varphi$$

By developing the sum of the squares of the four matrix elements, we obtain $(g_A{}^2 + g_B{}^2)(\beta B_1)^2/2$.

7.6. From position $B = 335.5$ mT at the centre of the spectrum, we deduce $g = 2.006$. The line splitting is 0.377 mT, which gives $A_{iso} = 10.5$ MHz. The concept of *normalised spin density* (appendix 5) can be used to interpret the value of A_{iso}:

- In a nitroxide radical, the spin population of the $2p_z$ orbital of the nitrogen atom is $\rho_N \approx 0.5$ (exercise 2.4).

- In a nitronyl-nitroxide radical, the two nitrogen atoms are equivalent and $\rho_N \approx 0.25$ (exercise 2.6).
- In a bis(nitronyl-nitroxide) derivative in the triplet state, the spin population on each nitronyl-nitroxide entity is equal to $K_A = K_B = ½$ (appendix 5), and $\rho_N = 0.125$.

7.7. From the position of the "centre of symmetry" for the spectrum $B = 333$ mT, we deduce $g = 2.01$. The separation between the extreme lines, which is $15a/g\beta$ in figure 7.12, equals 120 mT. From this value, we deduce $a = 220$ MHz. The anisotropy of the hyperfine matrix for Mn^{3+} and higher-order effects of perturbation theory explain the differences compared to the simplified diagram shown in figure 7.12.

Chapter 8

EPR spectrum for complexes of rare earth and actinide ions

8.1 – Rare earth ions

In the periodic table of the elements, the atoms with an atomic number between $Z = 58$ and $Z = 71$:

cerium, praseodymium, neodymium, promethium, samarium, europium, gadolinium, terbium, dysprosium, holmium, erbium, thulium, ytterbium and lutetium (see the etymology in complement 1),

occupy the same cell as lanthanum ($Z = 57$). For this reason they are called the lanthanides. The configuration of lanthanum is [...] $5s^2\ 5p^6\ 5d^1\ 6s^2$ and that of lanthanide cations is [...] $4f^n\ 5s^2\ 5p^6$, where n varies between 1 and 14. These elements were first discovered in the 19th century in rare ores, and they were therefore qualified as "rare", although it has since been realised that their abundance in the terrestrial crust is comparable to the abundance of elements from the first transition series. Rare earth atoms all have the same valence electrons, and as a result their chemical properties are very similar and for a long time it was difficult to separate them. Around twenty years ago, effective separation techniques were developed and applications of rare earth elements have expanded ever since, in particular in high-impact technologies such as electronics, imagery, permanent magnets, catalysis, phosphors, quantum computers. The paramagnetism of rare earth ions is due to electrons in the $4f$ subshell, whereas it is due to d electrons in transition ions. This difference impacts the magnetic properties of these two classes of cations:

P. Bertrand, *Electron Paramagnetic Resonance Spectroscopy*,
https://doi.org/10.1007/978-3-030-39663-3_8

- ▹ The $4f$ subshell has seven orbitals characterised by $\ell = 3$, the numbers L and S can therefore take larger values than with the five d orbitals characterised by $\ell = 2$.
- ▹ The $4f$ subshell is *internal*, its electrons are thus exposed to a high effective nuclear charge which increases the spin-orbit and hyperfine interactions. In contrast, they interact much more weakly with the ligands than the electrons in an *external d* subshell when the cation is part of a complex.

8.1.1 – Magnetic moment of free rare earth ions

Although spin-orbit coupling is strong in rare earth ions, the spectroscopic terms remain sufficiently separated for the magnetic characteristics of the ground state to be well described by the (L, S) coupling scheme (appendix 1). For example, consider the Nd^{3+} ion which has a $4f^3$ configuration. We construct the microstate which produces the maximal values of M_S, then of M_L:

–3	–2	–1	0	1	2	3
				↑	↑	↑

This microstate corresponds to $M_L = 6$, $M_S = 3/2$, and Hund's rule indicates that the *ground term* is ^4I $(L = 6, S = 3/2)$. The degeneracy of this term is partially removed by spin-orbit coupling, which results in energy levels identified by the number J associated with the total angular momentum $\mathbf{J} = \mathbf{L} + \mathbf{S}$ of the ion. This number varies from $J_{min} = |L - S| = 9/2$ to $J_{max} = L + S = 15/2$, and as the $4f$ subshell is less than half-full, the ground state is characterised by $J_{min} = 9/2$. Each value of J is associated with an (L, S, J) *multiplet* which has $(2J + 1)$ states, within which the magnetic moment is written

$$\boldsymbol{\mu} = -g_{Landé}\, \beta\, \mathbf{J}$$

where $g_{Landé}$ is given by equation [7] from appendix 1. In the case of Nd^{3+}, $g_{Landé} = 1 - 3/11\,(g_e - 1) \approx 8/11$ for the ground multiplet $^4I_{9/2}$. By proceeding in this way for all the rare earth cations, the results presented in table 8.1 are obtained.

Table 8.1 – Characteristics of the ground multiplet of trivalent rare earth cations. The Landé factor is calculated for $g_e = 2$. (Isoelectronic cations are indicated in brackets).

Z	conf.	Ion	L, S	J	$g_{Landé}$
58	$4f^1$	$Ce^{3+}(Pr^{4+})$	3, ½	5/2	6/7
59	$4f^2$	Pr^{3+}	5, 1	4	4/5
60	$4f^3$	Nd^{3+}	6, 3/2	9/2	8/11
61	$4f^4$	Pm^{3+}	6, 2	4	3/5
62	$4f^5$	Sm^{3+}	5, 5/2	5/2	2/7
63	$4f^6$	$Eu^{3+}(Sm^{2+})$	3, 3	0	0
64	$4f^7$	$Gd^{3+}(Eu^{2+}, Tb^{4+})$	0, 7/2	7/2	2
65	$4f^8$	Tb^{3+}	3, 3	6	3/2
66	$4f^9$	Dy^{3+}	5, 5/2	15/2	4/3
67	$4f^{10}$	$Ho^{3+}(Dy^{2+})$	6, 2	8	5/4
68	$4f^{11}$	$Er^{3+}(Ho^{2+})$	6, 3/2	15/2	6/5
69	$4f^{12}$	Tm^{3+}	5, 1	6	7/6
70	$4f^{13}$	$Yb^{3+}(Tm^{2+})$	3, ½	7/2	8/7

When a rare earth ion is placed in a magnetic field, the interaction of its unpaired electrons with **B** can therefore be written:

$$\hat{H}_{Zeeman} = g_{Landé}\, \beta\, \mathbf{B} \cdot \mathbf{J} \qquad [8.1]$$

8.1.2 – Hyperfine interaction with the nucleus

Numerous nuclei of rare earth atoms are paramagnetic, and the characteristics of the more abundant of these are indicated in table 8.2. The value of the nuclear spin is often high and isotopes of the same nucleus are generally characterised by quite different g_N. The hyperfine structure of the spectra is therefore very rich and it has been very useful to identify rare earth cations at a time when it was difficult to separate them by chemical methods.

The magnetic interactions between the nucleus and the unpaired electrons in the $4f$ subshell are described by 3 operators (section 4.4.1):

1. Polarisation of the electrons in the s orbitals by the f electrons creates a non-null spin density at the nucleus, which produces a term $A_s \mathbf{S} \cdot \mathbf{I}$.
2. The dipolar interactions between the moment $g_N \beta_N \mathbf{I}$ of the nucleus and the *orbital* magnetic moments of the electrons are described within an (L, S) term by the equivalent operator (appendix 4):

$$(\hat{H}_{dip}^{orb})_{eq} = P < r^{-3} > \mathbf{L} \cdot \mathbf{I} \text{ (assimilating 2 to } g_e\text{)}$$

$$P = \frac{\mu_0}{4\pi} g_e g_N \beta \beta_N \; ; \; < r^{-3} > = \int_0^\infty r^{-3} R_{n,\ell}^2(r) r^2 \, dr$$

$R_{n,\ell}(r)$ is the radial function common to all the orbitals in the incomplete subshell (n, ℓ) of the cation ($n = 4$ and $\ell = 3$ for a $4f$ orbital), and it depends on the nature of the cation.

Table 8.2 – Natural abundance and magnetic characteristics of rare earth nuclei [Lide, 2010].

nucleus	natural abundance %	I	g_N
^{141}Pr	100	5/2	1.71
^{143}Nd	12.2	7/2	−0.306
^{145}Nd	8.3	7/2	−0.188
^{147}Sm	15.02	7/2	−0.233
^{149}Sm	13.83	7/2	−0.192
^{151}Eu	47.8	5/2	1.389
^{153}Eu	52.2	5/2	0.6132
^{155}Gd	14.8	3/2	−0.172
^{157}Gd	15.7	3/2	−0.227
^{159}Tb	100	3/2	1.343
^{161}Dy	18.9	5/2	−0.192
^{163}Dy	24.9	5/2	0.2692
^{165}Ho	100	7/2	1.192
^{167}Er	22.9	7/2	−0.1611
^{169}Tm	100	1/2	−0.462
^{171}Yb	14.1	1/2	0.9873
^{173}Yb	16.1	5/2	−0.2719

3. The dipolar interactions with the *spin* magnetic moments of the electrons are described within an (L, S) term by the operator (appendix 4):

$$(\hat{H}_{dip}^{spin})_{eq} = P \xi < r^{-3} > [L(L+1)\, \mathbf{I} \cdot \mathbf{S} - \tfrac{3}{2} (\mathbf{L} \cdot \mathbf{I})(\mathbf{L} \cdot \mathbf{S}) - \tfrac{3}{2} (\mathbf{L} \cdot \mathbf{S}) \cdot (\mathbf{L} \cdot \mathbf{I})]$$

$$\xi = \frac{7 - 4S}{45 S(2L - 1)}$$

All these expressions can be simplified within an (L, S, J) multiplet. Indeed, in this multiplet we can replace $\mathbf{L} \cdot \mathbf{S}$ by $\tfrac{1}{2} [J(J+1) - L(L+1) - S(S+1)]$, $\mathbf{L}$ by $K_L \mathbf{J}$ and $\mathbf{S}$ by $K_S \mathbf{J}$, where K_L and K_S are the *spin coupling coefficients* defined in appendix 4. The sum of the three operators can then be written in the following compact form:

$$\hat{H}_{hyperfine} = A_J\,\mathbf{J}\cdot\mathbf{I} \qquad [8.2]$$

where the hyperfine coupling constant A_J is given by:

$$A_J = K_S A_s + P\langle r^{-3}\rangle\,\{K_L + \xi\,[K_S L(L+1) - \tfrac{3}{2}K_L\,(J(J+1) - L(L+1) - S(S+1))]\}$$

The quantity in curly brackets, which only depends on the numbers (S, L, J), has been tabulated [Abragam and Bleaney, 1986].

Note – For cations with a $4f^7$ configuration like Gd^{3+}, Eu^{2+} and Tb^{4+}, the ground term is $^8S(L = 0)$. We therefore have $g_{Landé} = g_e$, $K_L = 0$ and $K_S = 1$, and as a result $A_J = A_s$.

8.2 – Complexes of rare earth ions: effect of interaction with ligands

Complexes of rare earth cations can include up to 12 ligands in solids. They are often highly coloured due to the *f-f* transitions which are allowed in these complexes thanks to interaction with ligands which admixes configurations of opposite parities. The UV-visible and fluorescence absorption spectra are characterised by particularly narrow lines, which provide detailed information on the energy level diagram [Judd, 1988].

As the unpaired electrons in rare earth ions occupy *internal* orbitals, the splitting created by electrostatic interactions with the ligands is weak: it varies from a few tens to a few hundreds of cm^{-1}. In contrast, the splitting is around $10^3 - 10^4\ cm^{-1}$ in transition ion complexes.

8.2.1 – Expression describing the interaction of electrons in the 4*f* subshell with ligands

Given the mainly ionic nature of the bonds between the cation and the ligands, as a first approximation the effects of the electrostatic interactions between the 4*f* electrons and the ligands can be assimilated to those of an electric field created by point charges. To perform a detailed calculation requires integration over the coordinates of the electrons, but this integration can be avoided by describing these effects by an equivalent operator $\hat{H}_{ligand}$ within an (L, S) term, defined from the components of **L** and **S** [Stevens, 1952]. To construct the matrix representing $\hat{H}_{ligand}$ in the (L, S, J) ground multiplet, the equivalent operator $\hat{H}_{ligand}$ is expressed as a function of the components of **J**, which produces an expression with the form [Abragam and Bleaney, 1986]:

$$\hat{H}_{ligand} = \sum_{k} \sum_{q=-k}^{k} A_k^q < r^k > a_k(J) \hat{O}_k^q \qquad [8.3]$$

- ▷ The A_k^q parameters depend on the *geometry* and *nature* of the ligands. For electrons in the *f* orbitals, the number *k* can take the values 2, 4, 6. These parameters can be determined using a point charge model or semi-empirical methods [Newmann and Ng, 1989], or by considering them to be adjustable parameters.
- ▷ In $<r^k>$, *r* is the distance between the electron and the nucleus of the cation. Calculation of the mean involves the radial part of the 4*f* orbitals, which depends on the nature of the cation.
- ▷ The coefficients $a_k(J)$ depend on the numbers (L, S, J) and they have been tabulated [Abragam and Bleaney, 1986].
- ▷ The operators $\hat{O}_k^q$ are Hermitian and have a null trace. These operators are k^{th} degree polynomial functions of the components of **J** in a Cartesian reference frame. These operators and their matrix elements have also been tabulated [Abragam and Bleaney, 1986].

On the second side of equation [8.3], the number of terms increases when the symmetry of the complex is lower. Hereafter, we will more particularly consider complexes which have a 3-fold axis of symmetry denoted *z*, and a plane of symmetry perpendicular to this axis (C_{3h} or D_{3h} symmetry). This symmetry is close to that found in rare-earth ethyl sulfates which have been extensively studied by EPR. For these complexes, the $\hat{H}_{ligand}$ operator includes four terms:

$$\begin{aligned} \hat{H}_{ligand} &= A_2^0 < r^2 > a_2(J)\hat{O}_2^0 + A_4^0 < r^4 > a_4(J)\hat{O}_4^0 \\ &+ A_6^0 < r^6 > a_6(J)\hat{O}_6^0 + A_6^6 < r^6 > a_6(J)\hat{O}_6^6 \end{aligned} \qquad [8.4]$$

$$\begin{aligned} \hat{O}_2^0 &= 3\hat{J}_z^2 - J(J+1) \\ \hat{O}_4^0 &= 35\hat{J}_z^4 - [30J(J+1) - 25]J_z^2 - 6J(J+1) + 3J^2(J+1)^2 \\ \hat{O}_6^0 &= 231\hat{J}_z^6 - 105[3J(J+1) - 7]J_z^4 \\ &+ [105J^2(J+1)^2 - 525J(J+1) + 294]J_z^2 \\ &- 5J^3(J+1)^3 + 40J^2(J+1)^2 - 60J(J+1) \\ \hat{O}_6^6 &= \tfrac{1}{2}(\hat{J}_+^6 + \hat{J}_-^6) \end{aligned}$$

The operators $\hat{J}_+$ and $\hat{J}_-$ are defined in appendix 7.

Construction of the matrix representing $\hat{H}_{ligand}$ in the (L, S, J) multiplet is facilitated if we choose as basis the $(2J + 1)$ kets $\{|J, M_J\rangle\}$, eigenvectors shared by $(\hat{\mathbf{J}}^2, \hat{J}_z)$, where z is the symmetry axis of the complex. Indeed:

The $\hat{J}_z^n$ operators produce *diagonal* elements:

$$\langle J, M_J | \hat{J}_z^n | J, M_J \rangle = (M_J)^n$$

and the operator $\hat{O}_6^6$ produces off-diagonal elements with the form:

$$\begin{aligned} \langle J, M_J | \hat{O}_6^6 | J, M_J - 6\rangle &= \langle J, -M_J | \hat{O}_6^6 | J, -M_J + 6\rangle \\ \langle J, M_J | \hat{O}_6^6 | J, M_J + 6\rangle &= \langle J, -M_J | \hat{O}_6^6 | J, -M_J - 6\rangle \end{aligned} \qquad [8.5]$$

8.2.2 – Effects of interaction with ligands on the ground multiplet

These effects are similar to those produced by the zero-field splitting term $\hat{H}_{ZFS}$ on the lowest energy level for transition ion complexes in the low-field situation (sections 6.6 and 6.7). Here too, it is important to distinguish between two cases, depending on whether J is integer or half-integer.

◇ *Case of cations with half-integer J values*

▹ First, consider the simple case of a complex of the Ce^{3+} ion with C_{3h} symmetry for which the ground multiplet is $^2F_{5/2}$ ($L = 3$, $S = 1/2$, $J = 5/2$) (table 8.1). For this cation, the matrix representing the $\hat{H}_{ligand}$ operator in the basis $\{|5/2, M_J\rangle\}$ is *diagonal*. Indeed, since no $(|5/2, M_J\rangle, |5/2, M_J \pm 6\rangle)$ pair exists in this basis, the off-diagonal elements of equations [8.5] are not involved. The eigenvalues are therefore the diagonal elements obtained by replacing $\hat{J}_z$ by M_J in the $\hat{O}_2^0$, $\hat{O}_4^0$ and $\hat{O}_6^0$ operators. As $\hat{J}_z$ is present with *even powers* in these operators, the diagonal elements corresponding to opposite values of M_J are *equal*. There therefore exist three *doubly-degenerate* energy levels, the values of which are determined by the parameters included in the first three terms of $\hat{H}_{ligand}$, and for each level there is a corresponding *Kramers doublet* $\{|5/2, M_J\rangle, |5/2, -M_J\rangle\}$ (figure 8.1).

▹ The same method can be applied to a Yb^{3+} complex with C_{3h} symmetry, selecting the eight kets $\{|7/2, M_J\rangle\}$, eigenvectors shared by $(\hat{\mathbf{J}}^2, \hat{J}_z)$, as the basis for the ground multiplet $^2F_{7/2}$ ($L = 3, S = 1/2, J = 7/2$). The operator $\hat{O}_6^6$ now gives off-diagonal elements which verify relations [8.5]. The matrix representing $\hat{H}_{ligand}$ has the following form (the non-indicated elements are null):

	$\lvert 7/2, 7/2\rangle$	$\lvert 7/2, -5/2\rangle$	$\lvert 7/2, -7/2\rangle$	$\lvert 7/2, 5/2\rangle$	$\lvert 7/2, 3/2\rangle$	$\lvert 7/2, -3/2\rangle$	$\lvert 7/2, 1/2\rangle$	$\lvert 7/2, -1/2\rangle$
$\langle 7/2, 7/2\rvert$	A	a						
$\langle 7/2, -5/2\rvert$	a	B						
$\langle 7/2, -7/2\rvert$			A	a				
$\langle 7/2, 5/2\rvert$			a	B				
$\langle 7/2, 3/2\rvert$					C			
$\langle 7/2, -3/2\rvert$						C		
$\langle 7/2, 1/2\rvert$							D	
$\langle 7/2, -1/2\rvert$								D

Examination of the different shaded blocks shows that four energy levels exist:

- an energy level C associated with the doublet $\{\lvert 7/2, 3/2\rangle, \lvert 7/2, -3/2\rangle\}$
- an energy level D associated with the $\{\lvert 7/2, 1/2\rangle, \lvert 7/2, -1/2\rangle\}$ doublet
- two energy levels associated with doublets which are admixtures with the following form:

$$\{\cos\gamma \lvert 7/2, 7/2\rangle + \sin\gamma \lvert 7/2, -5/2\rangle, \cos\gamma \lvert 7/2, -7/2\rangle + \sin\gamma \lvert 7/2, 5/2\rangle\}$$

$$\{\sin\gamma \lvert 7/2, 7/2\rangle - \cos\gamma \lvert 7/2, -5/2\rangle, \sin\gamma \lvert 7/2, -7/2\rangle - \cos\gamma \lvert 7/2, 5/2\rangle\}$$

where the angle γ $(0 \leq \gamma \leq \pi/2)$ is determined by the matrix elements denoted A, B, a.

▷ In general, when J is half-integer, interaction with the ligands produces $(J + 1/2)$ doubly-degenerate energy levels. Each level corresponds to one (or two in cubic symmetry) Kramers doublets made up of two kets $\{\lvert u\rangle, \lvert\bar{u}\rangle\}$ conjugated by time-reversal [Abragam and Bleaney, 1986], which have the form:

$$\lvert u\rangle = \sum_{M_J} c(M_J)\lvert J, M_J\rangle \ ; \ \lvert\bar{u}\rangle = \sum_{M_J} (-1)^{J-M_J} c^*(M_J)\lvert J, -M_J\rangle$$

When the complex has C_{3h} or D_{3h} symmetry, the admixtures are due to the $\hat{O}_6^6$ term in $\hat{H}_{ligand}$ (equation [8.5]) and the doublets are given by the following expressions:

$$\{\alpha_1 \lvert J, 15/2\rangle + \beta_1\lvert J, 3/2\rangle + \gamma_1 \lvert J, -9/2\rangle, \alpha_1' \lvert J, -15/2\rangle + \beta_1' \lvert J, -3/2\rangle + \gamma_1' \lvert J, 9/2\rangle\} \quad [8.6]$$

$$\{\alpha_2 \lvert J, 13/2\rangle + \beta_2\lvert J, 1/2\rangle + \gamma_2\lvert J, -11/2\rangle, \alpha_2' \lvert J, -13/2\rangle + \beta_2' \lvert J, -1/2\rangle + \gamma_2' \lvert J, 11/2\rangle\} \quad [8.7]$$

$$\{\cos\gamma \lvert J, 7/2\rangle + \sin\gamma \lvert J, -5/2\rangle, \cos\gamma \lvert J, -7/2\rangle + \sin\gamma \lvert J, 5/2\rangle\} \quad [8.8]$$

In writing these expressions, we took into account the fact that the maximal value of J is $15/2$ (table 8.1).

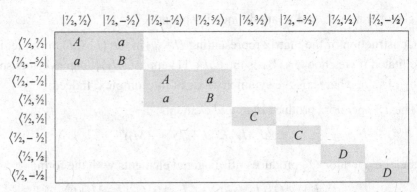

Note – For cations with a $4f^7$ configuration, for which the ground term is not orbitally degenerate, interaction with ligands modifies the energy of the ground multiplet but causes no splitting to the first order.

◇ *Case of cations with integer J values*

▹ We will first consider a complex of the Pr^{3+} ion with C_{3h} symmetry, for which the ground multiplet is $^3H_4$ ($L = 5, S = 1, J = 4$) (table 8.1). According to equation [8.5], the matrix representing the $\hat{H}_{ligand}$ operator in the basis made of the nine eigenvectors $\{|4, M_J\rangle\}$ shared by $(\hat{\mathbf{J}}^2, \hat{J}_z)$ has the following form:

	$\|4,0\rangle$	$\|4,1\rangle$	$\|4,-1\rangle$	$\|4,3\rangle$	$\|4,-3\rangle$	$\|4,4\rangle$	$\|4,-2\rangle$	$\|4,-4\rangle$	$\|4,2\rangle$
$\langle 4,0\|$	A								
$\langle 4,1\|$		B							
$\langle 4,-1\|$			B						
$\langle 4,3\|$				C	c				
$\langle 4,-3\|$				c	C				
$\langle 4,4\|$						D	d		
$\langle 4,-2\|$						d	E		
$\langle 4,-4\|$								D	d
$\langle 4,2\|$								d	E

- There exist three non-degenerate energy levels associated with "singlets": an energy level A associated with the ket $|4,0\rangle$ and two other levels determined by C and c, associated with admixtures of $|4,3\rangle$ and $|4,-3\rangle$.
- Three degenerate levels associated with Kramers doublets are found. Energy level B is associated with the $(|4,1\rangle, |4,-1\rangle)$ doublet. The two others are determined by D, E, d and are associated with kets with the form:

$$\{\sin\gamma\,|4,2\rangle + \cos\gamma\,|4,-4\rangle, \sin\gamma\,|4,-2\rangle + \cos\gamma\,|4,4\rangle\} \quad [8.9]$$

$$\{\cos\gamma\,|4,2\rangle - \sin\gamma\,|4,-4\rangle, \cos\gamma\,|4,-2\rangle - \sin\gamma\,|4,4\rangle\} \quad [8.10]$$

▹ In general, for complexes with C_{3h} or D_{3h} symmetry, taking the fact that the maximal value of J is 8 into account (table 8.1), the following conclusions can be drawn:

- The singlets are admixtures of the kets $\{|J,3\rangle, |J,-3\rangle\}$ or $\{|J,-6\rangle, |J,0\rangle, |J,6\rangle\}$.
- The Kramers doublets take the form

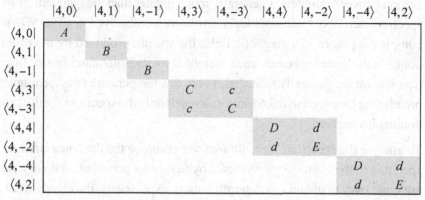

$$\{\alpha_1 |J,8\rangle + \beta_1 |J,2\rangle + \gamma_1 |J,-4\rangle, \alpha_1' |J,-8\rangle + \beta_1' |J,-2\rangle + \gamma_1' |J,4\rangle\}$$
$$\{\alpha_2 |J,7\rangle + \beta_2 |J,1\rangle + \gamma_2 |J,-5\rangle, \alpha_2' |J,-7\rangle + \beta_2' |J,-1\rangle + \gamma_2' |J,5\rangle\} \quad [8.11]$$

8.3 – The EPR spectrum for complexes of rare earth ions with half-integer *J* values

8.3.1 – Introduction

We saw that the electrostatic interaction of the $4f$ electrons with the ligands causes the ground multiplet to split into (J + ½) Kramers doublets. As the separation between doublets is generally much greater than the quantum $h\nu$ of the spectrometer, only "intradoublet" transitions are observed in EPR. When the complex is placed in a magnetic field, the spectra produced by the different doublets are added together, each weighted by its Boltzmann factor. In practice, spectra are generally recorded at very low temperature to avoid relaxation broadening (see section 8.3.5), and as a result only the spectrum for the ground doublet is observed.

To analyse the spectrum for a doublet, we compare the data measured on the spectrum to the expressions predicted by first order perturbation theory, which are quite easy to obtain. To determine these expressions, the eigenvalues and eigenvectors for the matrix representing the operator

$$\hat{H}_P = \hat{H}_{Zeeman} + \hat{H}_{hyperfine} \quad [8.12]$$

must be found, where $\hat{H}_{Zeeman}$ and $\hat{H}_{hyperfine}$ for each doublet are given by equations [8.1] and [8.2]. In the frequent situation where the nuclear spin is greater than ½ (table 8.2), a term due to interaction between the quadrupole moment of the nucleus and the asymmetric electric field created by the f electrons is also involved. For simplicity, we will ignore this term even though it sometimes produces visible effects on the spectrum [Guillot-Noël *et al.*, 2006]. In these conditions, the problem is identical to that in section 6.6, where we showed that $\hat{H}_{Zeeman}$ and $\hat{H}_{hyperfine}$ could be replaced by equivalent operators constructed from "effective" parameters and the components of an angular momentum of spin ½. By applying this method, we can use the results obtained for the position (chapter 4) and intensity (chapter 5) of the resonance lines for centres of spin ½. As always, the symmetry properties of the complex are manifested in the effective parameters:

▹ In cubic symmetry, we have $g_x^{\text{eff}} = g_y^{\text{eff}} = g_z^{\text{eff}}$ and $A_x^{\text{eff}} = A_y^{\text{eff}} = A_z^{\text{eff}}$.

▹ When the complex has a symmetry axis z of order greater than or equal to 3 (axial symmetry), we have $g_z^{\text{eff}} = g_{//}^{\text{eff}}$ and $g_x^{\text{eff}} = g_y^{\text{eff}} = g_\perp^{\text{eff}}$. The same can be said for the effective A values. The intensity factor and position for the resonance lines are determined by equations [6.34] to [6.36]:

$$(g')^2 = (g_\perp^{\text{eff}})^2 \sin^2\theta + (g_{//}^{\text{eff}})^2 \cos^2\theta$$

$$(A')^2 = (A_\perp^{\text{eff}})^2 (g_\perp^{\text{eff}}/g')^2 \sin^2\theta + (A_{//}^{\text{eff}})^2 (g_{//}^{\text{eff}}/g')^2 \cos^2\theta$$

$$g_P = (g_\perp^{\text{eff}})^2 \,[1 + (g_{//}^{\text{eff}}/g')^2]/8g'$$

where θ is the angle between the symmetry axis z and the field $\mathbf{B}$. When a transition is allowed ($g_\perp^{\text{eff}} \neq 0$), it produces an EPR spectrum similar to that for a centre of spin ½ with axial symmetry, in which the notable features are determined by the effective parameters.

▹ When the symmetry is lower, three distinct effective parameters exist. When the sample is a single crystal, these parameters can be determined by studying how the position of the line shifts when the direction of **B** is varied in three perpendicular planes.

8.3.2 – Expression for the effective parameters

This expression depends on the nature of the Kramers doublet. Consider, for example, the doublets described in section 8.2.2 (C_{3h} or D_{3h}-type axial symmetry).

▹ For doublets with the form $\{|J, M_J\rangle, |J, -M_J\rangle\}$, the expressions from section 6.6 can be used directly:

- Doublet $\{|J, ½\rangle, |J, -½\rangle\}$ (equations [6.33] and [6.37]):

$$g_{//}^{\text{eff}} = g_{Landé}\,; \quad g_\perp^{\text{eff}} = (J + ½)g_{Landé} \qquad [8.13]$$

$$A_{//}^{\text{eff}} = A_J\,; \quad A_\perp^{\text{eff}} = (J + ½)A_J \qquad [8.14]$$

- Doublets $\{|J, M_J\rangle, |J, -M_J\rangle\}$ with $M_J \neq \pm½$ (equation [6.40]):

$$g_{//}^{\text{eff}} = 2|M_J|g_{Landé}\,; \quad g_\perp^{\text{eff}} = 0 \qquad [8.15]$$

$$A_{//}^{\text{eff}} = 2|M_J|A_J\,; \quad A_\perp^{\text{eff}} = 0 \qquad [8.16]$$

▹ We will now examine the doublets described by equations [8.6] to [8.8]. Transition between the kets of the doublet described by equation [8.6] is not allowed. In contrast, it is allowed for the doublet described by [8.8], and the effective parameters are given by (exercise 8.1):

$$g_{//}^{\text{eff}} = g_{Landé}\,(1 + 6\cos 2\gamma)\,;\quad g_{\perp}^{\text{eff}} = g_{Landé}\,[J(J+1) - {}^{35}\!/\!_{4}]^{1/2}\sin 2\gamma \qquad [8.17]$$

$$A_{//}^{\text{eff}} = A_J(1 + 6\cos 2\gamma)\,;\quad A_{\perp}^{\text{eff}} = A_J\,[J(J+1) - {}^{35}\!/\!_{4}]^{1/2}\sin 2\gamma \qquad [8.18]$$

The variations of the $g_{//}^{\text{eff}}/g_{Landé}$ and $g_{\perp}^{\text{eff}}/g_{Landé}$ ratios as a function of γ are shown in figure 8.2 for $J = 9/2$ and $J = 15/2$. They can be observed to vary over a broad range, and to be very sensitive to the value of γ, which itself is determined by the $\hat{H}_{ligand}$ parameters (equation [8.4]).

The transition is allowed for the doublet described by equation [8.7], but the expression for the effective parameters is complex and will not be used in the examples presented here.

For all these doublets, the effective parameters given by the first order approximation of perturbation theory satisfy the relation (equations [6.33] and [6.37]):

$$A_{\perp}^{\text{eff}}/A_{//}^{\text{eff}} = g_{\perp}^{\text{eff}}/g_{//}^{\text{eff}} \qquad [8.19]$$

This relation has a very visible impact on the spectrum. Indeed, it implies that the $A_{\perp}^{\text{eff}}/g_{\perp}^{\text{eff}}$ and $A_{//}^{\text{eff}}/g_{//}^{\text{eff}}$ ratios are equal. As a result, the splitting of the hyperfine lines is independent of the direction of **B** relative to the complex. Their common value is the $A_J/g_{Landé}$ ratio which is characteristic of the cation (exercise 8.2). These properties remain valid in rhombic symmetry.

8.3.3 – The case of cations in an *S* state

The magnetic properties of cations with a $4f^7$ configuration, such as Eu^{2+}, Gd^{3+} and Tb^{4+}, which have a ground term ^8S ($L = 0$), are unusual. Indeed, $g_{Landé} = g_e$ and the constant A_J characterising the hyperfine interaction with the nucleus is reduced to the A_s term (see note at the end of section 8.1.2). In addition, in these complexes, the interaction of the seven unpaired electrons with the ligands causes no splitting to the first order (section 8.2.2). This situation is similar to that of transition ions with a ^6S ground term, such as Fe^{3+} and Mn^{2+} (section 3.3 of appendix 2). The EPR spectrum produced by an isotope of these cations should thus be described by the spin Hamiltonian:

$$\hat{H}_J = g_e\,\beta\mathbf{B}\cdot\mathbf{J} + A_s\,\mathbf{J}\cdot\mathbf{I}$$

where $J = 7/2$. A simple pattern of $2I + 1$ lines separated by $A_s/g_e\beta$, centred at $h\nu/g_e\beta$, is expected regardless of the molecules' orientation relative to the magnetic field. In fact, interactions with ligands and spin-orbit coupling produce higher order effects, and the spin Hamiltonian becomes:

$$\hat{H}_J = g\beta\mathbf{B}\cdot\mathbf{J} + A\,\mathbf{J}\cdot\mathbf{I} + D\,[\hat{J}_z^2 - J(J+1)/3] + E\,(\hat{J}_x^2 - \hat{J}_y^2) \quad [8.20]$$

The g and A parameters are slightly different from g_e and A_s, and the D and E terms cause zero-field splitting and anisotropic effects. 4th and 6th degree polynomial operators in $\hat{J}_x$, $\hat{J}_y$, $\hat{J}_z$ also exist; they are generally weaker. But in cubic symmetry, where the D and E parameters are null, these operators will determine the zero-field splitting.

8.3.4 – Application: analysing the data obtained for ethyl sulfates

Due to the strong anisotropy of the effective parameters and the high values of the hyperfine constants, powder spectra for complexes of rare earth ions spread over a very broad field range and are often difficult to observe. Therefore single crystals are generally studied. Classical experiments were performed on the homologous series of ethyl sulfates $Ln(C_2H_5SO_4^-)_3$, $9H_2O$ where Ln represents a trivalent cation of a rare earth element, and on cations substituted into crystals such as lanthanum chloride $LaCl_3$. In this type of study, the paramagnetic cations are "magnetically diluted" so as to reduce the line broadening due to unresolved splitting caused by intercentre interactions (chapter 7).

The data obtained with crystals of lanthanide ethyl sulfates, in which the cations are coordinated with nine H_2O molecules and the site symmetry is C_{3h}, provide a good illustration of the diversity of spectra produced by rare earth complexes. Most of these spectra were recorded at very low temperatures ($T < 20$ K) to avoid line broadening due to spin-lattice relaxation (section 8.3.3). The original references to these studies can be found in [Abragam and Bleaney, 1986]. Below, we will compare these data to the expressions predicted by first order perturbation theory (section 8.3.2). When a hyperfine interaction exists, relation [8.19] can be used to test the validity of this approximation.

◇ Ce^{3+} ($4f^1$, $J = 5/2$, $g_{Landé} = 6/7$) and Sm^{3+} ($4f^5$, $J = 5/2$, $g_{Landé} = 2/7$)

The nucleus of the cerium atom is not paramagnetic. We saw in section 8.2.2 that in C_{3h} symmetry the three Kramers doublets take the form $\{|5/2, M_J\rangle, |5/2, -M_J\rangle\}$. Equations [8.13] and [8.15] give the following effective parameters (figure 8.1):

$$\{|5/2, 1/2\rangle, |5/2, -1/2\rangle\}: \quad g_{//}^{\text{eff}} = 6/7 \approx 0.86; \qquad g_{\perp}^{\text{eff}} = 18/7 \approx 2.57$$

$$\{|5/2, 3/2\rangle, |5/2, -3/2\rangle\}: \quad g_{//}^{\text{eff}} = 18/7 \approx 2.57; \qquad g_{\perp}^{\text{eff}} = 0$$

$$\{|5/2, 5/2\rangle, |5/2, -5/2\rangle\}: \quad g_{//}^{\text{eff}} = 30/7 \approx 4.29; \qquad g_{\perp}^{\text{eff}} = 0$$

At 4.2 K, the EPR spectrum produced by Ce^{3+} in an ethyl sulfate crystal exhibits two components characterised by $g_{//} = 0.955$, $g_{\perp} = 2.185$ and $g_{//} = 3.72$, $g_{\perp} = 0.20$.

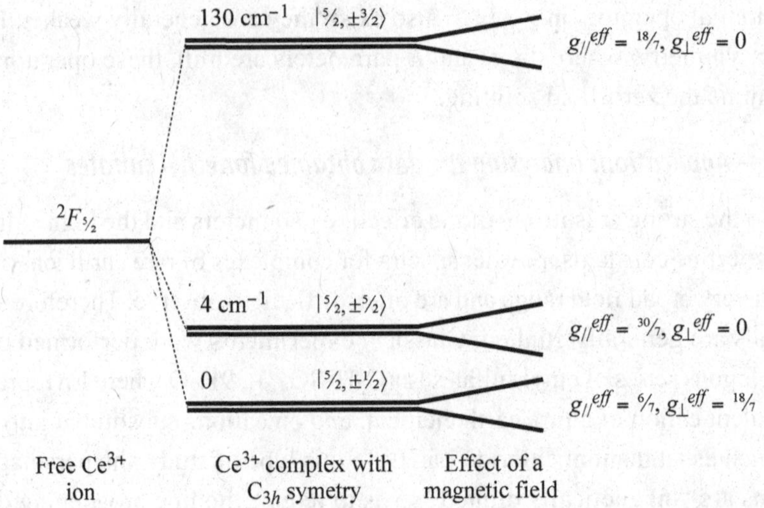

Figure 8.1 – Energy levels and spin states of the Ce^{3+} ion in lanthanum ethyl sulfate. The energies determined by UV-visible absorption spectroscopy are not to scale. The effective g values predicted by first order perturbation theory are indicated. The energy of the first excited multiplet $^2F_{7/2}$ is equal to 2240 cm^{-1}.

Their temperature-dependence shows that the first component arises from the ground doublet and that the second is due to an excited doublet at around 3 cm^{-1}. Comparison with the values predicted above suggests that the ground doublet is $\{|5/2, 1/2\rangle, |5/2, -1/2\rangle\}$ and that the first excited doublet is $\{|5/2, 5/2\rangle, |5/2, -5/2\rangle\}$ (figure 8.1). The deviations from the calculated values are due to weak admixture with the kets from other doublets, an effect which has been attributed to a lowering of the complex's symmetry due to vibrations in the crystal.

The Sm^{3+} ion is also characterised by $J = 5/2$ and its nucleus has two paramagnetic isotopes (table 8.2). Equations [8.13] to [8.16] predict:

$\{|5/2, 1/2\rangle, |5/2, -1/2\rangle\}$: $g_{//}^{\text{eff}} = 2/7 \approx 0.286$; $g_{\perp}^{\text{eff}} = 6/7 \approx 0.857$; $A_{//}^{\text{eff}} = A_J$; $A_{\perp}^{\text{eff}} = 3A_J$

$\{|5/2, 3/2\rangle, |5/2, -3/2\rangle\}$: $g_{//}^{\text{eff}} = 6/7 \approx 0.857$; $g_{\perp}^{\text{eff}} = 0$; $A_{//}^{\text{eff}} = 3A_J$; $A_{\perp}^{\text{eff}} = 0$

$\{|5/2, 5/2\rangle, |5/2, -5/2\rangle\}$: $g_{//}^{\text{eff}} = 10/7 \approx 1.43$; $g_{\perp}^{\text{eff}} = 0$; $A_{//}^{\text{eff}} = 5A_J$; $A_{\perp}^{\text{eff}} = 0$

At 4.2 K, the EPR spectrum given by a crystal of Sm^{3+} ethyl sulfate has a single component characterised by $g_{//} = 0.596$, $g_{\perp} = 0.604$ and $A_{//} = 180$ MHz,

$A_\perp = 752$ MHz for ^{147}Sm, and $A_{//} = 147$ MHz, $A_\perp = 615$ MHz for ^{149}Sm. The ratio of hyperfine constants is close to that of the g_N factors indicated in table 8.2 (exercise 8.3). From these data, we calculate:

$$A_\perp/A_{//} = 4.18; \quad g_\perp/g_{//} = 1.01$$

Relation [8.19] is far from being satisfied, and the experimental $g_{//}$ and $g_\perp$ values are very different from those predicted for the $\{|5/2, 1/2\rangle, |5/2, -1/2\rangle\}$ doublet. First order perturbation theory is therefore not applicable in this case. Admixing with the kets $\{|7/2, 1/2\rangle, |7/2, -1/2\rangle\}$ of the first excited multiplet ($J = 7/2$), which is brought by the second order correction, accounts for the experimental values.

◇ *Nd^{3+} ($4f^3$, $J = 9/2$, $g_{Landé} = 8/11$)*

For this value of J, the Kramers doublets which produce allowed transitions are defined by equations [8.7] and [8.8].

- ▹ The doublet defined by [8.7] in this case is reduced to $\{|9/2, 1/2\rangle, |9/2, -1/2\rangle\}$, and equations [8.13] and [8.14] produce:

$$g_{//}^{\text{eff}} = 8/11 \approx 0.727; \quad g_\perp^{\text{eff}} = 40/11 \approx 3.64$$
$$A_{//}^{\text{eff}} = A_J; \quad A_\perp^{\text{eff}} = 5A_J$$

- ▹ For doublets with the form [8.8], equations [8.17] and [8.18] become

$$g_{//}^{\text{eff}} = 8/11\,(1 + 6\cos 2\gamma); \quad g_\perp^{\text{eff}} = 32/11\,\sin 2\gamma \qquad [8.21]$$
$$A_{//}^{\text{eff}} = A_J\,(1 + 6\cos 2\gamma); \quad A_\perp^{\text{eff}} = A_J\,(4\sin 2\gamma)$$

The spectrum produced by a crystal of Nd^{3+} ethyl sulfate recorded at 20 K has a single component characterised by $g_{//} = 3.535$, $g_\perp = 2.072$ and hyperfine constants $A_{//} = 1140$ MHz, $A_\perp = 596$ MHz for the ^{143}Nd isotope and $A_{//} = 709$ MHz, $A_\perp = 371$ MHz for ^{145}Nd. Within experimental errors, the ratio of hyperfine constants is close to that of the g_N factors (exercice 8.3). In addition, we have:

$$A_\perp/A_{//} = 0.523; \quad g_\perp/g_{//} = 0.586$$

Relation [8.19] is quite well satisfied, suggesting that the first order approximation is good. The spectrum observed cannot be attributed to the $\{|9/2, 1/2\rangle, |9/2, -1/2\rangle\}$ doublet, but equation [8.21] gives $g_{//}^{\text{eff}} = 3.65$ and $g_\perp^{\text{eff}} = 2.16$ for $\gamma = 24$ °, which is in good agreement with the experimental data (figure 8.2).

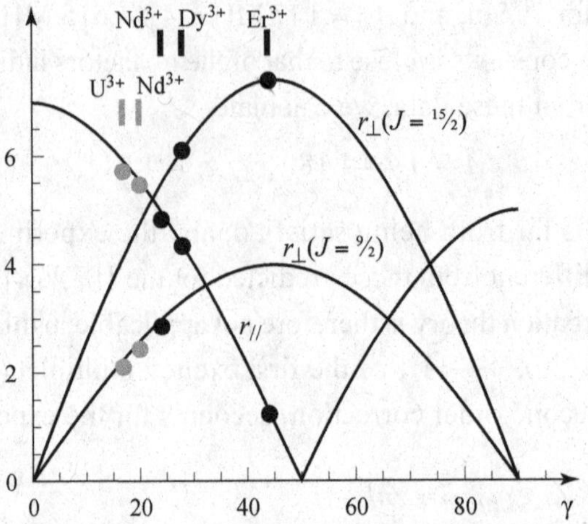

Figure 8.2 – Variation of the ratios $r_{//} = g_{//}^{eff}/g_{Landé}$ and $r_{\perp} = g_{\perp}^{eff}/g_{Landé}$ as a function of γ, as predicted by equations [8.17]. The black and grey circles represent the data obtained for ethyl sulfates and $LaCl_3$, respectively.

◇ *Er^{3+} ($4f^{11}$, $J = 15/2$, $g_{Landé} = 6/5$)*

For this value of *J*, the effective parameters of the Kramers doublets defined by [8.8] are given by (equations [8.17] and [8.18]):

$$g_{//}^{eff} = 6/5\,(1 + 6\cos 2\gamma);\quad g_{\perp}^{eff} = 6/5\sqrt{55}\,\sin 2\gamma \qquad [8.22]$$

$$A_{//}^{eff} = A_J(1 + 6\cos 2\gamma);\quad A_{\perp}^{eff} = A_J(\sqrt{55}\,\sin 2\gamma)$$

On the EPR spectrum recorded at 4.2 K, one can measure $g_{//} = 1.47$, $g_{\perp} = 8.85$ and $A_{//} = 156$ MHz, $A_{\perp} = 941$ MHz for the ^{167}Er isotope. From these values, we deduce

$$A_{\perp}/A_{//} = 6.04;\quad g_{\perp}/g_{//} = 6.02$$

The first order approximation is therefore excellent. Indeed, equation [8.22] gives $g_{//}^{eff} = 1.45$ and $g_{\perp}^{eff} = 8.89$ for γ = 44 ° (figure 8.2).

◇ *Dy^{3+} ($4f^{9}$, $J = 15/2$, $g_{Landé} = 4/3$)*

The ground doublet produces no signal, but a spectrum characterised by $g_{//} = 5.86$ and $g_{\perp} = 8.4$ can be attributed to an excited doublet with the form [8.8]. Indeed, equation [8.17] which is written here as

$$g_{//}^{eff} = 4/3\,(1 + 6\cos 2\gamma);\quad g_{\perp}^{eff} = 4/3\sqrt{55}\,\sin 2\gamma$$

predicts $g_{//}^{eff} = 5.80$ and $g_{\perp}^{eff} = 8.2$ for $\gamma = 28°$ (figure 8.2). For this complex, data obtained by EPR, UV-visible and infrared spectroscopies could be combined to determine all the ligand field parameters present in equation [8.4] [Powell and Orbach, 1961].

◇ *Gd^{3+} ($4f^7$, $J = 7/2$, $g_{Landé} = 2$)*

At 90 K, the Gd^{3+} ions incorporated into a lanthanum ethyl sulfate crystal produce a spectrum with a fine structure composed of seven lines centred at $g = 1.990$ [Bleaney *et al.*, 1954]. As a first approximation, this structure can be described with the help of the Hamiltonian [8.20] with $D = 0.02$ cm^{-1} and $E = 0$. A detailed study shows that the 4th degree polynomial terms in $\hat{J}_x$, $\hat{J}_y$, $\hat{J}_z$ are two orders of magnitude smaller than D.

The different examples presented above show that the expressions for the g^{eff} and A^{eff} parameters given by first order perturbation theory are a good starting point when attempting to interpret spectra given by rare earth ion complexes. Significant differences relative to these expressions appear when the separation between the doublets is small and/or the excited multiplets are relatively close. In this case, the spectrum can only be interpreted if complementary information supplied by other spectroscopic techniques is available.

8.3.5 – Spin-lattice relaxation for complexes with half-integer spin

We recall that in the solid state, in particular in crystals, the most efficient relaxation processes are due to coupling between the paramagnetic centres and the vibrational modes: the vibrations modulate the geometry of the complex composed of the cation and its ligands, which produces a time-dependent interaction between the unpaired electrons and the ligands, inducing transitions between the spin states (section 5.4.1).

- ▹ In transition ion complexes, interaction between the ligands and the orbital angular momenta of the electrons is strong, but it only affects the spin states through the intermediary of spin-orbit coupling in the very frequent situation where the orbital momentum is quenched.
- ▹ In rare earth complexes, interaction with the ligands is weak, but it directly affects the $\{|J, M_J\rangle\}$ states of the multiplets. In addition, we saw that the excited doublets are close (figure 8.1), which promotes the existence of very efficient Orbach-type relaxation processes (section 5.4.1). As a result, spin-lattice relaxation is generally fast in rare earth complexes. Due to the

combined effects of relaxation broadening and populating the excited doublets, the spectrum disappears at cryogenic temperatures. A large volume of data was obtained in the 1960s on the temperature-dependence of the spin-lattice relaxation time T_1 in crystals of rare earth elements. From these data, models describing the relaxation processes in solids were developed [Manenkov and Orbach, 1966]. The direct process dominates at very low temperatures, then 2-phonon processes take over. Raman processes are often masked by the Orbach process, of which the T_1 temperature-dependence can be used to determine the energy of the first excited level. The values obtained in this way are in good agreement with those deduced from UV-visible absorption and infrared spectroscopy experiments [Orbach and Stapleton, 1972].

For cations with a $4f^7$ configuration such as Gd^{3+} and Eu^{2+}, with a ground term characterised by $L = 0$, the coupling with the lattice is weak, the relaxation is slower and the EPR spectrum can be observed up to room temperature.

8.4 – The EPR spectrum for complexes of rare earth ions with integer J values

The same difficulties are encountered as in the case of transition ion complexes with integer spin in the low-field situation (section 6.7). Once again, the intensity of the signal increases when the $\mathbf{b}_1(t)$ component of the electromagnetic field is parallel to $\mathbf{B}$. We will simply indicate a few general characteristics and illustrate them with some examples taken from [Baker and Bleaney, 1958].

Consider once again the case of complexes with C_{3h} or D_{3h} symmetry. We saw that the interaction between the $4f$ electrons and the ligands creates degenerate levels (doublets) and non-degenerate levels (singlets) (section 8.2.2). When the complex is placed in a magnetic field, transitions can take place within doublets or between singlets.

8.4.1 – Intradoublet transitions

The doublets are defined by equations [8.11]. As the matrices representing the $\hat{J}_x$ and $\hat{J}_y$ operators within these doublets are null, intradoublet transitions are *forbidden*. In fact, crystal faults often cause distortions which locally reduce the symmetry, resulting in some admixing of the kets and creating a weak zero-field splitting. In this case the transitions are allowed, but the position and intensity of the lines are determined by the distribution of the distortions within the crystal, which produces a broad and often asymmetric spectrum.

An example of this type of transition is provided by Pr^{3+} in lanthanum ethyl sulfate. For the Pr^{3+} ion in the $4f^2$ configuration, we have $J = 4$ and $g_{Landé} = 4/5$ (table 8.1), and the ^{141}Pr nucleus is characterised by $I = 5/2$ (table 8.2). With a parallel-mode cavity, a broad asymmetric line is observed at X- and Q-bands, from which $g_{//} = 1.52$ and $A_{//} = 2260$ MHz can be deduced. The $A_{//}/g_{//}$ ratio is close to $A_J/g_{Landé}$ (exercise 8.2), which shows that the line is effectively produced by a doublet (section 8.3.2)..As it is not $\{|4,1\rangle, |4,-1\rangle\}$, it must be one of the doublets defined by equations [8.9] and [8.10]. For example, equation [8.10] produces (exercise 8.4):

$$g_{//}^{\text{eff}} = 2g_{Landé}\,(3\cos 2\gamma - 1);\; g_{\perp}^{\text{eff}} = 0$$

This expression gives $g_{//}^{\text{eff}} = 1.48$ for $\gamma = 25°$.

8.4.2 – Transitions between singlets

Transitions are possible between singlets when their separation is slightly less than the quantum for the spectrometer, $h\nu$. The situation is similar to that for transition ion complexes with an integer spin in the low-field situation (section 6.7 and complement 2 to chapter 6), and the same treatment can be used. In axial symmetry, these transitions are allowed when **B** is parallel to z if $\mathbf{B}_1$ is also parallel to z. However, transitions are allowed when $\mathbf{B}_1$ is perpendicular to **B** if the local symmetry of the crystal is lower, or when the centre has a *dipolar electric* moment perpendicular to the axis of symmetry, z, and the electric component $\mathbf{E}_1$ of the radiation is perpendicular to z [Abragam and Bleaney, 1986].

This type of transition is observed for Tb^{3+} ($4f^8$) in lanthanum ethyl sulfate. The ground multiplet is characterised by $J = 6$, $g_{Landé} = 3/2$ and the spin for the ^{159}Tb nucleus is $I = 3/2$. With a parallel-mode cavity, a pattern of narrow symmetric lines is observed at low field, characterised by $g_{//} \approx 18, A_{//} = 6300$ MHz, which has been attributed to a transition between singlets which are admixtures of $\{|J,-6\rangle, |J,0\rangle, |J,6\rangle\}$.

8.5 – Actinide complexes

8.5.1 – Introduction

Elements for which the atomic number is between 90 and 103:

thorium, protactinium, uranium, neptunium, plutonium, americium, curium, berkelium, californium, einsteinium, fermium, mendelevium, nobelium and lawrencium (etymology in complement 1),

occupy the same cell as actinium ($Z = 89$) in the periodic table, and are therefore called the actinides. The configuration of actinium is [...] $6s^2\ 6p^6\ 6d^1\ 7s^2$ and that of actinide cations is [...] $5f^n\ 6s^2\ 6p^6$, where n varies between 1 and 14. Among these 14 elements, only thorium and uranium exist naturally and their abundance in the earth's crust is comparable, for example, to that of molybdenum. Traces of protactinium are present in uranium deposits. The other actinides are produced in nuclear reactors, and they have numerous radioactive isotopes (α radioactivity) for which the half-life varies between a few days and several million years.

The magnetic characteristics of complexes of trivalent cations of actinides are often quite similar to those of lanthanides. However, the $5f$ orbitals are less internal than the $4f$ orbitals as they are more "screened" from the nucleus by the core electrons. As a result actinides can more easily shed electrons, so that actinide cations also exist in a 4+, 5+ and even 6+ state in some complexes (table 8.3).

Table 8.3 – Electronic configuration of the main actinide cations

Z	trivalent ion		Other cations
90	$5f^1$	Th^{3+}	Th^{4+}
91	$5f^2$	Pa^{3+}	Pa^{4+}, Pa^{5+}
92	$5f^3$	U^{3+}	U^{4+}, U^{5+}, U^{6+}
93	$5f^4$	Np^{3+}	Np^{4+}, Np^{5+}, Np^{6+}
94	$5f^5$	Pu^{3+}	Pu^{4+}, Pu^{5+}, Pu^{6+}
95	$5f^6$	Am^{3+}	Am^{4+}, Am^{5+}, Am^{6+}
96	$5f^7$	Cm^{3+}	

Another consequence is that the unpaired electrons interact more strongly with the ligands, and the spin density is sometimes sufficiently delocalised to produce a superhyperfine structure on the EPR spectrum. It even happens that the $5f$ electrons are involved in chemical bonds. The magnetic properties of high valence actinide complexes therefore present some analogies with those of transition ion complexes.

The strong α radioactivity of most actinides makes spectroscopic studies difficult: samples must be manipulated in appropriate glove boxes and they are more or less rapidly destroyed. These problems disappear if it is possible to work on samples strongly enriched in isotopes with a very long half-life, obtained by exploiting specific nuclear reactions.

8.5.2 – Comparison of spectra for complexes of trivalent rare earth and actinide cations

We saw in section 8.3.1 that to the first order approximation, the effective g values for complexes of rare earth cations depend on J, on $g_{Landé}$ and on the relative values of the $\hat{H}_{ligand}$ parameters (section 8.3.1). As the J and $g_{Landé}$ values are the same for the $4f^n$ and $5f^n$ configurations, homologous cations of rare earth elements and actinides are expected to produce similar spectra when they occupy the same crystal site. This is often, but not always, observed.

▹ Consider, for example, the Nd^{3+} ($4f^3$) and U^{3+} ($5f^3$) cations substituted into $LaCl_3$, where they are coordinated by nine chloride ions in a site with C_{3h} symmetry. As the nucleus of the most abundant uranium isotope, ^{238}U (natural abundance 99.7 %, half-life 4.5 × 10^9 years), is non-paramagnetic, experiments were performed with samples enriched in ^{235}U or ^{233}U. Table 8.4 lists the g and A values deduced from the experimental spectra [Abragam and Bleaney, 1986].

Table 8.4 – Comparison of the parameters measured for Nd^{3+} and U^{3+} in $LaCl_3$.

Nd^{3+}			U^{3+}		
	^{143}Nd	^{145}Nd		^{233}U	^{235}U
	$I = 7/2$	$I = 7/2$		$I = 5/2$	$I = 7/2$
$g_{//} = 3.996$	$A_{//} = 1270$ MHz	$A_{//} = 790$ MHz	$g_{//} = 4.153$	$A_{//} = 1130$ MHz	$A_{//} = 528$ MHz
$g_{\perp} = 1.763$	$A_{\perp} = 500$ MHz	$A_{\perp} = 310$ MHz	$g_{\perp} = 1.520$	$A_{\perp} = 369$ MHz	$A_{\perp} = 175$ MHz

The parameters of Nd^{3+} are close to those measured in ethyl sulfate (section 8.3.4). We calculate:

$$A_{\perp}/A_{//} = 0.39; \ g_{\perp}/g_{//} = 0.44$$

These ratios are similar, which indicates that the first order approximation of perturbation theory is correct. Indeed, equation [8.21] reproduces the $g_{//}$ and $g_{\perp}$ values quite well with $\gamma = 19 - 20°$ (figure 8.2).

The g values measured for U^{3+} are close to those for Nd^{3+}. The ratios of $A_{//}$ and $A_{\perp}$ measured for the two isotopes are equal to those of their g_N factors, which are equal to $(g_N)_{233} = +0.236$ and $(g_N)_{235} = -0.109$. From these values, we deduce that:

$$A_{\perp}/A_{//} = 0.33; \; g_{\perp}/g_{//} = 0.37$$

The experimental g values can be approximately reproduced using equation [8.21] with $\gamma = 17 - 18°$ (figure 8.2). It should be noted that there is no reason for the hyperfine constants to be equal for the two cations: in the expression for A_J (equation [8.2]), the various contributions depend on the nature of the cation and of the isotope considered.

▹ We will now examine the case of the Gd^{3+} ($4f^7$) and Cm^{3+} ($5f^7$) cations which have the same ground multiplet, $^8S_{7/2}$. The spectra produced by Gd^{3+} in lanthanum ethyl sulfate and lanthanum trichloride are characterised by $g \approx g_e$ and a fine structure due to weak zero-field splitting terms (section 8.3.4). Curium is radioactive, with a half-life of 160 days. When Cm^{3+} is incorporated into these crystals, an axial spectrum is observed, characterised by $g_{//} = 1.926$ and $g_{\perp} = 7.73$ with no fine structure [Abraham *et al.*, 1962]. These values are very different from g_e, which shows that the spectrum is produced by a Kramers doublet, and the $g_{\perp}/g_{//} = 4.0$ ratio is characteristic of the $\{|7/2, 1/2\rangle, |7/2, -1/2\rangle\}$ doublet (equation [8.13]). In the case of Cm^{3+}, it is therefore the zero-field splitting term that dominates the Hamiltonian [8.20]: the separation of the doublets is greater than the quantum $h\nu$ of the spectrometer, and only the spectrum for the ground doublet is observed. A similar situation is encountered with another cation in an S state, the Fe^{3+} ion, for which the zero-field splitting term is often much greater than the Zeeman term (figures 6.10 and 6.11).

8.5.3 – Complexes of high valence actinides: example of cations with a $5f^1$ configuration

In general, the first order approximation, which is based on the hypothesis that interaction with the ligands is much smaller than the separation of the multiplets, cannot be used to interpret the spectrum for complexes containing high valence actinides. But it is sometimes a good starting point for a more precise calculation. The different possible situations are well illustrated by complexes of cations with a $5f^1$ configuration, such as Pa^{4+}, U^{5+}, Np^{6+}, for which the ground multiplet is $^2F_{5/2}$ with $J = 5/2$ and $g_{Landé} = 6/7$ (table 8.1).

We will first examine the case of the bis (cycloheptatrienyl)-uranium $[U(\eta^7 - C_7H_7)_2]^-$ sandwich compound in which the uranium is formally in the 5+ valence state [Gourier *et al.*, 1998]. At low temperature, a frozen solution produces the X-band spectrum shown in figure 8.3, characterised by $g_{//} = 1.244$ and $g_{\perp} = 2.365$. As the symmetry of the complex is D_{7d} or D_7, the $\hat{H}_{ligand}$ operator (equation [8.3]) only includes the $\hat{O}_2^0$ and $\hat{O}_4^0$ terms which produce diagonal elements in the basis $\{|J, M_J\rangle\}$ (section 8.2.1). This operator is therefore represented by a diagonal matrix in the basis $\{|5/2, M_J\rangle\}$ for the ground multiplet of the U^{5+} ion, like in the case of the Ce^{3+} $(4f^1)$ ion in C_{3h} symmetry (section 8.2.2). As a result, the expressions for the effective g for the three Kramers doublets given by the first order approximation are identical to those calculated for Ce^{3+} in section 8.3.4. The experimental g values do not differ much from the $g_{//}^{eff} = 0.86$, $g_{\perp}^{eff} = 2.57$ values predicted for the $\{|5/2, 1/2\rangle, |5/2, -1/2\rangle\}$ doublet. A weak admixing of the doublets due to a reduction of the symmetry to C_{2h} suffices to reproduce the experimental values.

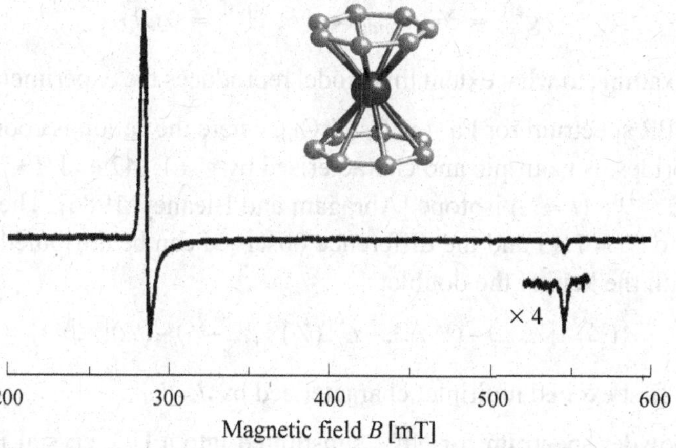

Figure 8.3 – X-band spectrum of the sandwich compound $[U(\eta^7\text{-}C_7H_7)_2]^-$ at 15 K. Microwave frequency: 9.44 GHz, power 4 mW. Modulation frequency: 100 kHz, peak-to-peak amplitude 1 mT. [Reproduced with permission from Gourier D. *et al.*, *Journal of the American Chemical Society*, **120**, 6084–6092 © 1998, American Chemical Society]

Proton ENDOR experiments showed that the spin population of the $2p\pi$ orbitals for the carbon atoms in the heptatrienyl cycles is greater than or equal to 4×10^{-2}, thus demonstrating the covalent nature of the orbital occupied by the unpaired electron in this complex [Gourier *et al.*, 1998].

We will now consider complexes of Pa^{4+}, U^{5+}, Np^{6+} with octahedral symmetry. For this symmetry, the $\hat{H}_{ligand}$ operator is written:

$$\hat{H}_{ligand} = A_4^0 \langle r^4 \rangle a_4(J)\,(\hat{O}_4^0 + 5\hat{O}_4^4) + A_6^0 \langle r^6 \rangle a_6(J)\,(\hat{O}_6^0 - 21\hat{O}_6^4)$$

where $\hat{O}_4^0$ and $\hat{O}_6^0$ are defined in section 8.2.1 and $\hat{O}_4^4$, $\hat{O}_6^4$ can be found in [Abragam and Bleaney, 1986].

When the unpaired electron from a cation with configuration $5f^1$ interacts with the ligands, the $^2F_{5/2}$ multiplet is split and three doublets form. The ground doublet takes the form [Abragam and Bleaney, 1986]:

$$\{1/\sqrt{6}\,(|5/2, 5/2\rangle - \sqrt{5}\,|5/2, -3/2\rangle), 1/\sqrt{6}\,(|5/2, -5/2\rangle - \sqrt{5}\,|5/2, 3/2\rangle)\} \qquad [8.23]$$

To first order in perturbation theory, the energy levels of this ground doublet are described by the spin Hamiltonian for a centre with effective spin $J = 1/2$ (exercise 8.5):

$$\hat{H}_J = g^{eff}\,\beta \mathbf{B}\cdot\mathbf{J} + A^{eff}\,\mathbf{J}\cdot\mathbf{I} \qquad [8.24]$$

$$g^{eff} = 5/3\, g_{Landé} = 10/7; \quad A^{eff} = 5/3\, A_J$$

We will examine to what extent this model reproduces the experimental results.

- The EPR spectrum for Pa^{4+} in Cs_2ZrCl_6, where the cation is coordinated to 6 chlorides, is isotropic and characterised by $g = 1.142$ and $A = 1580$ MHz for the ^{231}Pa ($I = 3/2$) isotope [Abragam and Bleaney, 1986]. The g value is close to $10/7 = 1.43$ and the difference observed can be attributed to admixing with the kets of the doublet

$$\{(3/4)^{1/2}\,|7/2, 5/2\rangle - (1/2)|7/2, -3/2\rangle, (3/4)^{1/2}\,|7/2, -5/2\rangle - (1/2)|7/2, 3/2\rangle\}$$

  of the first excited multiplet characterised by $J = 7/2$.

- The powder spectrum for Np^{6+} substituted into a UF_6 crystal is isotropic and characterised by $g = 0.604$ and $A = 2000$ MHz for the ^{237}Np isotope ($I = 5/2$). The superhyperfine structure due to the six fluoride ligands is clearly visible on the spectrum. The g value observed is very different from 1.43 and it can only be explained by giving up a description based on the ground multiplet and by simultaneously considering the spin-orbit coupling and the interaction with the ligands [Hutchinson and Weinstock, 1960].

- The difference relative to what is predicted by the Hamiltonian [8.24] increases further for U^{5+} in LiF. When the Li^+ ions are replaced by U^{6+}, the large charge difference between the two cations results in significant reorganisation within the crystal. X or γ irradiation of the crystal creates U^{5+} ions which produce several EPR signals corresponding to sites with different symmetries. Among these, an isotropic signal without superhyperfine structure characterised by $g = 0.333$ has been attributed to an octahedral UO_6 centre [Lupei and Lupei, 1979]. Another signal characterised by $g_{//} = 0.2526$, $g_\perp = 0.4716$, a hyperfine structure due to ^{235}U ($I = ½$) (in a strongly enriched sample), and a superhyperfine structure due to an F^- ion ($I = ½$), was attributed to a site with tetragonal symmetry where the sixth ligand is a fluoride ion [Lupei *et al.*, 1976]. In the latter case, analysis of the spectrum showed that the quadrupole interaction between the 5*f* electron and the ^{235}U nucleus is of the same order of magnitude as the hyperfine interaction.

8.6 – Points to consider in applications

8.6.1 – Comparison of the EPR characteristics of transition ion and rare earth ion complexes with a half-integer spin.

◇ *Transition ion complexes*

1. In numerous complexes, it is the interaction between the unpaired electrons and the field **B** which mainly determines the splitting of the ground level. This is the case in complexes of spin ½ (chapter 4) and complexes with spin greater than ½ in the high-field situation (sections 6.4 and 6.5). In these cases, the spectrum is centred at around $g \approx 2$ and its fine or hyperfine structure, more or less anisotropic, is relatively easy to interpret.
2. For complexes with spin greater than ½ in the low-field situation, the zero-field splitting term dominates and there exist 2 or 3 Kramers doublets depending on whether S is equal to $3/2$ or $5/2$. In principle, the features of the spectrum can be used to measure the effective g values which depend on the spin, S, the E/D ratio and the principal values of the $\tilde{\mathbf{g}}$ matrix (section 6.6.2). As these principal values are always close to g_e, S and E/D can be determined by comparing the effective g values to simplified charts calculated with $g_x = g_y = g_z = 2$ (section 6.6.2 and figure 6.12). In practice, the spreading of the powder spectrum is often significant and not all of its features are always visible.

◇ *Rare earth complexes*

In these complexes, the zero-field splitting is caused by the electrostatic interaction between the 4*f* electrons and the ligands, which is generally larger than the interaction with the magnetic field. The situation is therefore similar to case number 2 above, with strong anisotropy of the resonance lines and features which are difficult to observe on a powder spectrum. Interpretation of the spectrum is complicated by the fact that, as J can take the values $5/2$, $7/2$, $9/2$, $15/2$, the number $(J + 1/2)$ of Kramers doublets is high. The effective g values depend on J, on the numerous $\hat{H}_{ligand}$ parameters, and on $g_{Landé}$ which varies between $2/7 \approx 0.3$ and $4/3 = 1.33$ depending on the cation considered (table 8.1). Unlike with transition ion complexes, it is not possible to construct charts and the attribution of the experimental g values to a specific doublet requires additional information deduced by analysing the hyperfine structure and the UV-visible absorption spectrum.

8.6.2 – Interpreting spectra for rare earth complexes

When the EPR lines are narrow, studies performed on single crystals at low temperature provide very detailed information on the principal values and principal axes of the $\tilde{\mathbf{g}}$ and $\tilde{\mathbf{A}}$ matrices defined for an effective spin $1/2$, and on the quadrupole $\tilde{\mathbf{P}}$ matrix which sometimes contributes to the shape of the spectrum [Guillot-Noël *et al.*, 2006]. This type of study revealed the existence of exchange interactions between rare earth ions; the value of the exchange parameter, J, deduced from the EPR spectrum or the fluorescence spectrum can vary from a fraction of a cm^{-1} to several cm^{-1} [Guillot-Noël *et al.*, 2000; Guillot-Noël *et al.*, 2003].

Interpretation of the parameters deduced from the spectrum is rarely simple. To determine the identity of the ground doublet, the principal values of the $\tilde{\mathbf{g}}$ and $\tilde{\mathbf{A}}$ matrices can first be compared to the expressions given by the first order approximation of perturbation theory, taking the known $g_{Landé}$ and A_J values into consideration. Calculations performed on rare earth ion complexes are facilitated by the fact that the matrix elements have been tabulated for the free ions [Abragam and Bleaney, 1986]. An example of calculation of the effective g values based on a crystal field model is presented in [Guillot-Noël *et al.*, 1998].

8.6.3 – Actinide complexes

The nature of the ground state for actinide complexes is generally more difficult to determine than for complexes of rare earth ions. For high valence cations, the notion of multiplet must often be abandoned in favour of a description similar to that of transition ion complexes, and additional information is required to correctly interpret the spectra. Unfortunately, the UV-visible absorption spectra for actinide complexes are much less resolved that those for rare earth complexes. Historically, EPR played a significant role in determining the spin values, I, and the values of the g_N parameter for the different paramagnetic isotopes of actinides. Given their current importance in industrial and military fields, as well as their potential catalytic properties, it is to be expected that EPR spectroscopy applications specific for actinide complexes will be developed.

Complement 1 – Rare earth elements and actinides: etymological considerations

The aim of this complement is not to reduce the mystery surrounding the names of some elements, but to help readers to memorise them. It is inspired from the small book *La découverte de l'atome* [atom discovery], by Alfred Romer (*Petite Bibliothèque Payot*, 1962). The dates for discovery of the oldest known elements are approximate.

Rare earths

Cerium	**Ce**	1804	from the Ceres asteroid discovered in 1801.
Dysprosium	**Dy**	1886	from the Greek *dysprositos*, "difficult to obtain".
Erbium	**Er**	1843	from Ytterby, Swedish village where several rare earth ores were discovered.
Europium	**Eu**	1900	from Europe.
Gadolinium	**Gd**	1886	for Johan Gandolin, Finnish chemist who discovered yttrium.
Holmium	**Ho**	1879	from Stockholm.
Neodymium	**Nd**	1885	from the Greek *neos*, new, and *didymos*, twin.
Praseodymium	**Pr**	1885	from the Greek *praseos*, green, and *didymos*, twin.
Promethium	**Pm**	1947	for Prometheus.
Samarium	**Sm**	1879	from samarskite, a mineral discovered by the Russian engineer W. V. Von Samarski.
Terbium	**Tb**	1843	from Ytterby, Swedish village.
Thulium	**Tm**	1879	from *Thule,* an Ancient Greek place-name associated with Scandinavia or Iceland.
Ytterbium	**Yb**	1905	from Ytterby, a Swedish village.
Lutetium	**Lu**	1907	from Lutetia, the Latin name for Paris. Name attributed by Georges Urbain, who was the first to isolate this element.

Actinides

Americium	**Am**	1944	from America.
Berkelium	**Bk**	1949	from Berkeley, in California.
Californium	**Cf**	1950	from California.
Curium	**Cm**	1914	for Pierre and Marie Curie.
Einsteinium	**Es**	1952	for Albert Einstein.
Fermium	**Fm**	1953	for Enrico Fermi.
Lawrencium	**Lr**	1961	for E. Lawrence, inventor of the cyclotron.
Mendelevium	**Md**	1955	for Dimitri Mendeleev, father of the periodic table of the elements.
Neptunium	**Np**	1940	from the planet Neptune.
Nobelium	**No**	1957	for Alfred Nobel.
Plutonium	**Pu**	1940	from the dwarf planet Pluto.
Protactinium	**Pa**	1917	from the Greek *protos*, first, and *actinium* (disintegrates to form actinium).
Thorium	**Th**	1929	from *Thor*, the god in Scandinavian mythology.
Uranium	**U**	1789	from the planet Uranus, discovered in 1781.

References

ABRAGAM A. & BLEANEY B. (1986) *Electron Paramagnetic Resonance of transition ions*, Dover, New York.

ABRAHAM M, JUDD B. & WICKMAN H. (1962) *Physical Review* **130**: 611-612.

BAKER J.M. (1993) "EPR and ENDOR of the lanthanides" in *Electron Spin Resonance*, volume 13B, Royal Society of Chemistry, Athenaeum Press, Newcastle upon Tyne.

BAKER J.M. & BLEANEY B. (1958) *Proceeding of the Royal Society of London* A **245**: 156-174.

BLEANEY B., SCOVIL H. & TRENAM R. (1954) *Proceeding of the Royal Society* A **223**: 15-29.

GOURIER D. *et al.* (1998) *Journal of the American Chemical Society* **120**; 6084-6092.

GUILLOT-NOËL O. *et al.* (1998) *Journal of Physics: Condensed Matter* **10**: 6491-6503.

GUILLOT-NOËL O. *et al.* (2000) *Physical Review* B **61**: 15338-15346.

GUILLOT-NOËL O. *et al.* (2003) *Chemical Physics Letters* **380**: 563-568.

GUILLOT-NOËL O. *et al.* (2006) *Physical Review* B **74**: 214409-1 - 214409-8.

HUTCHINSON C.A. & WEINSTOCK B. (1960) *Journal of Chemical Physics* **32**: 56-59.

JUDD B.R.(1988) "Atomic Theory and Optical Spectroscopy" in *Handbook on the physics of the Rare Earths*, GSCHNEIDER K.A. & EYRING L. eds, Elsevier, Amsterdam.

LIDE D.R. (2010) *Handbook of Chemistry and Physics*, section 1, CRC Press, Boca Raton.

LUPEI V. & LUPEI A. (1979) *Physica Status Solidi* B **94**: 301- 307.

LUPEI V. *et al.* (1976) *Journal of Physics* C: *Solid State Physics* **9**: 2619-2626.

MANENKOV A.A. & ORBACH R. (1966) *Spin-Lattice Relaxation in Ionic Solids*, Harper & Row, New York.

NEWMANN D.J. & NG B. (1989) *Report on Progress in Physics* **52**: 699- 763.

ORBACH R. & STAPLETON H.J (1972) "Electron Spin-Lattice Relaxation" in *Electron Paramagnetic Resonance*, GESCHWIND S. ed. Plenum Press, New York.

POWELL M.J.D. & ORBACH R. (1961) *Proceeding of the Physical Society* **78**: 753-758.

STEVENS K.W.H. (1952) *Proceeding of the Physical Society* (London) A **65**: 209-215.

Exercises

8.1. Consider the Kramers doublet defined by equation [8.8]. Using the matrix elements from appendix 7, construct the matrices representing the $(\hat{J}_x, \hat{J}_y, \hat{J}_z)$ operators for this doublet. Show that they can be written:

$$\tilde{\mathbf{J}}_\mathbf{x} = [J(J+1) - 35/4]^{1/2} \sin 2\gamma \; \tilde{\mathbf{J}}'_\mathbf{x}$$

$$\tilde{\mathbf{J}}_\mathbf{y} = [J(J+1) - 35/4]^{1/2} \sin 2\gamma \; \tilde{\mathbf{J}}'_\mathbf{y}$$

$$\tilde{\mathbf{J}}_\mathbf{z} = (1 + 6\cos 2\gamma) \; \tilde{\mathbf{J}}'_\mathbf{z}$$

where $(\tilde{\mathbf{J}}'_\mathbf{x}, \tilde{\mathbf{J}}'_\mathbf{y}, \tilde{\mathbf{J}}'_\mathbf{z})$ are the matrices representing the $(\hat{J}'_x, \hat{J}'_y, \hat{J}'_z)$ operators associated with an angular momentum $J' = 1/2$ in the basis $\{|1/2, 1/2\rangle, |1/2, -1/2\rangle\}$ (section 3.2.5). From this, deduce that the matrix representing the $\hat{H}_P$ operator (equation [8.12]) can be replaced by the equivalent operator.

$$\hat{H}'_P = \beta(g_\perp{}^{\text{eff}} B_x \hat{J}'_x + g_\perp{}^{\text{eff}} B_y \hat{J}'_y + g_{//}{}^{\text{eff}} B_z \hat{J}'_z)$$
$$+ (A_\perp{}^{\text{eff}} \hat{J}'_x \hat{I}_x + A_\perp{}^{\text{eff}} \hat{J}'_y \hat{I}_y + A_{//}{}^{\text{eff}} \hat{J}'_z \hat{I}_z)$$

where the effective parameters are given by equations [8.17] and [8.18].

8.2. The relations given in section 8.3.2 can be used to deduce the hyperfine constant A_J from the effective parameters. Using the results obtained with rare earth ethyl sulfates (sections 8.3.4 and 8.4.1), determine A_J for the trivalent cations $^{143}Nd^{3+}$, $^{167}Er^{3+}$ and $^{141}Pr^{3+}$. Compare the results to the values tabulated in the literature: –220 MHz, –125 MHz, +1093 MHz [Baker, 1993].

8.3. Verify that the ratio of the hyperfine constants measured for the isotopes of Sm^{3+} in ethyl sulfate is equal to the ratio of the g_N. Same question for the two isotopes of Nd^{3+}.

8.4. Consider the Kramers doublet defined by equation [8.10]. Construct the matrices representing the $(\hat{J}_x, \hat{J}_y, \hat{J}_z)$ operators for this doublet and deduce the expressions for the effective parameters.

8.5. Same questions as the previous exercise for the doublet defined by equation [8.23].

Answers to exercises

8.1.

$$\tilde{\mathbf{J}}_{\mathbf{x}} = \begin{bmatrix} 0 & \sin\gamma\cos\gamma\left[J(J+1) - {}^{35}\!/_{4}\right]^{1/2} \\ \sin\gamma\cos\gamma\left[J(J+1) - {}^{35}\!/_{4}\right]^{1/2} & 0 \end{bmatrix}$$

$$\tilde{\mathbf{J}}_{\mathbf{y}} = \begin{bmatrix} 0 & -\mathrm{i}\sin\gamma\cos\gamma\left[J(J+1) - {}^{35}\!/_{4}\right]^{1/2} \\ \mathrm{i}\sin\gamma\cos\gamma\left[J(J+1) - {}^{35}\!/_{4}\right]^{1/2} & 0 \end{bmatrix}$$

$$\tilde{\mathbf{J}}_{\mathbf{z}} = \begin{bmatrix} \dfrac{7\cos^2\gamma - 5\sin^2\gamma}{2} & 0 \\ 0 & -\dfrac{7\cos^2\gamma - 5\sin^2\gamma}{2} \end{bmatrix}$$

Comparison with the matrices in section 3.2.5 produces the predicted results.

8.2. From the relations given in section 8.3.2, we can deduce:

$$A_J = A_{//}^{\mathrm{eff}}\,(g_{Land\acute{e}}\,/\,g_{//}^{\mathrm{eff}}) = A_{\perp}^{\mathrm{eff}}\,(g_{Land\acute{e}}\,/\,g_{\perp}^{\mathrm{eff}})$$

- For $^{143}Nd^{3+}$, the calculation based on $(A_{//}^{\mathrm{eff}}, g_{//}^{\mathrm{eff}})$ gives $A_J = 235$ MHz, and that based on $(A_{\perp}^{\mathrm{eff}}, g_{\perp}^{\mathrm{eff}})$ gives $A_J = 209$ MHz. The mean, 222 MHz, is in good agreement with the tabulated value.
- For $^{167}Er^{3+}$, the two calculations produce $A_J = 127$ MHz.
- For $^{141}Pr^{3+}$, we obtain $A_J = 1190$ MHz, which is in reasonable agreement with the tabulated value and confirms the attribution of this line to the doublet.

8.3. For Sm^{3+} we calculate $\dfrac{(A_{//})_{147}}{(A_{//})_{149}} = \dfrac{(A_{\perp})_{147}}{(A_{\perp})_{149}} = 1.22$ which must be compared to $\dfrac{(g_N)_{147}}{(g_N)_{149}} = 1.21$.

For Nd^{3+}: $\dfrac{(A_{//})_{143}}{(A_{//})_{145}} = \dfrac{(A_{\perp})_{143}}{(A_{\perp})_{145}} = 1.61$ which should be compared to $\dfrac{(g_N)_{143}}{(g_N)_{145}} = 1.63$.

8.4. The $\tilde{\mathbf{J}}_{\mathbf{x}}$ and $\tilde{\mathbf{J}}_{\mathbf{y}}$ matrices are null and $\tilde{\mathbf{J}}_{\mathbf{z}}$ is given by:

$$\tilde{\mathbf{J}}_{\mathbf{z}} = \begin{bmatrix} 2\cos^2\gamma - 4\sin^2\gamma & 0 \\ 0 & -2\cos^2\gamma + 4\sin^2\gamma \end{bmatrix}$$

We deduce $g_{\perp}^{\mathrm{eff}} = 0$; $g_{//}^{\mathrm{eff}} = 2g_{Land\acute{e}}\ (3\cos 2\gamma - 1)$ $A_{\perp}^{\mathrm{eff}} = 0$, $A_{//}^{\mathrm{eff}} = 2A_J\,(3\cos 2\gamma - 1)$.

8.5. We obtain:

$$\tilde{\mathbf{J}}_{\mathbf{x}} = \begin{bmatrix} 0 & -5/6 \\ -5/6 & 0 \end{bmatrix}, \quad \tilde{\mathbf{J}}_{\mathbf{y}} = \begin{bmatrix} 0 & 5i/6 \\ -5i/6 & 0 \end{bmatrix}, \quad \tilde{\mathbf{J}}_{\mathbf{z}} = \begin{bmatrix} -5/6 & 0 \\ 0 & 5/6 \end{bmatrix}$$

Comparison to the matrices for $J = ½$ produces the expression [8.24].

Chapter 9

How instrumental parameters affect the shape and intensity of the spectrum. Introduction to simulation methods

9.1 – Introduction

By performing measurements on the EPR spectrum, either directly or using numerical simulations, information can be obtained on the paramagnetic centres contained in a sample. The measurements that can be performed relate to the position and amplitude of the notable features of the spectrum, its total intensity or the spin-lattice relaxation time, T_1. The quality of the spectral information depends on the sample studied (spectrum complexity, concentration) and on factors linked to the researcher's skill:

1. The instrumental parameters used during recording determine phenomena which take place within the sample and in the detection system (figure 9.1). Poorly selected parameters can produce a *deformed* spectrum.
2. To measure the position of the spectral features, the magnetic field at the level of the sample and the microwave frequency must be known. The values indicated by the spectrometer are not always sufficiently precise and calibration should be performed regularly using carefully selected standard samples. The choice of reference samples is also important when performing intensity measurements.
3. With a continuous wave spectrometer, the spin-lattice relaxation time, T_1, can be measured using methods based on saturation and spectral broadening. However, the values obtained are only reliable if these phenomena are appropriately modelled.

P. Bertrand, *Electron Paramagnetic Resonance Spectroscopy*,
https://doi.org/10.1007/978-3-030-39663-3_9

4. When the spectral features are unresolved or when they are difficult to interpret, numerical simulation can be used to determine the value of some parameters. Simulation software are necessarily based on simplified models, and it is best to be aware of their limitations.

In this chapter, we have gathered together information on these different points, limiting ourselves to a simple description of the phenomena and emphasising practical considerations. Point 2 is dealt with in complement 4 and the technical elements related to point 3 can be found in complements 2 and 3. The main simulation software suites are described in complement 5.

9.2 – How field sweep and field modulation affect the shape of the spectrum

In previous chapters, the position and intensity of the resonance lines were calculated assuming that the paramagnetic centres were exposed to a constant magnetic field. However, while the spectrum is being recorded, the total magnetic field in the cavity is the sum of the field created by the electromagnet – which varies linearly with time due to the field *sweep* – and the field created by the *modulation coils* – which varies sinusoidally:

$$B_{total}(t) = B(t) + b_{\mathrm{m}}(t)$$

$$B(t) = B_{min} + vt \quad [9.1]$$

$$b_{\mathrm{m}}(t) = (B_{\mathrm{m}}/2)\cos\omega_{\mathrm{m}}t \quad [9.2]$$

In these expressions, v is the scan rate, B_{m} is the "peak-to-peak" amplitude of field modulation and $\omega_{\mathrm{m}} = 2\pi\nu_{\mathrm{m}}$ is its angular frequency. The signal recorded in these conditions only reproduces the shape of the spectrum if $B_{total}(t)$ varies "sufficiently slowly". We will see what this means below. The *amplitude of the modulation* and the *scan rate* also contribute to the shape of the signal produced by the detection system (figure 9.1).

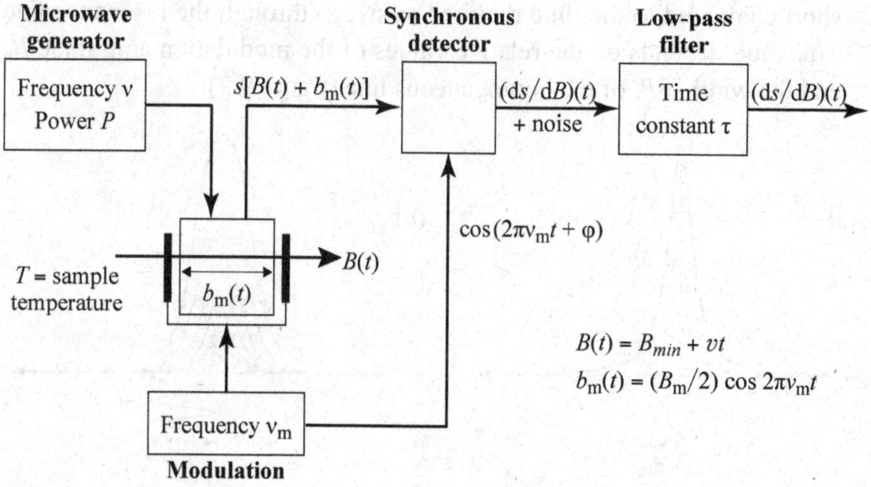

Figure 9.1 – Instrumental parameters for which the effects are discussed in this chapter.

9.2.1 – Effects of modulation at the level of the sample

We will first compare the speed of variation of the field sweep and field modulation.

- ▹ The scan rate is generally less than 1 mT s^{-1}.
- ▹ According to equation [9.2], the instantaneous speed of field modulation is $(B_m\omega_m/2)\sin\omega_m t$, and its maximum value is $B_m\omega_m/2$. For $B_m = 0.1$ mT, this quantity is 3×10^4 mT s^{-1} at the standard frequency $\nu_m = 100$ kHz, and 30 mT s^{-1} at the smallest modulation frequency available on spectrometers, $\nu_m = 100$ Hz.

Thus, the field modulation varies more rapidly.

- ▹ In the non-saturated regime, modulation reduces the lifetime of the states between which transitions takes place, this results in *broadening* of the resonance line δB_m given by:

$$\hbar\omega_m = g'\beta\delta B_m$$

Thus, $\delta B_m = \omega_m/\gamma$ when we set $\gamma = g'\beta/\hbar$. For $\nu_m = 100$ kHz and $g' = 2$, $\delta B_m = 3.6 \times 10^{-3}$ mT. This type of broadening is only observable on very narrow lines, and it can easily be eliminated by reducing ν_m.

- ▹ Complex phenomena occur in the saturated regime. In practice, we can consider expression [5.24], which gives the absorption signal in the saturated regime, to be valid as long as the spin-lattice relaxation time T_1 is short compared to the time required to sweep through the resonance line. This time depends on the relative values of the modulation amplitude B_m and the width, δB, of the homogeneous line (figure 9.2):

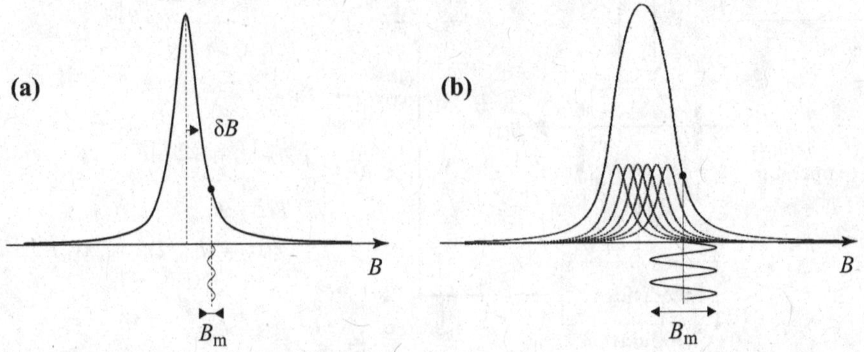

Figure 9.2 – Modulation of lines: (**a**) homogeneous, (**b**) inhomogeneous.

- For $B_m \leq \delta B$ (figure 9.2a), the characteristic time is $1/\omega_m$ and the condition given above is written:

$$\omega_m T_1 < 1 \qquad [9.3]$$

- For $B_m > \delta B$, which generally occurs when the line is *inhomogeneous* (figure 9.2b), the sweep time for a homogeneous line is shortened by a factor $B_m/\delta B$, and the condition becomes

$$(B_m/\delta B)\,\omega_m T_1 < 1 \qquad [9.4]$$

By noting that $\delta B = 1/\gamma T_2$ (equation [5.11]), this relation can be written:

$$\omega_m B_m < 1/\gamma T_1 T_2$$

At the frequency $\nu_m = 100$ kHz, inequality [9.3] is satisfied provided $T_1 < 1.6 \times 10^{-6}$ s. At very low temperatures, where T_1 is long (section 5.4), it is necessary to use a smaller modulation frequency to record the spectrum in the saturated regime, e.g. when a progressive saturation experiment is performed to measure T_1 (section 9.4).

9.2.2 – Effects of magnetic field modulation and sweep at the level of the detection chain

◇ *Overmodulation*

We will start by describing the principle of synchronous detection, also known as "phase-sensitive detection" (figure 9.1). If $s(B)$ is the absorption signal, the input signal at the detector when modulation occurs is $s[B + b_m(t)]$, where $b_m(t)$ is given by equation [9.2]. If the amplitude, B_m, is small, we can write:

$$s[B + b_m(t)] = s(B) + b_m(t)\, s'(B) + \ldots + [(b_m(t))^n/n!]\, s^{(n)}(B) + \ldots$$

Thus:

$$s(B, t) = s(B) + (B_m/2)\, s'(B) \cos\omega_m t + \ldots + [(B_m/2)^n/n!]\, s^{(n)}(B)\, (\cos\omega_m t)^n + \ldots$$

The quantities $(\cos\omega_m t)^n$ can be linearised using the following identity:

$$(\cos\omega_m t)^{2p+1} = (\tfrac{1}{2})^{2p}\left[\cos(2p+1)\omega_m t + C^1_{2p+1}\cos(2p-1)\omega_m t + \ldots + C^p_{2p+1}\cos\omega_m t\right] \quad [9.5]$$

and a similar identity for $(\cos\omega_m t)^{2p}$ [Zwillinger, 2003]. We then obtain an expression with the form:

$$s(B, t) = a_0(B) + a_1(B)\cos\omega_m t + a_2(B)\cos 2\omega_m t + \ldots$$

Synchronous detection consists in multiplying $s(B,t)$ by a signal *with the same frequency* (figure 9.1), which produces:

$$s(B, t)\cos(\omega_m t + \varphi) = \tfrac{1}{2}\, a_1(B)\cos\varphi + A_1(B)\cos(\omega_m t + \varphi_1) + A_2(B)\cos(2\omega_m t + \varphi_2) + \ldots$$

A "low-pass" filter is used to conserve only the first term, which varies slowly over time due to the field sweep and can be optimised by adjusting the phase φ. The expression for $a_1(B)$ is deduced from the identity [9.5]:

$$a_1(B) = (B_m/2)\, s'(B) + C^1_3/(3!\,2^2)\,(B_m/2)^3 s^{(3)}(B) + C^2_5/(5!\,2^4)\,(B_m/2)^5 s^{(5)}(B) + \ldots \quad [9.6]$$

The successive numerical coefficients for this development decrease rapidly. As a result, the signal delivered by the detector is proportional to $B_m s'(B)$ if the amplitude of the modulation is small. Otherwise, the shape of $a_1(B)$ differs from $s'(B)$, depending on the derivatives of $s(B)$, and consequently on the *shape* of the line. *Overmodulation* is then said to occur.

In practice, the following points must be retained:

1. Overmodulation creates line *broadening*. To demonstrate this effect, in figure 9.3 we compared the first term (continuous lines) and the sum of the first and second terms (dashes) from development [9.6] for $B_m = \delta B_{pp}$, where δB_{pp} is the peak-to-peak width of $s'(B)$, when $s(B)$ is a Lorentzian or a Gaussian. Broadening of around 20 % for the Lorentzian and 15 % for the Gaussian is observed.

2. At the start of overmodulation, the amplitude of the line increases with B_m up to a maximum, which is reached when [Poole, 1967]:

$$B_m \approx 3\,\delta B_{pp} \text{ for a Lorentzian line}$$

$$B_m \approx 2\,\delta B_{pp} \text{ for a Gaussian line}$$

3. Overmodulation deforms the resonance lines but has no effect on their intensity (exercise 9.1).

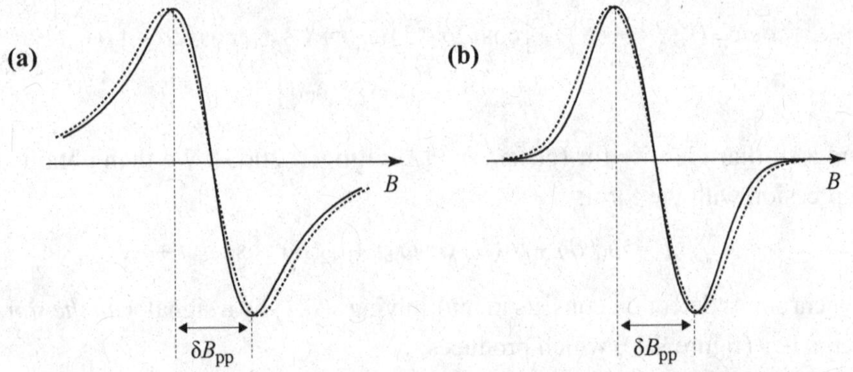

Figure 9.3 – Broadening of a line upon detection when the amplitude modulation B_m is equal to the line's peak-to-peak width, δB_{pp}. (**a**) Lorentzian line, (**b**) Gaussian line. The line amplitudes are normalised.

◇ *Scan rate and filtering*

In the absence of overmodulation, the synchronous detector delivers a signal proportional to $B_m s'(B)$, where B varies linearly with time due to the field sweep (equation [9.1]). An undesirable random signal of various origins, known as "noise", is often superimposed on this signal. Noise can be reduced by using a low-pass filter characterised by an adjustable time constant τ (figure 9.1). This filter reduces the components with a frequency exceeding the "cutoff frequency" $\nu_C = 1/(2\pi\tau)$, and has no effect on the components with a frequency less than ν_C. To eliminate as much noise as possible *without deforming* the signal $s'(B)$, the maximum frequency of the signal components must be known. This maximum frequency can be estimated as follows: if v is the scan rate, the duration of a spectral feature of width δB is $\delta t \approx \delta B/v$. In the frequency domain, such a feature is represented by components with a maximal frequency of around $1/\delta t$. The cutoff frequency ν_C must be greater than $1/\delta t$, i.e., $\tau < 0.16\,\delta t$, to avoid deforming the feature. In practice, the narrowest feature on the spectrum is considered, and τ is selected such that $\tau < 0.1\ \delta t$ [Poole, 1967]. This condition can be satisfied either by adjusting τ, or the scan rate (exercise 9.2).

When this condition is not satisfied, the spectrum is deformed. For example, a symmetric peak becomes asymmetric and its maximum is shifted in the direction of the sweep.

When recording the spectrum, the operator does not directly set the scan rate v, rather he sets the sweep width ΔB, the number of sampling points N, and the "conversion time" t_C – the time interval between two points. The scan rate in this case is equal to:

$$v = \Delta B/(N\, t_C)$$

9.3 – How the power and frequency of the radiation affect the spectrum. The temperature parameter

9.3.1 – Effect of the power and frequency of the radiation

◇ *Microwave power*

Readers are reminded that in the *unsaturated* regime, the absorption signal $s(B)$ and its derivative $s'(B)$ are proportional to the amplitude B_1 of the magnetic component of the radiation, and consequently to $\sqrt{P}$ (equation [5.13]). In the *saturated* regime, at first the absorption signal increases more slowly than $\sqrt{P}$, then it passes through a maximum and tends towards zero at very high powers (figure 5.8a). However, this behaviour is influenced by a number of factors: the homogeneous or inhomogeneous nature of the resonance lines, the line's shape and width, and, of course, the value of the spin-lattice relaxation time T_1 which is itself very sensitive to temperature. This behaviour is different for the absorption signal and its derivative (see section 9.4).

◇ *Frequency*

The intensity of a resonance line is proportional to the microwave frequency (section 5.2.1). But spectrometers with a higher frequency have a smaller resonance cavity volume, and as a result the sample volume also generally decreases. There may therefore be no net gain in signal-to-noise ratio. In fact, the major advantage of high-frequency spectrometers is that they allow the application of *high magnetic fields* to the paramagnetic centres. Indeed, the maximum value for the magnetic field B_{max} available with a spectrometer of frequency ν is such that the $\beta B_{max}/h\nu$ ratio is close to one. How changing the frequency affects the shape of the spectrum is closely linked to the nature of the paramagnetic centres:

▹ For a centre of spin ½ in which the unpaired electron only interacts with the magnetic field, the spin Hamiltonian is reduced to the Zeeman term and the positions of the notable features on the powder spectrum are given by $B_i = h\nu/g_i\beta$, where $g_i (i = x, y, z)$ are the principal values for the $\tilde{\mathbf{g}}$ matrix. The features on spectra recorded at different frequencies are therefore related by simple *homothetic transformations* (figure 9.4), but the *width* of the resonance lines varies differently depending on the broadening mechanism (complement 3 to chapter 5):

- When caused by "g-strain", the linewidth is proportional to the frequency. In this case, the homothetic transformation is applied to the whole spectrum (figure 9.4a).

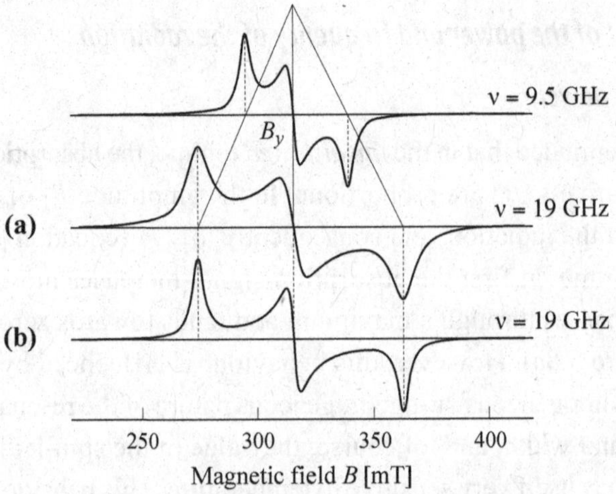

Figure 9.4 – How doubling the frequency affects the shape of the spectrum when the linewidth is determined by (**a**) the *g-strain* mechanism, (**b**) unresolved hyperfine interactions. The spectra are calculated with $g = (2.00; 2.15; 2.30)$. For the needs of the figure, the spectra shown in (**a**) and (**b**) were shifted towards weak fields with a value equal to B_y.

- When the lines are *homogeneous,* or *inhomogeneous* with a broadening mainly due to unresolved hyperfine patterns, the linewidth is not affected by the frequency. In this case, the spectrum is "better resolved" at high frequency (figure 9.4b). This is the case, for example, for the spectrum for semiquinone Q_A from photosystem II (figure 9.5); whereas the structures corresponding to the canonical directions of the field are unresolved at X-band, they are clearly visible at 285 GHz.

▷ If the unpaired electron interacts with nuclei and/or if the spin S is greater than ½, the spin Hamiltonian includes hyperfine terms (chapters 2 and 4) or zero-field splitting terms (chapter 6). Increasing the magnetic field selectively amplifies the Zeeman term, altering the shape of the spectrum. The same effect is produced when two paramagnetic centres are coupled by exchange and dipolar interactions (chapter 7). The frequency dependence of the shape of the spectrum can be studied by plotting spectra recorded with different frequencies on a common "g-value scale", obtained by replacing B by the dimensionless quantity $h\nu/\beta B$ (example shown in figure 7.11).

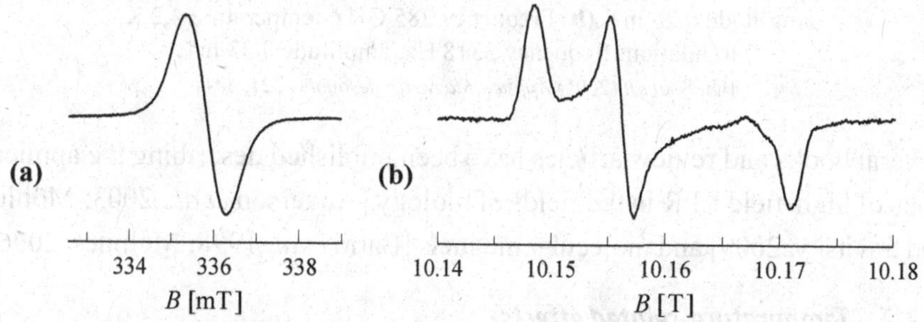

Figure 9.5 – Comparison of the spectra for semiquinone Q_A from photosystem II recorded at two frequencies: (**a**) X-band: microwave frequency 9.4204 GHz, temperature 100 K, power 0.25 mW. Modulation: frequency 100 kHz, amplitude 0.25 mT. (**b**) Frequency 285 GHz, temperature 4.2 K. Modulation: frequency 3318 Hz, amplitude 2 mT. [Un S. *et al.* (2001) Applied Magnetic Resonance, **21**, 341]

Comparison of spectra recorded at different frequencies makes their interpretation much easier. For example, figure 9.6 shows the spectra for the radical form of tyrosine D from photosystem II recorded at X-band and at 285 GHz. Although the hyperfine structure of the spectrum is clearly visible at X-band, its interpretation is much easier at 285 GHz where the patterns corresponding to the canonical directions of the field are clearly separated.

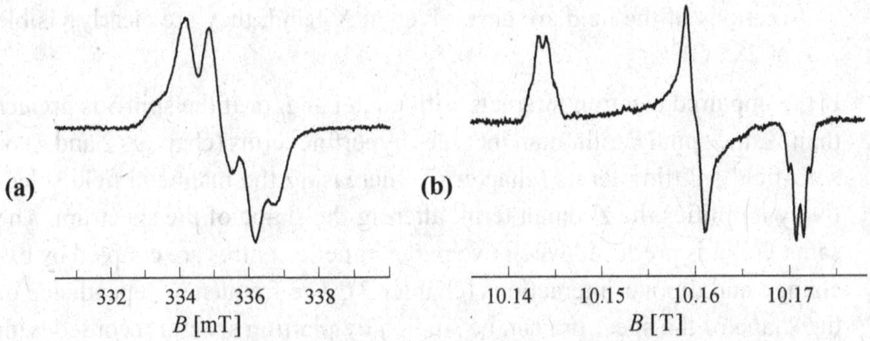

Figure 9.6 – Comparison of the spectra for the tyrosyl D radical from photosystem II recorded at two frequencies: (**a**) X-band: microwave frequency 9.4060 GHz, temperature 100 K, power 0.25 mW. Modulation: frequency 100 kHz, amplitude 0.25 mT. (**b**) Frequency 285 GHz, temperature 4.2 K. Modulation: frequency 3318 Hz, amplitude 0.33 mT.
[Un S. *et al.* (2001) *Applied Magnetic Resonance*, **21**, 341]

Several books and review articles have been published describing the applications of high-field EPR in the fields of biology [Anderson *et al.*, 2003; Möbius and Savitsky, 2009] and molecular magnets [Barra *et al.*,1998; McInnes, 2006].

9.3.2 – Temperature-related effects

The temperature of the paramagnetic centres can be varied over a broad range when the samples are in a solid state, but the temperature range where a spectrum can effectively be observed is determined by the temperature-dependence of the spectrum intensity and by spin-lattice relaxation processes.

◇ Temperature-dependence of the intensity

- ▹ In some cases, the amplitude and total intensity of the spectrum are proportional to $\tanh(h\nu/2k_BT)$ which can be assimilated to $(h\nu/2k_BT)$ with an error of less than 2 % when $k_BT \geq 2h\nu$. The variation inversely proportional to T corresponds to Curie's law. It applies to centres of spin ½ which have no low-lying excited states, but also to centres of spin greater than ½ in the high-field situation within the "high temperature" limit (section 6.4.2 and complement 1 to chapter 6).

- ▹ In other cases, increasing the temperature causes the intensity to decrease faster than predicted by Curie's law. This phenomenon is due to depopulation of the energy levels involved in the EPR transitions in favour of excited states, and is very evident when the product IT is plotted as a function of T. It occurs for centres with a half-integer spin ($S > ½$) in the low-field situation, when the temperature is high enough for several Kramers doublets to be populated (section 6.6.1). It is also observed with biradicals, polynuclear transition ion complexes (section 7.5.1) and rare earth complexes (section 8.3.4), which often have low-lying excited states.

The temperature-dependence of the intensity must, naturally, be taken into account when using the intensity of the spectrum to determine the number of centres contained in a sample (see exercise 6.6).

◇ *Spin-lattice relaxation*

We have already underlined the importance of relaxation-related phenomena in practical EPR spectroscopy (section 5.4.3). Here, we will simply recall that spin-lattice relaxation processes are not very effective at very low temperatures. This leads to *signal saturation* when the microwave power is too great. As the temperature increases, relaxation speeds up and in some cases results in *broadening* of the resonance lines. This broadening does not alter the total intensity of the spectrum, but it reduces its amplitude, and may make it undetectable. Saturation and relaxation broadening of the spectrum can be quantitatively studied to determine the spin-lattice relaxation time, T_1 (section 9.4 and complement 3).

When the paramagnetic centres have low-lying excited states, relaxation processes are particularly efficient and the spectrum sometimes disappears at a relatively low temperature due to the combined effects of the reduced intensity and relaxation broadening.

9.3.3 – Case study: seeking the origin of line splitting in an EPR spectrum

To illustrate the importance of the frequency, power and temperature parameters and the advantage of measuring spectral intensity, we will show how they can be used to determine the origin of the splitting visible on a spectrum. Assume that a frozen solution containing a known number of molecules N produces one of the two spectra shown in figure 9.7 at very low temperature. These spectra present features that we wish to interpret.

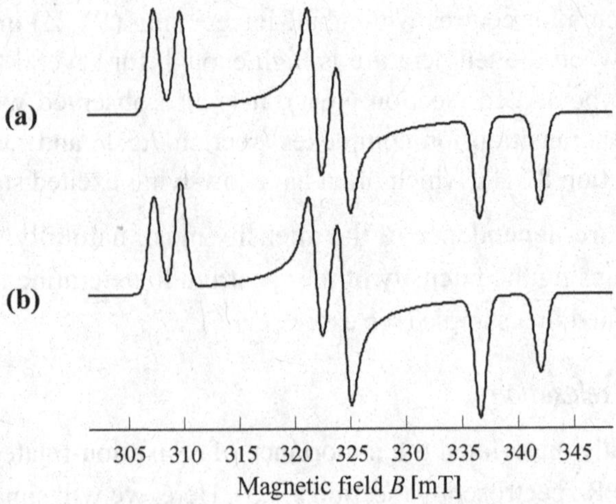

Figure 9.7 – EPR spectra showing line splitting of different origins. Microwave frequency: 9.5 GHz.

▹ Calculation of the numbers $h\nu/\beta B$ representing the positions of the barycentres for the split lines gives 2.00, 2.10, 2.20. As these values are close to $g_e = 2.0023$, we can deduce that the paramagnetic centres have a spin $S = ½$.

▹ The first step consists in ensuring that the features visible in figure 9.7 are not the result of superposition of two spectra produced by two paramagnetic centres with slightly different $\tilde{\mathbf{g}}$ matrices. Measuring the total intensity of the spectrum provides the answer: if the intensity corresponds to a single paramagnetic centre per molecule, this hypothesis can be rejected. In addition, if the spectrum were due to the presence of two types of centres, their spin-lattice relaxation times would probably be different. Some features of the spectrum would therefore saturate more readily than others at low temperature, while others would broaden earlier at high temperatures. The most general method consists in comparing spectra recorded at two different frequencies: the differences between the features due to the Zeeman terms are proportional to the frequency, whereas the splitting produced by other terms in the Hamiltonian are frequency-independent (section 9.3.1).

▹ We assume that line splitting has been established. In this case, the numbers (2.00, 2.10, 2.20) represent the principal values of the $\tilde{\mathbf{g}}$ matrix, and the splittings correspond to the canonical directions of **B**. If the splitting is due to hyperfine interaction with a nucleus of spin $I = ½$, the intensity of the spectrum corresponds to a single centre per molecule. The two lines in a hyperfine pattern have the same intensity (section 5.2.4) and the splitting is not temperature-dependent.

▷ Splitting can also be caused by weak interactions between two paramagnetic centres. As these produce splitting on the spectra for *the two centres* (section 7.3), there are two possibilities:

- The spectrum visible in the figure is that of a single centre. Its intensity then corresponds to one centre per molecule and the spectrum for the other centre must be observed outside the field range shown in the figure.
- The principal values of the $\tilde{\mathbf{g}}_A$ and $\tilde{\mathbf{g}}_B$ matrices are identical. In this case, the intensity of the spectrum corresponds to two centres per molecule.

In both cases, increasing the temperature will result in shorter spin-lattice relaxation times, causing "collapse" of the split lines and the appearance of a spectrum from which the effects of the interactions have completely disappeared (section 7.5.2).

▷ In fact, the *shape of the spectrum* is not the same depending on whether it is the result of hyperfine or intercentre interactions: the lines in a hyperfine pattern are always of equal intensity (section 5.2.4), whereas lines produced by intercentre interactions are only equally intense when the interaction is very weak. For example, the lines in figure 7.5 are of different intensities. The spectra shown in figure 9.7 illustrate these differences:

- The spectrum in figure 9.7a was calculated with ($g_x = 2.00, g_y = 2.10, g_z = 2.20$) and ($A_x = 150$ MHz, $A_y = A_z = 75$ MHz).
- That in figure 9.7b is the result of magnetic dipolar interactions between two centres A and B of spin ½, for which the principal values of the $\tilde{\mathbf{g}}$ matrices are equal ($g_x = 2.00, g_y = 2.10, g_z = 2.20$) and the principal axes are parallel (section 7.3.1). The intercentre axis is directed along the principal axis x and the intercentre distance $r = 10$ Å is such that $\frac{3}{2}(\mu_0/4\pi)g^2\beta^2/r^3 = 75$ MHz.

9.4 – Simulating spectrum saturation

The spin-lattice relaxation time T_1 can be measured by simulating a saturation curve (figure 5.8). This "progressive saturation" method requires quantitative analysis based on a realistic model of the resonance lines and of the spectrum, and we will describe the main elements of this approach.

9.4.1 – Simulating saturation of a homogeneous Lorentzian line

▹ Remember that a resonance line is *homogeneous* when its width is determined by the lifetime of the states between which the transition takes place. This lifetime is the spin-spin relaxation time T_2 (complement 3 to chapter 5). The shape of a homogeneous line centred at B_0 is described by the normalised Lorentzian function (figure 9.8a)

$$f(B - B_0) = \frac{\frac{\delta B}{\pi}}{(B - B_0)^2 + \delta B^2} \qquad [9.7]$$

Its half width at half maximum, δB, is linked to T_2 by equation [5.11]. The absorption signal produced by the spectrometer is given by equation [5.24]:

$$s(B - B_0) = g_P \Delta n_e \frac{B_1 f(B - B_0)}{1 + \pi \gamma' B_1^2 T_1 f(B - B_0)} \qquad [9.8]$$

g_P is the intensity factor for the line (section 5.2.1), Δn_e is the population difference at thermal equilibrium and we have set $\gamma' = g_P \beta / \hbar$. We have omitted the $(\pi\beta/2\hbar)$ factor which is not involved here. By replacing $f(B - B_0)$ by its expression, we obtain:

$$s(B - B_0) = (g_P \Delta n_e B_1 \sqrt{\sigma}) \frac{\frac{\delta B}{\pi\sqrt{\sigma}}}{(B - B_0)^2 + \frac{\delta B^2}{\sigma}} \qquad [9.9]$$

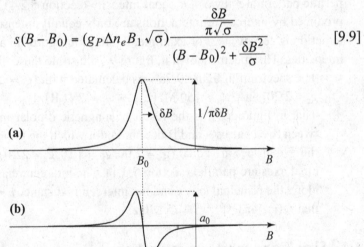

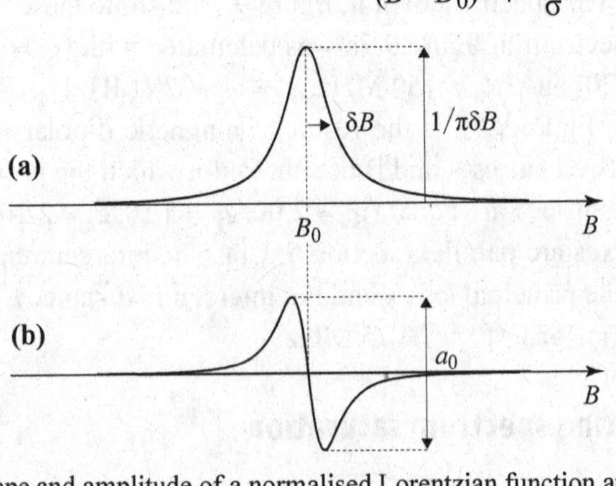

Figure 9.8 – Shape and amplitude of a normalised Lorentzian function and its derivative.

The quantity

$$\sigma = \frac{1}{1 + \frac{\gamma' B_1^2 T_1}{\delta B}}$$

is the “saturation factor” which is 1 in the absence of saturation and tends towards zero in the limit of strong saturation. Comparison of equations [9.9] and [9.7] shows that $s(B - B_0)$ is a Lorentzian characterised by the following parameters:

- half width at half maximum: $\delta B/\sqrt{\sigma}$ (saturation *broadens* the homogeneous line)
- intensity: $g_P\, \Delta n_e\, B_1 \sqrt{\sigma}$
- amplitude at the centre of the line $s(0) = g_P\, \Delta n_e\, B_1\, \sigma/(\pi\delta B)$

In figure 9.9a we show how the normalised amplitude

$$\frac{s(0)}{g_P \Delta n_e\, B_1\, f(0)} = \sigma = \frac{1}{1 + \dfrac{\gamma' B_1^{\,2} T_1}{\delta B}} \qquad [9.10]$$

where $f(0) = 1/(\pi\delta B)$, varies as a function of B_1. It reduces 2-fold when $B_1 = (\delta B/\gamma' T_1)^{1/2}$, when it is said to be at “half-saturation”.

▹ We will now examine how *the derivative* of the absorption signal saturates. The peak-to-peak amplitude of the derivative of a normalised Lorentzian is $a_0 = 9/(4\pi\sqrt{3}\,\delta B^2)$ (figure 9.8b). According to equation [9.9], the peak-to-peak amplitude a_{pp} of the derivative of the absorption signal is obtained by multiplying a_0 by $(g_P\, \Delta n_e\, B_1 \sqrt{\sigma})$ and by replacing δB by $\delta B/\sqrt{\sigma}$, which produces the normalised amplitude:

$$\frac{a_{pp}}{g_P \Delta n_e\, B_1\, a_0} = \sigma^{3/2} = \frac{1}{\left(1 + \dfrac{\gamma' B_1^{\,2} T_1}{\delta B}\right)^{3/2}} \qquad [9.11]$$

This normalised amplitude is represented as a function of B_1 in figure 9.9b. The amplitude of the derivative of the signal is seen to be easier to saturate than the amplitude at the centre of the absorption signal. Half-saturation occurs when $B_1 = (2^{2/3} - 1)^{1/2}\, (\delta B/\gamma' T_1)^{1/2}$ (exercise 9.3).

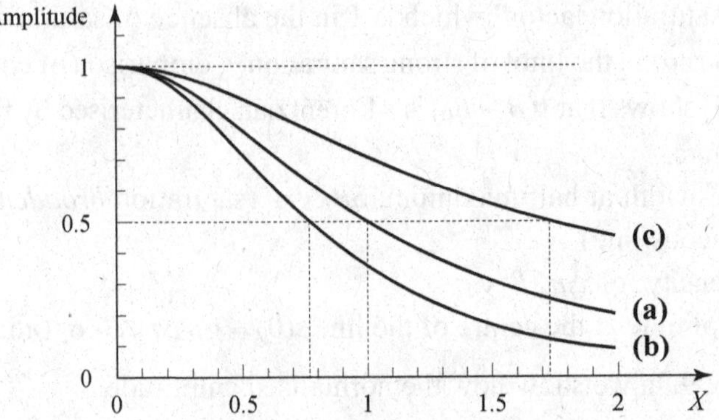

Figure 9.9 – Saturation behaviour of the amplitude of different lines. (**a**) Homogeneous Lorentzian line. (**b**) Derivative of a homogeneous Lorentzian line. (**c**) Completely inhomogeneous line composed of Lorentzian packets. The abscissa is the dimensionless quantity $X = (\gamma' T_1/\delta B)^{1/2} B_1$, the ordinate is normalised to 1 in the absence of saturation.

Homogeneous lines, which are usually very narrow (complement 3 to chapter 5), only form in very specific conditions. A classical example is that of free radicals in the isotropic regime (section 5.4.1). We can use equations [9.10] and [9.11] to simulate the saturation of these lines and determine T_1. The parameters involved are known: as the intensity factor is $g_P = g_{iso}/4$ (section 5.2.2), γ' equals $g_{iso}\beta/4\hbar$ and the width δB can be directly measured on the spectrum. To determine T_1, the amplitude of the magnetic component of the radiation, B_1, must also be known. This amplitude is proportional to $\sqrt{P}$, where P is the microwave power, but the proportionality coefficient depends on the geometry of the cavity. At the centre of a standard rectangular cavity which operates in the TE_{102} mode, we basically have [Poole, 1967]:

$$B_1\,[\text{mT}] \approx 0.2\,\sqrt{P}\,[\text{watt}]$$

The precise value of the proportionality coefficient depends on what is placed in the cavity (e.g. EPR tube or flat cell).

9.4.2 – Simulating saturation of an inhomogeneous line

In general, the molecules oriented in the same way relative to the magnetic field **B** produce *an inhomogeneous line* due to the superposition of a set of homogeneous lines. The source of this line "dispersion" was discussed in complement 3 to chapter 5. The homogeneous lines making up the inhomogeneous line are often quite "closely spaced" so that their position, B_0, varies almost continuously around a value B_r. We therefore define their *density* $h(B_0 - B_r)$: the quantity $h(B_0 - B_r)\mathrm{d}B_0$ represents the fraction of lines with a position between B_0 and $B_0 + \mathrm{d}B_0$, and the absorption signal can be written:

$$S(B - B_r) = \int_0^{+\infty} h(B_0 - B_r)\, s(B - B_0)\,\mathrm{d}B_0 \qquad [9.12]$$

where $s(B - B_0)$ is given by equation [9.9]. All the lines making up an inhomogeneous line can be considered to be characterised by the same g_P and T_1 values. The second side of this equation thus represents the *convolution product* (complement 1) of $h(B_0 - B_r)$ known as the "envelope" of the inhomogeneous line, and of $s(B - B_0)$ which is called a "spin packet" in this context.

Saturation of the inhomogeneous line described by equation [9.12] depends on the shape of the envelope and on the characteristics of the spin packets, i.e., the width δB, the intensity factor g_P, and T_1. In the frequent situation where the inhomogeneous line is considerably broader than the homogeneous lines, the shape of the envelope can be approximated based on the shape of the experimental line. The intensity factor g_P is calculated using equation [5.15] and is very close to 0.5 as long as the anisotropy of the $\tilde{\mathbf{g}}$ matrix is moderate (table 1 in chapter 5). We can then determine the unknown parameters δB and T_1 by comparing the shape of the experimental saturation curve to a set of curves calculated for different values of these parameters (complement 2).

Equation [9.12] is simplified when the width of the envelope is much greater than that of the saturated packets, $\delta B/\sqrt{\sigma}$ ("completely inhomogeneous" line). In this case, the normalised Lorentzian $s(B - B_0)/(g_P\,\Delta n_e\,B_1\sqrt{\sigma}\,)$ behaves like a Dirac function in the convolution product and we obtain (complement 1):

$$S(B - B_r) \;=\; g_P\,\Delta n_e\,B_1\sqrt{\sigma}\;h(B - B_r)$$

The shape of the saturated line is the shape of the envelope, whatever the value of B_1, and its amplitude at any point saturates as $\sqrt{\sigma}$. The same is true for its derivative. The normalised value of the amplitude $S(0)$ at the centre of the line is given by:

$$S(0)/[g_P\,\Delta n_e\,B_1\,h(0)] \;=\; \sqrt{\sigma} \qquad [9.13]$$

It is more difficult to saturate than a homogeneous line (figure 9.9c) and half-saturation occurs when $B_1 = \sqrt{3}\,(\delta B/\gamma' T_1)^{1/2}$ (exercise 9.3).

These results can be used to simulate the saturation curves given by single crystal samples. For a given orientation of the crystal relative to $\mathbf{B}$, the equivalent centres in the crystal generally produce an inhomogeneous line, characterised by the parameters g_P and T_1. When the orientation of the crystal relative to the magnetic field is changed, these parameters vary and the saturation curve is altered.

9.4.3 – Simulating saturation of a powder spectrum

A powder spectrum can be considered to be structured in the following way (figure 9.10):

1. The centres oriented in the same way relative to **B** resonate for a field value B_r. If all possible orientations of the sample are considered, B_r varies almost continuously between B_{min} and B_{max} and its distribution is described by a density $D(B_r)$.
2. Centres with the same orientation relative to **B** produce an *inhomogeneous* line centred at B_r, composed of homogeneous lines with a distribution described by the density $h(B_0 - B_r)$. The shape and width of this distribution depend on the orientation of the centres relative to **B**.
3. In each inhomogeneous line, the saturated packets are described by $s(B - B_0)$ (equation [9.9]). The spin-spin relaxation time, T_2, is independent of the orientation of the centres relative to **B**, in contrast to the width $\delta B = \hbar / g'\beta T_2$, the intensity factor g_P and the spin-lattice relaxation time T_1 which are all influenced by it.

This general scheme produces the following expression for the absorption signal at B:

$$sp(B) = \int_0^{+\infty} D(B_r)\, S(B - B_r)\, \mathrm{d}B_r \qquad [9.14]$$

where $S(B - B_r)$ is given by equation [9.12]. Contrary to appearances, the second side of equation [9.14] is not a convolution product as the shape, width and intensity of $S(B - B_r)$ depend on the orientation of the centres relative to **B**, and therefore on the B_r integration variable.

The amplitude at any point on the spectrum is the result of superposition of inhomogeneous lines which do not saturate in the same way, and the saturation curve is difficult to simulate. The difficulty is overcome for lines forming at the extremities of the spectrum, which are produced by centres with the same orientation (canonical directions of **B**). For example, we will examine the shape of the spectrum near the field B_{min} which corresponds to a certain canonical direction (figure 9.10). In this part of the spectrum, $sp(B)$ is the result of superposition of the inhomogeneous lines produced by centres with an orientation close to the canonical orientation (complement 2 to chapter 4) and which therefore saturate in the same way. Near to B_{min}, equation [9.14] is therefore equivalent to a convolution product and the derivative of $sp(B)$ is obtained by replacing $D(B_r)$ by its derivative in the integral (complement 1). As the density $D(B_r)$ takes the form of a step function near to B_{min}, this derivative is a very narrow function which behaves like a Dirac function centred at B_{min} (complement 1). As a result, the derivative of $sp(B)$ reproduces *the shape of the inhomogeneous line* $S(B - B_{min})$ near to B_{min}. We can therefore use equation [9.12] to simulate the saturation of the low- and high-field peaks in a powder spectrum.

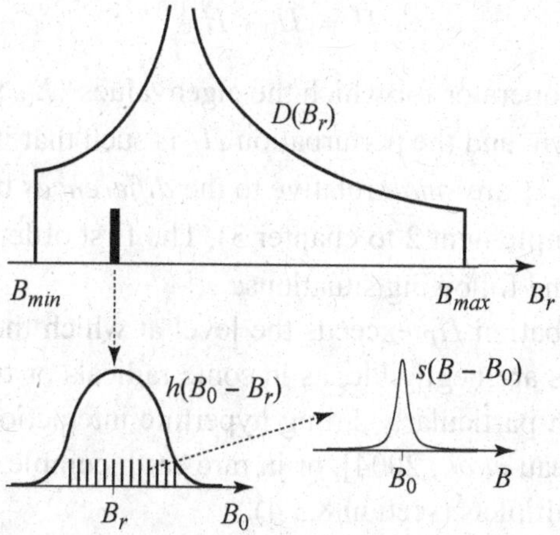

Figure 9.10 – Various levels of structuration of a powder EPR spectrum. $D(B_r)$ is the density of the resonance fields, $h(B_0 - B_r)$ describes the distribution of the lines in the inhomogeneous line, and $s(B - B_0)$ describes the homogeneous line.

9.5 – Introduction to numerical simulation of the EPR spectrum

9.5.1 – Why simulate a spectrum?

Numerical simulation is used whenever it is impossible to analyse the features of a spectrum using closed-form expressions such as those obtained in the previous chapters. These expressions may be unusable due to the complexity of the spectrum or because they are not valid.

1. Overlap between hyperfine patterns can produce complex spectra with a difficult-to-analyse structure. A typical example is that of free radicals where the unpaired electron interacts with a number of non-equivalent nuclei (section 2.4.2). The spectral features may also be masked by the broad width of the inhomogeneous lines. This occurs, for example, with transition ion complexes of integer spin in the low-field situation (section 6.7 and complement 2 to chapter 6), and with rare earth complexes of integer spin (section 8.4).

2. Most of the expressions that we have obtained are based on the first order approximation of perturbation theory. It must be recalled that this calculation is only possible when the spin Hamiltonian has the form:

$$\hat{H} = \hat{H}_0 + \hat{H}_P \qquad [9.15]$$

where $\hat{H}_0$ is an operator for which the eigenvalues $\{E_{0k}\}$ and eigenvectors $\{|u_k\rangle\}$ are known, and the perturbation $\hat{H}_P$ is such that its matrix elements in the basis $\{|u_k\rangle\}$ are *small* relative to the *differences* between the eigenvalues E_{0k} (complement 2 to chapter 3). The first order approximation is insufficient in the following situations:

- The perturbation $\hat{H}_P$ exceeds the level at which the effects of higher order terms are negligible, as in some radicals or transition ion complexes with particularly strong hyperfine interactions [Marque *et al.*, 1997; Hureau *et al.*, 2004], or in rare earth complexes with low-lying excited multiplets (section 8.3.4).
- The spin Hamiltonian cannot be written in the form given by equation [9.15]. For example, in chapter 6 we examined the effect of the zero-field splitting term in the high- and low-field limits where one of the terms in the Hamiltonian is much larger than the other, but we were unable to treat the intermediate case where they are comparable [Blanchard *et al.*, 2003].

Second order perturbation theory can be used and closed-form expressions of the resonance field can be obtained for more general situations than those we have considered, but these expressions become complex and difficult to use when analysing the spectrum. In these conditions, it is better to perform a complete numerical calculation and determine the values of the unknown parameters by reasoning.

9.5.2 – How can the spectrum be numerically calculated?

The spectrum is constructed as shown in figure 9.10:

1. The position of the lines produced by all the paramagnetic centres in the sample, B_r, and their intensity factor, g_P are calculated. This step is based on the spin Hamiltonian which describes the magnetic interactions.
2. For each value of B_r, the dispersion due to the various causes of inhomogeneity are described. Several models may be used depending on the nature of the centres (section 9.5.3).
3. When necessary, the narrow width of the homogeneous lines is taken into account.

Hereafter, we will describe the principle of the calculations performed in the first step in the case of a powder spectrum; the questions relating to the linewidth are dealt with in the following section.

To determine the characteristics (B_r, g_P) of the resonance lines for all possible orientations of the centres relative to **B**, the direction of **B** is varied relative to the principal axes of the $\tilde{\mathbf{g}}$ matrix (figure 5.1), and for each direction we seek a value B satisfying the resonance condition in a previously-defined range ΔB. If such a value is found, the intensity factor is calculated for the line, the direction of **B** is incremented and the calculation is repeated. The procedure is therefore as follows:

▹ Numerical values are given to the spin Hamiltonian parameters: principal values of the $\tilde{\mathbf{g}}$, $\tilde{\mathbf{D}}$ and $\tilde{\mathbf{A}}$ matrices and Euler angles defining the relative orientation of their principal axes.

▹ For the **B** direction defined by the angles (θ, φ), the energy levels of the centre are calculated for each B value in the ΔB range. To do so, the matrix representing the Hamiltonian must be constructed in a certain basis and its eigenvalues must be determined. In simulation software, the Hamiltonian and the basis are predefined and all the matrix elements likely to be involved in the calculation are tabulated in advance to speed up the calculation. From these elements and the values of the parameters supplied by the user, a purely numerical matrix is constructed and its eigenvalues and eigenvectors are calculated. If the splitting between two energy levels is equal to the quantum $h\nu$ of the spectrometer (within the limit of a predefined value ε), resonance is possible and the intensity factor or the transition is calculated. Otherwise, B is incremented and the energy levels are re-calculated. When the whole ΔB range has been explored, the position and intensity of the line or lines for the direction of **B** defined by (θ, φ) are known. The angles are incremented and the calculation continues until all possible directions have been screened.

9.5.3 – The linewidth problem

It is often more difficult to reproduce the *relative amplitudes* of the spectral features than their *positions*. Indeed, some authors mask this difficulty by discreetly shifting the experimental and calculated spectra vertically rather than superposing them. This problem is often linked to poor modelling of the *linewidth*, as the amplitude of the derivative of a line is inversely proportional to the square of its width (section 9.4.1). The shape and width of EPR lines are difficult to model, but it can be done in some cases. We already indicated in section 5.4.1 that models exist to describe the width of homogeneous lines produced in the isotropic regime, and in this section we will examine the case of inhomogeneous lines.

▹ *g*-strain is a frequent cause of inhomogeneous broadening of lines. This phenomenon occurs when the principal values and the principal axes of the $\tilde{\mathbf{g}}$ matrix vary slightly from one paramagnetic centre to another. It can be modelled thanks to a general mathematical treatment which consists in writing the $\tilde{\mathbf{g}}$ matrix in the form [Hagen *et al.*, 1985]

$$\tilde{\mathbf{g}} = \tilde{\mathbf{g}}_0 + \tilde{\mathbf{p}}$$

where $\tilde{\mathbf{g}}_0$ is a "mean" matrix and $\tilde{\mathbf{p}}$ is a matrix for which the principal values $\{p_i\}$ are centred *random normal variables* characterised by their standard deviations $\{\sigma_i\}$ and coefficients of correlation $\{c_{ij}\}$. The orientation of the principal axes of the $\tilde{\mathbf{p}}$ matrix relative to those of the $\tilde{\mathbf{g}}_0$ matrix is defined by a set of three Euler angles. This model therefore uses 12 independent parameters. For a given orientation of **B** relative to the principal axes of $\tilde{\mathbf{g}}_0$, the energy ΔE of a transition depends on the principal values and principal axes of the $\tilde{\mathbf{g}}$ matrix. It is therefore a random variable for which the mean value and the standard deviation can be expressed as a function of the parameters defined above and the numerically-calculable partial derivatives $\partial(\Delta E)/\partial p_i$ and $\partial^2(\Delta E)/\partial p_i^2$. From these partial derivatives, we can deduce the mean value of the resonance field and its standard deviation, i.e., the position and width of the inhomogeneous line. This treatment gives excellent simulations of the spectra produced by metal centres present in biological macromolecules, where broadening is often due to "g-strain" [Hagen *et al.*, 1985; Hearshen *et al.*, 1986]. The efficacy of the method is partly linked to the large number of adjustable parameters, but the quality of the simulations remains good if the number is restricted to six by assuming that the principal axes of the $\tilde{\mathbf{g}}_0$ and $\tilde{\mathbf{p}}$ matrices are parallel and that the random variables $\{p_i\}$ are fully correlated. The six adjustable parameters are then the mean values $\{g_{0i}\}$ and the standard deviations $\{\sigma_i\}$. A detailed description of this treatment and a discussion of its limitations can be found in [More, 1998]. This treatment can also be applied to other parameters of the spin Hamiltonian, such as those appearing in the $\tilde{\mathbf{D}}$ and $\tilde{\mathbf{A}}$ matrices, and even the intercentre distance r and the exchange parameter J which are involved in intercentre interactions [Guigliarelli *et al.*, 1986; Bertrand *et al.*, 1994].

Another, more "physical" approach to the g-strain phenomenon consists in distributing the geometric parameters determining the principal values and the principal axes of the $\tilde{\mathbf{g}}$ matrix rather than the values themselves. This type of approach, based on ligand field-type models, produces good results in the case of haeme-iron centres and [2Fe – 2S] centres in proteins [More et *al.*, 1987; More *et al.*, 1990].

▹ The width of inhomogeneous lines is also influenced by unresolved hyperfine features due to weak interactions with distant paramagnetic nuclei. This contribution is very difficult to model, and it is generally taken into account by performing a simple convolution product by an adjustable-width Gaussian.

In complement 5 we will present some details on the software that is currently most frequently used.

9.6 – Points to consider in applications

9.6.1 – How should the modulation and sweep parameters be selected?

As the signal delivered by the synchronous detector is proportional to the amplitude of the modulation B_m (section 9.2.2), the largest amplitude compatible with non-deformation of the spectrum must be used. Narrow lines are the first to undergo overmodulation, when B_m exceeds their peak-to-peak width (figure 9.3). Beyond this value, the line broadens but its amplitude continues to increase. If deformation of a line is acceptable, overmodulation can be used to improve the signal-to-noise ratio. This strategy can be used to cause weak signals to emerge and even to determine the number of paramagnetic centres since overmodulation has no effect on the *intensity* of the line.

The scan rate to be selected is determined by the well-known slogan: "sweep slower to filter more". Indeed, the longer the time taken to sweep through a line, the lower the frequency of its components, and the easier it is to filter out noise without modifying these components. This is essential when dealing with narrow lines (exercise 9.2). An alternative is to sweep more rapidly with a relatively short time constant and to filter the noise by *accumulating* spectra. In this case, the signal-to-noise ratio increases as $\sqrt{N}$, where N is the number of spectra accumulated.

The foregoing considerations are very important when performing EPR spectroscopy experiments. Beginners are advised to familiarise themselves with them by recording the spectrum for a sample which produces a simple spectrum at room temperature, such as "*strong pitch*", varying the amplitude of the modulation, the scan rate (by adjusting the conversion time) and the time constant for the low-pass filter.

9.6.2 – How can a saturation curve be simulated?

The models described in section 9.4 can be used to simulate the saturation curves obtained in "slow passage" conditions. The inequality $(B_m/\delta B)\omega_m T_1 < 1$, which is enough to guarantee these conditions (section 9.2.1), is very restrictive at low temperature where the spin-lattice relaxation time T_1 is long. For example, for $B_m = \delta B$, it requires a modulation frequency $\nu_m < 150$ Hz for $T_1 = 1$ ms. In practice, the smallest values of ν_m and B_m compatible with a good signal-to-noise ratio are used.

To determine T_1, the shape of the experimental saturation curve, which gives the amplitude at a point on the spectrum as a function of $\sqrt{P}$, is compared to a set of theoretical curves calculated for different values of the width of the packets δB, which is unknown. This comparison is only reasonable if all the centres in the sample which contribute to the amplitude are characterised by the same T_1 value. In the case of a powder spectrum, this condition is satisfied for the extreme values of the field, B_{min} and B_{max} in figure 9.10. The peaks which form at these points in the spectrum are often much broader than the spin packets, and they can be considered to represent the *envelopes* of the inhomogeneous lines at B_{min} and B_{max}. Details on the construction of the set of theoretical curves and the procedure used to determine T_1 are given in complement 2.

9.6.3 – How can an EPR spectrum be simulated?

Numerical simulation of the spectrum is generally used when the spin Hamiltonian involves numerous parameters. Before undertaking the simulation *per se*, it is a good idea to perform a few experiments likely to guide the search for the most appropriate set of parameters: compare spectra recorded at different frequencies, examine the saturation behaviour of the different patterns on the spectrum, and how the shape changes as a function of temperature. In the case of organic molecules, isotopic substitution experiments can save significant time. The method used to describe the linewidth is not without influence, and it should not be surprising if we cannot reproduce the relative amplitudes of the features of the spectrum when this method is not model-based. In particular, when the linewidths are mainly determined by the *g*-strain, the mathematical treatment which describes this mechanism in a very general way should be used (section 9.5.3).

Complement 1 – Some properties of the convolution product

Here, we indicate the properties of the convolution product used in this chapter, without aiming for mathematical completeness.

◇ *Definition*

$f_1(x)$ and $f_2(x)$ are two functions of the real variable x. Their convolution product is the function:

$$g(x) = \int_{-\infty}^{+\infty} f_1(u)\, f_2(x-u)\, \mathrm{d}u$$

The signification of the product present in the integral is given in figure 9.11. The convolution product is *commutative*. Indeed, if we perform the $y = x - u$ change of variable, we obtain:

$$g(x) = \int_{+\infty}^{-\infty} f_1(x-y)\, f_2(y)(-\mathrm{d}y) = \int_{-\infty}^{+\infty} f_2(y)\, f_1(x-y)\, \mathrm{d}y$$

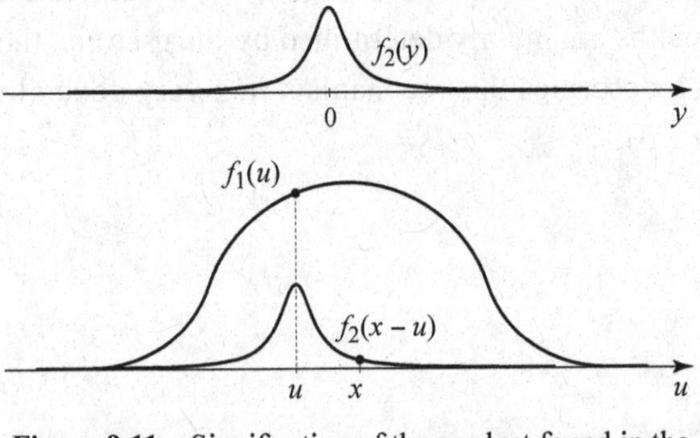

Figure 9.11 – Signification of the product found in the integral defining the convolution product of $f_1(x)$ and $f_2(x)$.

This property is expressed by writing:

$$g(x) = f_1(x) * f_2(x) = f_2(x) * f_1(x)$$

It can be shown that the area under the curve $g(x)$ is equal to the product of the areas under the curves $f_1(x)$ and $f_2(x)$.

◇ *Derivative of a convolution product*

If $\mathrm{d}f_1/\mathrm{d}x$ and $\mathrm{d}f_2/\mathrm{d}x$ are the derivatives of $f_1(x)$ and $f_2(x)$, it can be shown that

$$\mathrm{d}g/\mathrm{d}x = \mathrm{d}f_1/\mathrm{d}x * f_2(x) = f_1(x) * \mathrm{d}f_2/\mathrm{d}x$$

We therefore obtain the derivative of the convolution product by deriving one of the two functions in the integral.

◇ *Associativity of the convolution product*

The associativity property is expressed as follows:

$$[f_1(x) * f_2(x)] * f_3(x) = [f_2(x) * f_3(x)] * f_1(x) = [f_1(x) * f_3(x)] * f_2(x)$$

◇ *Definition and properties of the Dirac function*

The "unit step" function $H(x)$ is defined in figure 9.12a. Its derivative with respect to x is the limit when $\delta x \to 0$ of the "rectangular" function

$$f_{\delta x}(x) = \frac{H(x+\delta x) - H(x)}{\delta x}$$

which is represented in figure 9.12b. When $\delta x \to 0$, the height $1/\delta x$ of the rectangle becomes infinite, but its area remains equal to 1. This function is said to tend towards a "Dirac function" denoted $\delta(x)$.

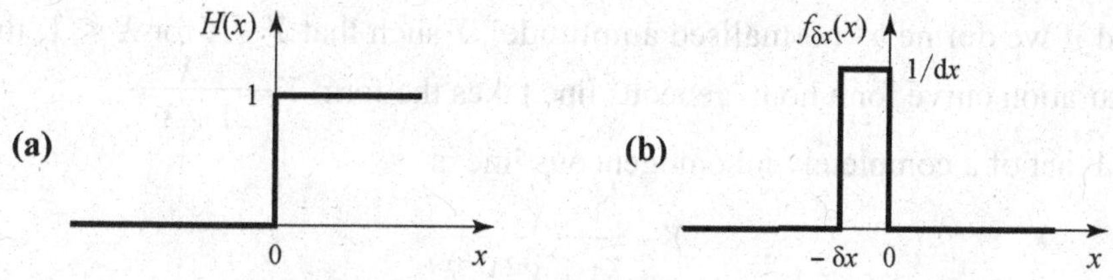

Figure 9.12 – Definition of the unit step function $H(x)$ and of the rectangular function $f_{\delta x}(x)$.

If we perform the convolution product of the rectangular function $f_{\delta x}(x)$ by any function $f_1(x)$:

$$g(x) = \int_{-\infty}^{+\infty} f_1(u)\, f_{\delta x}(x-u)\, \mathrm{d}u$$

The construction from figure 9.11 shows that only the values of u which verify $x \le u \le x + \delta x$ contribute to the integral. For these values, we have $f_{\delta x}(x-u) = 1/\delta x$. We then deduce:

$$g(x) = \frac{1}{\delta x} \int_{x}^{x+\delta x} f_1(u)\, \mathrm{d}u$$

When δx tends to zero, the integral tends to $\delta x\, f_1(x)$ and $g(x)$ tends to $f_1(x)$. This property can be generalised to any function with unit area for which the width tends towards zero, and is written in the form:

$$\int_{-\infty}^{+\infty} f_1(u)\, \delta(x-u)\, \mathrm{d}u = f_1(x)$$

where $\delta(x - u)$ is a "Dirac function" centred at x.

Complement 2 – Quantitative analysis of the saturation curve for an inhomogeneous line

According to equations [9.10] and [9.13], the amplitude of a saturated line can be written:

- For a homogeneous line: $\dfrac{s(0)}{g_P \Delta n_e f(0)} = \dfrac{B_1}{1 + \dfrac{\gamma' B_1^2 T_1}{\delta B}}$

- For a completely inhomogeneous line: $\dfrac{S(0)}{g_P \Delta n_e h(0)} = \dfrac{B_1}{\left(1 + \dfrac{\gamma' B_1^2 T_1}{\delta B}\right)^{1/2}}$

It is convenient to introduce dimensionless quantities. If we set

$$X = (\gamma' T_1/\delta B)^{1/2} B_1 \qquad [1]$$

and if we define a "normalised amplitude" Y such that $Y = X$ for $X \ll 1$, the saturation curve for a homogeneous line takes the form $Y = \dfrac{X}{1 + X^2}$ and that of a completely inhomogeneous line is

$$Y = \frac{X}{(1 + X^2)^{1/2}}.$$

These curves, which are shown in figure 9.13 on a log-log plot, describe saturation in two limit cases. To obtain the curves for intermediate situations, the quantity $S(0)$ defined by equation [9.12] must be numerically calculated by approximating the envelope using the spectrum's low-field or high-field peak and writing equation [9.9] in the form:

$$s(B - B_0) \propto \frac{X}{(B - B_0)^2 + \delta B^2 (1 + X^2)}$$

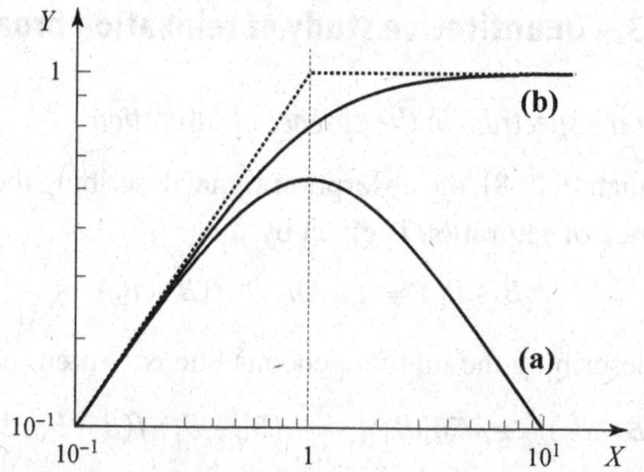

Figure 9.13 – Log-log plot showing saturation curves for two very different lines: (**a**) Homogeneous Lorentzian line. (**b**) Completely inhomogeneous line composed of Lorentzian packets. The abscissa is the dimensionless quantity $X = (\gamma' T_1/\delta B)^{\frac{1}{2}} B_1$, the ordinate is the normalised amplitude defined in the text.

By varying X and the width of the packets δB, and by normalising the amplitude $S(0)$ as indicated above, a set of "*theoretical*" saturation curves is obtained. The shapes of these curves depend on the parameter δB. They are intermediate between those of the curves presented in figure 9.13 which correspond to $\delta B \gg \delta B_e$ (homogeneous line) and $\delta B \ll \delta B_e$ (completely inhomogeneous line), where δB_e is the width of the envelope.

In practice, the experimental saturation curve is determined by plotting the amplitude at the centre of the peak as a function of $\sqrt{P}$, where P [watt] is the microwave power, also using a log-log plot. If the model is appropriate, horizontal and vertical translations should allow the experimental curve to be superposed on one of the theoretical curves. The selected theoretical curve thus supplies δB, and by making the X scale (theoretical curve) and $\sqrt{P}$ scale (experimental curve) to coincide, T_1 can be determined if λ, the "conversion factor" for the cavity defined by $B_1 = \lambda\sqrt{P}$, is known. Assume, for example, that $X = 10$ corresponds to $\sqrt{P} = 0.1\ \mathrm{W}^{\frac{1}{2}}$ ($P = 10$ mW). Equation [1] gives $(\gamma' T_1/\delta B)^{\frac{1}{2}} = 100/\lambda$. Given δB and λ we can deduce T_1. As indicated in section 9.4.1, λ is around $10^{-4}\ \mathrm{T\ W}^{-\frac{1}{2}}$ for a standard cavity which operates in the TE_{102} mode.

Complement 3 – Quantitative study of relaxation broadening

◇ *Expression for the spectrum in the absence of saturation*

According to equation [9.8], the absorption signal describing the homogeneous line in the absence of saturation is given by

$$s(B - B_0) = g_P \, \Delta n_e \, B_1 f(B - B_0)$$

and the signal describing the inhomogeneous line is written (equation [9.12])

$$S(B - B_r) = g_P \Delta n_e B_1 \int_0^{+\infty} h(B_0 - B_r) \, f(B - B_0) \, \mathrm{d}B_0$$

Equation [9.14], which gives the amplitude at any point of the spectrum, becomes:

$$sp(B) = \Delta n_e B_1 \int_0^{+\infty} g(B_0) \, f(B - B_0) \, \mathrm{d}B_0 \qquad [1]$$

with:

$$g(B_0) = \int_0^{+\infty} D(B_r) \, g_P(B_r) \, h(B_0 - B_r) \, \mathrm{d}B_r$$

The integral present on the second side of equation [1] is the convolution product of $f(B - B_0)$ by $g(B_0)$. The function $g(B_0)$ is determined by:

- ▹ The density $D(B_r)$ of the resonance lines and their intensity factor $g_P(B_r)$.
- ▹ The $h(B_0 - B_r)$ envelopes of the inhomogeneous lines.

◇ *Expression for the spectrum with relaxation broadening*

- ▹ At sufficiently low temperatures for T_1 to be long relative to T_2, the width of the Lorentzian function $f(B - B_0)$ present in equation [1] is determined by T_2 (equation [5.11]):

$$\delta B_2 = \hbar / g' \beta T_2$$

It is noted $f_2(B - B_0)$. The amplitude at any point of the spectrum then becomes:

$$sp_2(B) = \Delta n_e B_1 \int_0^{+\infty} g(B_0) \, f_2(B - B_0) \, \mathrm{d}B_0 \qquad [2]$$

- ▹ When the temperature is high enough for T_1 to become comparable to or shorter than T_2, $f(B - B_0)$ is the convolution product of $f_2(B - B_0)$ by a normalised Lorentzian function $f_1(B - B_0)$ of width (equation [5.28]):

$$\delta B_1 = \hbar / g' \beta T_1$$

Using the associativity of the convolution product, equation [1] can therefore be written in the form:

$$sp(B) = sp_2(B) * f_1(B - B_0) \quad [3]$$

Thus, the relaxation broadening of the spectrum is simulated by performing the convolution product of the experimental spectrum $sp_2(B)$ recorded at low temperature, by a *normalised* "relaxation Lorentzian" $f_1(B - B_0)$ for which the width is adjusted to reproduce the experimental spectrum [Bertrand *et al.*, 1980]. The property of the convolution product relating to the area (complement 1) shows that relaxation broadening does not alter the total intensity of the spectrum.

As the spin-lattice relaxation time T_1 is generally anisotropic, equation [3] is strictly valid only for B_{min} and B_{max} values.

Complement 4 – Using standard samples in EPR spectroscopy

◇ *Magnetic field standards*

During recording of the spectrum, the magnetic field is regulated by comparing the value of B set during calibration of the electromagnet to that measured by a sensor located in the gap of the magnet. This sensor is a Hall probe placed against one of the pole caps or an NMR probe located near to the sample. The magnetic field set in this way, the value of which is presented on the abscissa of the spectrum, is generally slightly different to that to which the sample is exposed. The offset depends on the experimental setup, but in the case of a 13″ magnet with a 56 mm gap, it can be up to several tenths of a mT at X-band and several mT at Q-band. It is therefore important to determine this difference if the positions of the spectral features of the sample being studied are to be precisely measured. To do so, a "standard" sample is used for which the spectrum has a specific feature identified by a well known $g = h\nu/\beta B$ number which is unaffected by the magnetic field. The standard is placed in the same position in the cavity as the sample being studied, and its spectrum is recorded. The field corresponding to the specific feature is calculated from g and the frequency determined by a frequency meter. A magnetic field standard must have several properties:

- ▹ It must be stable and produce narrow lines at room temperature.
- ▹ Its signal must be *isotropic*, i.e., its shape and position must not depend on the orientation of the sample relative to **B**.

These conditions are fulfilled by some free radicals in solid samples. Spectrometer manufacturers supply carbon-powder based standards called "*weak pitch*". The one supplied by Bruker has $g = 2.0028$ and a peak-to-peak width of 0.17 mT. A single crystal or polycrystalline powder of the diphenyl-picryl-hydrazyl (DPPH) radical can also be used, which produces a line centred at $g = 2.0036 \pm 0.0002$, with a full width at half maximum of 0.14 mT. More demanding operators can consult [Weil *et al.*, 1994] for a list of free radicals in solution for which the g_{iso} value is known with very good precision. Other standards are used to determine the principal values (g_x, g_y, g_z) of radicals determined by high-field EPR experiments [Un *et al.*, 2001].

◇ *Intensity standards*

As stated previously, the intensity of the spectrum is proportional to the amplitude of the modulation (section 9.2.2), the mean intensity factor (equation [5.15]) and, in the absence of saturation, to the amplitude B_1 of the magnetic component of the microwave radiation (section 5.2). When only the ground state is populated at the temperature of the experiment, the intensity varies as $1/T$ (Curie's law) even when the spectrum is broadened due to spin-lattice relaxation (complement 3). If excited states are populated, the expression for the intensity must be weighted by an appropriate Boltzmann factor (see section 9.3.2).

Naturally, all these factors must be considered when comparing the intensity of the spectra produced by a sample and a standard. But we must also take the conditions imposed by the variations of B_1 inside the cavity into account. For example, consider a rectangular cavity for X-band, with the dimensions $a \approx 23$ mm, $b < a$, $c \approx 2a$ (figure 9.14a). The z axis is parallel to the magnetic field **B**, and we are interested in the vertical median plane defined by $y = c/2$. In the TE_{102} mode generally used in EPR, the magnetic component $\mathbf{b}_1(t)$ is *parallel to the* x axis at any point in this plane and its amplitude only depends on x [Poole, 1967]:

$$B_1(x) = B_{1max} \sin \pi x/a$$

It is therefore maximal at the centre of the cavity and null at the extremities (figure 9.14b). The electric component $\mathbf{e}_1(t)$ is null at any point in this plane.

It is important to consider this distribution of the electromagnetic field when placing the sample in the cavity:

- ▹ When the sample is small (single crystal, polycrystalline powder), it should be placed at the centre of the cavity to avail of the maximum value of B_1, and its intensity should be quantified with a standard of small dimension placed in the same position.
- ▹ EPR tubes and capillaries are positioned along the vertical axis of the plane. To maximise the signal and facilitate comparisons of intensities, the part located inside the cavity must be completely filled (height $\geq a \approx 23$ mm). In these conditions, not all paramagnetic centres are subjected to the same value of B_1. This introduces a complication (which is generally ignored) in the interpretation of progressive saturation experiments, but does not affect intensity measurements if the sample and the standard are placed and filled in the same way.

▹ Flat cells used with solvents with a high dielectric constant, are placed in the median plane (figure 9.14a) and the same precautions as to their filling should be observed.

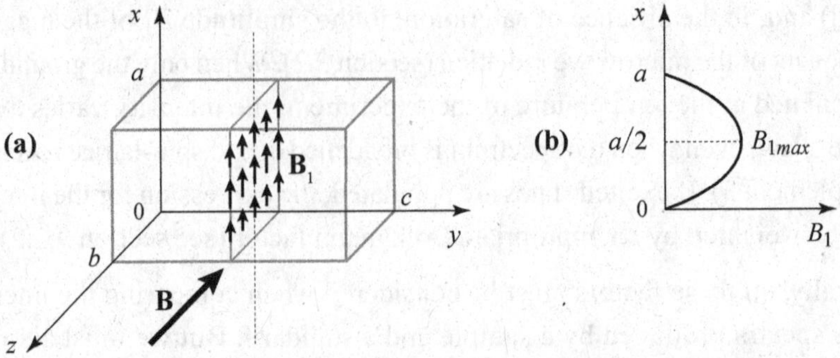

Figure 9.14 – (**a**) Rectangular cavity operating in the TE_{102} mode. **B** is the field produced by the electro-magnet. In the median plane, $\mathbf{B}_1$ is parallel to x. (**b**) Variation of the amplitude B_1 as a function of x in the median plane.

Readers should be aware that there exist "double" rectangular cavities (TE_{104} mode) which can simultaneously contain two samples. These cavities are useful when comparing the intensities of two spectra on the condition that the two places are exactly symmetric with respect to the values of B_1.

As intensity standards, samples are selected that are easy to prepare, stable, and for which the signal follows Curie's law over a broad temperature range while also being easy to integrate (relatively symmetric shape, relatively narrow lines). The most frequently used are the following:

▹ DPPH as a powder or in solution in benzene, at room temperature.

▹ Nitroxide radicals in solution, at room temperature.

▹ $MnSO_4$, H_2O; $CuSO_4$, $5H_2O$; $[Cu(EDTA)]^{2-}$ compounds in solution at room temperature or at low temperatures.

All these samples degrade more or less rapidly, and it is important to remember to prepare fresh samples frequently.

Complement 5 – Numerical simulation software

For a long time, only researchers in laboratories hosting programming experts could simulate EPR spectra. The programmes developed were generally dedicated to specific applications and they were not particularly user-friendly. Nowadays, researchers have wider access to easier-to-use software.

◇ *WinSim*

This freeware can be used to simulate spectra for free radicals in the isotropic regime. It was created by D.R. Duling [Duling, 1994] and further developed by D.O. O'Brien, D.R. Duling and Y.C. Fann. Its most recent version [WinSim 2002] is available from:

http://www.niehs.nih.gov/research/resources/software/tox-pharm/tools/index.cfm

The software can be used to calculate the spectrum for a radical from user-input characteristics: g_{iso}, spin I and hyperfine constants A_{iso} values, number of equivalent nuclei. The shape and linewidth can be specified. The spectra produced by a mixture of radicals in known proportions can also be calculated. An experimental spectrum can be loaded, the hyperfine splitting on the spectrum measured, and a simulation based on an automatic regression procedure launched for several parameters.

◇ *EasySpin*

The *general use* software EasySpin is a *toolbox* for the commercial Matlab calculation software. It uses the same syntax and formats and can avail of Matlab's resources [Stoll and Schweiger, 2006]. It can be downloaded from:

http://www.easyspin.org/

The spectrum is numerically calculated from the spin Hamiltonian for a set of paramagnetic centres, the parameters for which are user-defined. Matlab commands are used to view the result.

▹ Each centre is characterised by its spin S, its $\tilde{\mathbf{g}}$ matrix, its zero-field splitting matrix $\tilde{\mathbf{D}}$, the $\tilde{\mathbf{A}}$ matrices describing hyperfine interactions with paramagnetic nuclei, and the $\tilde{\mathbf{P}}$ matrices describing quadrupole electric interactions when the nuclei have a spin greater than ½. As the g_N factors and the abundances of the different isotopes are tabulated, users can simply indicate the name of the atom when isotopes are present in the sample at their natural abundance. Otherwise, their abundance must be specified. Each matrix is characterised by its principal values and principal axes, for which the orientation relative to a system of molecular axes is identified by three Euler angles (verify their definition).

▹ The spectrum is calculated for various regimes of molecular movement (see section 4.5):
- The limit of very fast motion, which corresponds to the isotropic regime.
- Fast motion, characterised by a correlation time.
- Slow motion, characterised by a diffusion matrix.
- The "solid-state" limit where the molecules are in fixed positions.

▹ Within the solid-state limit, calculation is performed for different types of sample (see section 4.3.2):
- A single crystal. The user indicates the orientation of the molecules in the unit cell relative to the axes of the crystal. The orientation of the crystal relative to the laboratory axes is defined, and, if necessary, the direction of the axis around which the crystal is rotated.
- A polycrystalline powder or frozen solution.
- A partially oriented system. The user describes the distribution of the molecular orientation relative to the support rotated in the magnetic field.

▹ The linewidth can be described using several options (see section 9.5.3):
- A simple convolution by a Lorentzian or a Gaussian.
- An unresolved hyperfine structure (constant width in MHz).
- Treatment based on the distribution of the parameters of the $\tilde{\mathbf{g}}$ (*g* strain), $\tilde{\mathbf{D}}$, and $\tilde{\mathbf{A}}$ matrices.

▹ The spectrum can be calculated for $\mathbf{B}_1$ perpendicular to $\mathbf{B}$ (standard cavity) or $\mathbf{B}_1$ parallel to $\mathbf{B}$ (dual-mode cavity, see section 6.7). Complete diagonalisation of the Hamiltonian is possible, but time-consuming. Calculations based on the expressions given by second order perturbation theory is quicker, but only provide an approximation. It is recommended to use this approximate calculation first and to verify the final result by performing the complete calculation. Various algorithms based on the least-squares method can be used to perform automatic regression on several parameters.

▹ EasySpin can be used to simulate ENDOR and ESEEM spectra (for a single centre of spin ½). The software also provides tables listing the physical constants, the Clebsch-Gordan coefficients (appendix 4) and various functions used in magnetic resonance.

Comments – To use this software it is necessary to have a license for, and be familiar with, Matlab. While offering numerous options, it remains pleasant and easy to use. It is regularly updated and its potential has considerably increased since the earliest versions. For example, the maximum number of paramagnetic nuclei in the sample is currently almost unlimited. It should be noted that the matrices corresponding to the different interactions are "black boxes": if the effect of a molecular parameter (such as the distance) on the spectrum is to be examined, a function which calculates the matrix elements as a function of this parameter must first be defined.

References

ANDERSON K.K. *et al.* (2003) *Journal of Biological Inorganic Chemistry* **8**: 235-247.

BARRA A.-L. *et al.* (1998) *Account of Chemical Research* **31**: 460-466.

BERTRAND P. *et al.* (1980) *Journal of Magnetic Resonance* **40**: 539-549.

BERTRAND P. *et al.* (1994) *Journal of the American Chemical Society* **116**: 3078-3086.

BLANCHARD S. *et al.* (2003) *Inorganic Chemistry* **42**: 4568-4578.

DULING D.R. (1994) *Journal of Magnetic Resonance* série B **104**: 105-110.

GUIGLIARELLI B. *et al* (1986) *Journal of Chemical Physics* **85**: 2774-2778.

HAGEN W.R. *et al.* (1985) *Journal of Magnetic Resonance* **61**: 220-232.

HAGEN W.R. *et al.* (1985) *Journal of Magnetic Resonance* **61**: 233-244.

HEARSHEN D.O. *et al.* (1986) *Journal of Magnetic Resonance* **69**: 440- 459.

HUREAU C. *et al.* (2004) *Chemistry-A European Journal* **10**: 1998-2010.

MARQUE S. *et al.* (1997) *Journal of Physical Chemistry A* **101**: 5640-5645.

MCINNES E.J.L. (2006) *Structure and Bonding* **122**: 69-102.

MÖBIUS K. & SAVITSKY A. (2009) *High-field EPR spectroscopy on proteins and their model systems*, RSC, Cambridge.

MORE C. (1998) Dossier d'Habilitation à Diriger les Recherches, Université de Provence, Marseille.

MORE C., BERTRAND P. & GAYDA J-P. (1987) *Journal of Magnetic Resonance* **73**: 13-22.

MORE C., GAYDA J-P. & BERTRAND P. (1990) *Journal of Magnetic Resonance* **90**: 486-499.

POOLE P.P. (1967) *Electron Spin Resonance. A Comprehensive Treatise on Experimental Techniques*. Intersciences Publishers, New York.

STOLL S. & SCHWEIGER A. (2006) *Journal of Magnetic Resonance* **178**: 42-45.

UN S., DORLET P. & RUTHERFORD A.W. (2001) *Applied Magnetic Resonance* **21**: 341-361.

WEIL J.A, BOLTON J.R. & WERTZ J.E. (1994) *Electron Paramagnetic Resonance. Elementary Theory and Practical Application*, John Wiley and Sons, New York.

ZWILLINGER D. (2003) *Standard Mathematical Tables and Formulae*, CRC Press, Boca Raton.

Exercises

9.1. The, possibly overmodulated, signal returned by the synchronous detector is proportional to the quantity $a_1(B)$ given by equation [9.6].

a) Give the expression for the primitive of $a_1(B)$.

b) Show that the area under this primitive is equal to $\frac{B_m}{2}\int_0^{+\infty} s(B)\,dB$, where $s(B)$ is the absorption signal.

9.2. A spectrum has lines with a width δB and noise with frequency ν_{noise} [in Hz]. To reduce the noise, a low-pass filter with time constant τ is used and we allow that the filtering is effective when $\nu_C < \nu_{noise}/10$ where ν_C is the cutoff frequency for the filter linked to τ by $\nu_C = 1/(2\pi\tau)$ (section 9.2.2).

a) Show that the scan rate v must be less than a specific value for the noise to be filtered out without deforming the signal.

b) Application: assume that $\nu_{noise} = 100$ Hz. How should the time constant and scan rate be chosen for $\delta B = 0.1$ mT and $\delta B = 5$ mT?

9.3. Determine for which value of B_1 half-saturation is achieved when we are interested in saturation of the:

a) peak-to-peak amplitude of a Lorentzian line,

b) amplitude of a completely inhomogeneous line.

Answers to exercices

9.1. a) The primitive for $a_1(B)$ is written

$$\int_0^B a_1(b)\,db = \frac{B_m}{2}\int_0^B s'(b)\,db + (\cdots)\int_0^B s^{(3)}(b)\,db + (\cdots)\int_0^B s^{(5)}(b)\,db + \cdots$$
$$= \frac{B_m}{2}[s(b)]_0^B + (\cdots)[s^{(2)}(b)]_0^B + (\cdots)[s^{(4)}(b)]_0^B + \cdots$$

The absorption signal $s(B)$ and all its derivatives vanish at $B = 0$. The primitive for $a_1(B)$ is therefore written:

$$\frac{B_m}{2}s(B) + (\cdots)s^{(2)}(B) + (\cdots) + s^{(4)}(B) + \cdots$$

b) The area under this signal is equal to:

$$\frac{B_m}{2}\int_0^{+\infty} s(B)\,dB + (\cdots)\int_0^{+\infty} s^{(2)}(B)\,dB + (\cdots)\int_0^{+\infty} s^{(4)}(B)\,dB + \cdots$$

Only the first term is different from zero, as all the derivatives of $s(B)$ vanish for $B = 0$ and when $B \to \infty$. From this result, we conclude that overmodulation does not alter the area of the line.

9.2. a) For a line not to be deformed, the following condition must be satisfied: $\tau < 0.1\ \delta t$, with $\delta t = \delta B/v$ (section 9.2.2). In addition, filtering is effective if $\nu_C = 1/(2\pi\tau) < \nu_{noise}/10$. From these relations, we deduce

$$v < (2\pi/100)\ \nu_{noise}\delta B$$

b) For $\nu_{noise} = 100$ Hz, we can choose $\tau = 15$ ms to efficiently filter the signal, and we must take $v < 6\delta B$ so as not to deform the line. This requires $v < 0.6$ mT s^{-1} for $\delta B = 0.1$ mT, and $v < 30$ mT s^{-1} for $\delta B = 5$ mT.

9.3. Saturation of the peak-to-peak amplitude of the derivative of a Lorentzian line is described by equation [9.11]. The second side is equal to ½ for $B_1 = (2^{2/3} - 1)^{1/2}\,(\delta B/\gamma' T_1)^{1/2}$ with $(2^{2/3} - 1)^{1/2} \approx 0.766$. Saturation of a completely inhomogeneous line is described by [9.13]. The second side is equal to ½ for $B_1 = \sqrt{3}\,(\delta B/\gamma' T_1)^{1/2}$.

Appendix 1

Expression of the magnetic moment of a free atom or ion

The expression for the moment of a paramagnetic centre can be used to express the interaction of its unpaired electrons with a magnetic field. This expression depends on the nature of the ground state of the centre, which is determined by all the interactions to which the electrons are subjected. The ground state can be determined by first considering the effect of the strongest interactions and by introducing additional interactions in decreasing order of magnitude. In the case of an atom or a "free" (i.e., isolated) ion, this method leads from the notion of *microstate* to that of *multiplet* and provides an expression producing a good approximation of the magnetic moment of the ground multiplet. In this appendix, we qualitatively describe the different steps in this calculation.

First step: electrostatic interactions to which electrons are subjected

These are electron-nucleus and electron-electron *electrostatic interactions*.

1 – Microstates

In a simplified description, we first assume that the electrons are *independent* of each other, each being exposed to a "central field" due to the attractive interaction with the nucleus and to the *mean* repulsive interaction with the other electrons. We can then define *orbitals* labelled with three numbers (n, ℓ, m_ℓ) where:

$$n = 1, 2, \ldots$$

$$\ell = 0, 1, \ldots n - 1$$

$$m_\ell = -\ell, -\ell + 1, \ldots \ell \qquad [1]$$

P. Bertrand, *Electron Paramagnetic Resonance Spectroscopy*,
https://doi.org/10.1007/978-3-030-39663-3

The lowercase letters *s*, *p*, *d*, *f*, etc. are used to indicate that $\ell = 0, 1, 2, 3, \ldots$ and an orbital characterised by ($n = 3, \ell = 2$), for example, is called a $3d$ orbital.

These orbitals are generally complex functions of the coordinates of an electron, for which the square of the modulus represents the electron probability density at a given point. The full physical state of an electron is described by a "spin orbital", obtained by multiplying an orbital by a function defining its spin state ($m_s = \pm ½$). When an electron "occupies" an orbital identified by the triplet (n, ℓ, m_ℓ), its energy depends only on the numbers (n, ℓ). All of the $(2\ell + 1)$ orbitals characterised by the same value of (n, ℓ) (equation [1]) make up a *subshell*. By attributing an orbital to each electron in order of increasing energy, with a maximum of two electrons characterised by $m_s = \pm ½$ per orbital, we can determine the ground *electronic configuration* from which the periodic table of the elements is established.

Let us consider an atom or ion which has N electrons in an *incomplete subshell* (n, ℓ). There exist $2(2\ell + 1)$ possible pairs (m_ℓ, m_s) for an electron, and the number of ways they can be attributed to N electrons is equal to the number of ways of sorting N indistinguishable objects into $2(2\ell + 1)$ boxes, i.e., $C_{2(2\ell+1)}^N$. For example, for the carbon atom which has 2 electrons in its $2p$ subshell ($\ell = 1$), there exist 6 possible (m_ℓ, m_s) pairs, and therefore $C_6^2 = 15$ different ways to attribute them to the 2 electrons. Each of these ways defines a "microstate" of the atom. Each microstate is represented by a product of N spin orbitals, from which an *antisymmetric* wave function (Slater determinant) can be developed to take the *indistinguishable* nature of the electrons into account (section 7.2.1).

The magnetic moment associated with a given microstate can be written as:

$$\boldsymbol{\mu} = -\beta \sum_{i=1}^{N} (\mathbf{l}_i + g_e \mathbf{s}_i) \qquad [2]$$

where the $\mathbf{l}_i$ and $\mathbf{s}_i$ represent the orbital and spin angular momenta of the electrons in the incomplete subshell. Its projection on any Z axis is given by:

$$\mu_Z = -\beta \sum_{i=1}^{N} (m_{\ell i} + g_e m_{si})$$

2 – (*L*, *S*) terms

To improve on the previous model, we retain the notion of orbital, but we calculate the effect of the repulsive interactions between electrons *taking the orbitals they occupy into account*. This effect depends on how the electrons are distributed among the orbitals in the subshell, and thus all microstates are no longer equivalent from an energy perspective. The microstates concept is thus ill-adapted to describe these new energy levels. It can be shown that the symmetry properties for all the electrostatic interactions in which electrons are involved are such that the energy levels can be identified by pairs of numbers (L, S) linked to the total orbital angular momentum **L** and total spin angular momentum **S** of the atom, defined by:

$$\mathbf{L} = \sum_{i=1}^{N} \mathbf{l}_i; \quad \mathbf{S} = \sum_{i=1}^{N} \mathbf{s}_i \qquad [3]$$

To determine all the possible (L,S) pairs, we project the two members of equation [3] on any Z axis:

$$L_Z = \sum_{i=1}^{N} l_{iZ}; \quad S_Z = \sum_{i=1}^{N} s_{iZ} \qquad [4]$$

As **L** and **S** are angular momenta, the components L_Z and S_Z can only take values of the following form:

$$M_L = -L, -L+1, \ldots, L; \quad M_S = -S, -S+1, \ldots, S \qquad [5]$$

However, according to equation [4], the numbers M_L and M_S are as follows:

$$M_L = \sum_{i=1}^{N} m_{\ell i}; \quad M_S = \sum_{i=1}^{N} m_{si}$$

From these relations, and considering all microstates, we can deduce possible values for (M_L, M_S), and consequently values for the numbers (L,S) defined by equations [5]. An ensemble of $(2L+1)(2S+1)$ states (M_L, M_S) correspond to energy level $E_{L,S}$; this ensemble is called an (L, S) term. The wave functions representing these states are linear combinations of Slater determinants representing microstates.

To designate the terms, we indicate the value of L with the help of a "letter code" similar to the code used for orbitals, but using uppercase letters: S, P, D, F, G, H… for $L = 0, 1, 2, 3, 4, 5$…. and we write the $(2S+1)$ value as a left superscript. A term characterised by $(L = 1, S = 1)$ is thus written ^3P.

The term of interest in EPR is the *ground term*, i.e., the term with the lowest energy. We can determine this term by applying "Hund's rule": we identify the microstate which produces the *maximum value* of M_S, i.e., $(M_S)_{max}$, then the *maximum value* for M_L, i.e., $(M_L)_{max}$. This microstate corresponds to a state $((M_L)_{max}, (M_S)_{max})$ which is part of the ground term characterised by:

$$L = (M_L)_{max}; \; S = (M_S)_{max}$$

If we return to the example of the carbon atom once again, it can be shown that the 15 microstates corresponding to configuration $2p^2$ produce the following three (L, S) terms:

$$(L = 0, S = 0)\ (^1S):\ 1 \text{ state } (M_L, M_S)$$

$$(L = 1, S = 1)\ (^3P):\ 9 \text{ states } (M_L, M_S)$$

$$(L = 2, S = 0)\ (^1D):\ 5 \text{ states } (M_L, M_S)$$

The total number of (M_L, M_S) states is observed to be equal to the number of microstates. This is a general property as the passage from microstates to (M_L, M_S) states corresponds mathematically to a change of basis in a vector space, in this case a 15-dimensional space. The ground term is obtained by applying Hund's rule. The microstate which gives the maximum value of M_S, followed by the maximum value of M_L is represented in the following table, where the two lines schematically indicate the values of (m_ℓ, m_s):

−1	0	1
	↑	↑

This corresponds to the $(M_L = 1, M_S = 1)$ state. The ground term for the carbon atom is therefore $(L = 1, S = 1)$, i.e., ^3P, and its degeneracy is equal to 9. For the nitrogen atom, with configuration $2p^3$, this method produces a ground term $(L = 0, S = 3/2)$ written ^4S, for which the degeneracy is equal to 4. For fluorine, with configuration $2p^5$, the following microstate is obtained:

−1	0	1
↑	↑↓	↑↓

which gives the ground term $(L = 1, S = 1/2)$ written ^2P, with degeneracy equal to 6. Within an (L, S) term, expression [2] of the magnetic moment is written:

$$\boldsymbol{\mu} = -\beta\,(\mathbf{L} + g_e\,\mathbf{S}) \tag{6}$$

For the (M_L, M_S) state, its projection on any axis Z is given by:

$$\mu_Z = -\beta\,(M_L + g_e\,M_S)$$

Second step: magnetic interactions to which electrons are subjected

1 – Multiplets

We now introduce the so-called "spin-orbit" interaction between the angular momenta $\mathbf{l}_i$ and $\mathbf{s}_i$ for each electron. For atoms for which the atomic number is not too large ($Z < 80$), these magnetic interactions of relativistic origin are smaller than the splitting between energy levels for the terms. These interactions partly remove the degeneracy of the terms and produce energy levels identified by the three numbers (L, S, J). The number J is linked to the total angular momentum of the atom, defined by:

$$\mathbf{J} = \mathbf{L} + \mathbf{S}$$

For fixed (L, S), the rules for addition of two angular momenta require the number J to take values between $J_{min} = |L - S|$ and $J_{max} = L + S$ in unit steps. An ensemble of $(2J + 1)$ states $\{(L, S, J, M_J), M_J = -J, -J+1, \ldots J\}$ corresponding to energy level $E_{L,S,J}$ is called a *multiplet*. This multiplet is written ${}^{2S+1}X_J$, where X is the "letter code" for L. The wave functions representing the states for a multiplet are linear combinations of functions of the (L, S) term.

In the case of the carbon atom, the three terms from configuration $2p^2$ produce the following multiplets:

(L,S) term	(L,S,J) multiplet	number of (L,S,J,M_J) states
$(0,0)$ $({}^1S)$	$(0,0,0)$ $({}^1S_0)$	1
$(1,1)$ $({}^3P)$	$(1,1,0)$ $({}^3P_0)$	1
	$(1,1,1)$ $({}^3P_1)$	3
	$(1,1,2)$ $({}^3P_2)$	5
$(2,0)$ $({}^1D)$	$(2,0,2)$ $({}^1D_2)$	5

The total number of (L, S, J, M_J) states produced by the multiplets remains equal to 15, for the reasons given above. The ground multiplet is obtained by applying the following rule, justified in appendix 4:

- ▹ If the incomplete subshell is less than half full, the ground multiplet is characterised by $J = J_{min}$.
- ▹ If it is more than half full, it is characterised by $J = J_{max}$.

2 – Expression for the magnetic moment

Thanks to the properties of angular momenta, equation [6] can be replaced within an (L,S,J) multiplet where J is different from zero, by the equivalent expression (appendix 4):

$$\boldsymbol{\mu} = -g_{Landé}\,\beta\,\mathbf{J} \qquad [7]$$

$$g_{Landé} = 1 + (g_e - 1)\frac{J(J+1) + S(S+1) - L(L+1)}{2J(J+1)}$$

This formula, known as the *Landé* formula, gives the magnetic moment associated with the multiplet of an atom or a free ion. When applied to the ground multiplets for C, N, F atoms, for example:

- ▹ The ground term for the carbon atom is $(L = 1, S = 1)$ (^3P), and J can take the values (0,1,2). As the $2p$ subshell is less than half full, the ground multiplet is characterised by $J_{min} = 0$: the ground state for the carbon atom is therefore *not paramagnetic*. This example illustrates how not all atoms with an incomplete subshell have a paramagnetic ground state.
- ▹ For the nitrogen atom, the ground term is $(L = 0, S = 3/2)$ (^4S). The only possible value for J is 3/2. The ground state is therefore paramagnetic and the Landé formula gives $g_{Landé} = g_e = 2.0023$.
- ▹ For the fluorine atom, the ground term is $(L = 1, S = 1/2)$ (^2P). The possible values for J are 1/2 and 3/2. As the subshell is more than half full, the ground multiplet is characterised by $J_{max} = 3/2$. Landé's formula gives $g_{Landé} = 1 + (g_e - 1)/3 = 1.3341$.

When a paramagnetic atom is placed in a field **B**, interaction with the magnetic moment $\boldsymbol{\mu}$ (equation [7]) removes the degeneracy of the multiplets: each multiplet (L,S,J) gives $(2J + 1)$ equidistant energy levels, split by $g_{Landé}\beta B$ (figure A1). This splitting induced by application of a magnetic field has been known since the end of the 19th century, thanks to UV-visible absorption spectroscopy experiments performed on atomic gases (Zeeman effect). This effect was studied in detail for a number of atoms, through which the validity of the Landé formula was confirmed for light- and medium-weight atoms. In atoms and ions with a high atomic number, like those of the third transition series (incomplete $5d$ subshell) and actinides (incomplete $5f$ subshell), the spin-orbit interaction is comparable to the difference between the energies of the terms and the procedure that we described in this appendix, known as "(L,S) coupling", does not always apply.

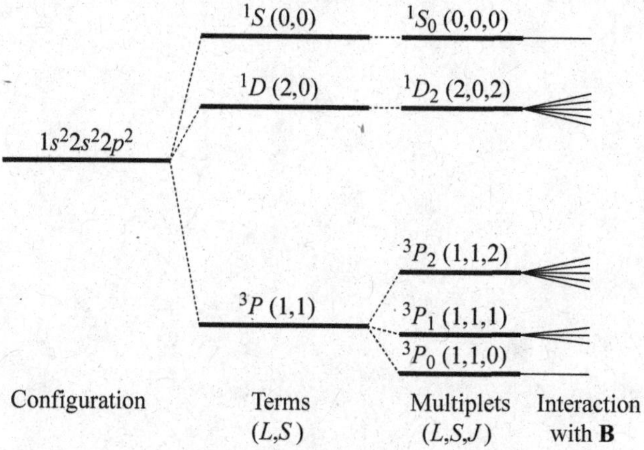

Figure A1 – Energy levels for the ground configuration of the carbon atom (splitting is not to scale).

Appendix 2

Expression of $\tilde{g}$ and $\tilde{A}$ matrices given by ligand field theory for a transition ion complex

We use *perturbation theory* to build the spin Hamiltonian describing the effect of magnetic interactions on the "orbital singlet" ground state of a mononuclear transition ion complex. This model and its limitations are illustrated by the example of a Cu^{2+} complex with square planar coordination.

1 – Electrostatic interactions in free ions

We will consider a "free" transition cation with an incomplete $3d$ subshell. The electrostatic interactions to which electrons are subjected produce energy levels identified by the numbers (L, S), each level corresponding to an ensemble of $(2L + 1)(2S + 1)$ states called a (L, S) *term* (appendix 1). We are interested in the *ground term* which is determined by applying Hund's rule. Its states can be represented by kets $|u\rangle|v\rangle$, where $|u\rangle$ and $|v\rangle$ are the kets for spaces $\mathcal{E}_L$ and $\mathcal{E}_S$ associated with the angular momenta **L** and **S**, representing the *total orbital state* and the *total spin state* of the free ion, respectively. Any direction Z of the Euclidean space defines $\hat{L}_Z$ and $\hat{S}_Z$ operators acting in the $\mathcal{E}_L$ and $\mathcal{E}_S$ spaces. If their eigenvectors are written as $\{|L, M_L\rangle\}$ and $\{|S, M_S\rangle\}$, the kets $\{|L, M_L\rangle|S, M_S\rangle\}$ form a basis for the (L, S) term. Z is completed by two axes, X and Y, to produce an orthonormal reference frame. The action of the operators $(\hat{L}_X, \hat{L}_Y, \hat{L}_Z)$ on the kets $\{|L, M_L\rangle\}$ and of the operators $(\hat{S}_X, \hat{S}_Y, \hat{S}_Z)$ on the kets $\{|S, M_S\rangle\}$ is described by equations [3.3] and [3.9]. This produces, for example:

P. Bertrand, *Electron Paramagnetic Resonance Spectroscopy*,
https://doi.org/10.1007/978-3-030-39663-3

$$\hat{L}_X|L, M_L\rangle = ½\,[L(L+1) - M_L(M_L+1)]^{½}\,|L, M_L+1\rangle + ½[L(L+1) - M_L(M_L-1)]^{½}\,|L, M_L-1\rangle$$

$$\hat{L}_Y|L, M_L\rangle = -½\,[L(L+1) - M_L(M_L+1)]^{½}\,|L, M_L+1\rangle + ½\,[L(L+1) - M_L(M_L-1)]^{½}\,|L, M_L-1\rangle$$

$$\hat{L}_Z|L, M_L\rangle = M_L|L, M_L\rangle \quad [1]$$

Remember that when the ion occupies a state of the (L, S) term, its magnetic moment can be written (Appendix 1):

$$\boldsymbol{\mu} = -\beta\,(\mathbf{L} + g_e\,\mathbf{S}) \quad [2]$$

2 – Electrostatic interactions with ligands

In "ligand field theory", interaction of the cation's electrons with the ligands is assimilated to the interaction with an electric field with the same symmetry as that of the complex. This interaction is represented by an operator $\hat{H}_{ligand}$. To use perturbation theory (see complement 2 in chapter 3), we assume that this interaction is weak relative to the differences between the terms (high-spin situation). As the ground term (L, S) is degenerate, we obtain the energy levels $\{E_k\}$ for the ion in the complex to the first order of perturbation theory, by adding to $E_{L,S}$ the eigenvalues for the matrix which represents $\hat{H}_{ligand}$ in the basis $\{|L, M_L\rangle\,|S, M_S\rangle\}$. Since this operator represents an electrostatic interaction, it acts only on the orbital states of the electrons and the spin states are not modified: the eigenvectors associated with the energy level E_k can be written $\{|\Phi_k\rangle\,|S, M_S\rangle\}$, where the orbital state $|\Phi_k\rangle$ is a linear combination of the kets $\{|L, M_L\rangle\}$ determined by the geometry of the complex. The expression for the orbital states and the degeneracy of the energy levels $\{E_k\}$ can be determined by using *Group Theory*, which describes the relation between the electronic properties and the symmetry properties of molecules in a very general way [Jean, 2005].

As a concrete example of this, we will consider a Cu^{2+} ion for which the ground configuration $3d^9$ gives a single term ^2D $(L = 2, S = ½)$. Any quantization axis defines a 10-ket basis $\{|2, M_L\rangle|½, M_S\rangle\}$ corresponding to the energy $E_{2,½}$. When the Cu^{2+} ion is coordinated by 4 ligands in a square planar geometry (figure A2.1a), the energies of the five $3d$ orbitals defined by the axes (x, y, z) are different (figure A2.1b and complement 1 in chapter 4). If we choose the four-fold symmetry axis z as the quantization axis, the 5 orbital states obtained by placing the unpaired electron in one of these orbitals can be written as follows:

$$|\Phi_{xz}\rangle = 1/\sqrt{2}\ (|2,-1\rangle - |2,1\rangle)$$
$$|\Phi_{yz}\rangle = \mathrm{i}/\sqrt{2}\ (|2,-1\rangle + |2,1\rangle)$$
$$|\Phi_{xy}\rangle = \mathrm{i}/\sqrt{2}\ (|2,-2\rangle - |2,2\rangle)$$
$$|\Phi_{z^2}\rangle = |2,0\rangle$$
$$|\Phi_{x^2-y^2}\rangle = 1/\sqrt{2}\ (|2,-2\rangle + |2,2\rangle) \qquad [3]$$

The subscript indicates which orbital is occupied by the unpaired electron. The ground state is the $|\Phi_{x^2-y^2}\rangle$ state and the energy level diagram is represented in figure A2.2b. The precise order and energies of the excited states (ranging from 12,000 to 20,000 cm^{-1}) depend on the nature of the ligands, but the energies E_{xz} and E_{yz} are always equal because of the equivalence of the x and y directions in the complex.

We will now return to the general case. At this level of description, the $(2S + 1)$ states $\{|\Phi_k\rangle|S, M_S\rangle, M_S = -S, -S+1, \ldots S\}$ are associated with the same energy E_k. The symmetry of transition ion complexes is generally low enough for their *ground state* to be an "orbital singlet", i.e., an orbitally non-degenerate state, written as $|\Phi_1\rangle$ ($|\Phi_{x^2-y^2}\rangle$ in the case illustrated in figure A2.2b). In these conditions, if Z is the quantization axis, the matrix elements of the operators $(\hat{L}_X, \hat{L}_Y, \hat{L}_Z)$ in state $|\Phi_1\rangle$ are *null*:

$$\langle \Phi_1|\hat{L}_X|\Phi_1\rangle = \langle \Phi_1|\hat{L}_Y|\Phi_1\rangle = \langle \Phi_1|\hat{L}_Z|\Phi_1\rangle = 0 \qquad [4]$$

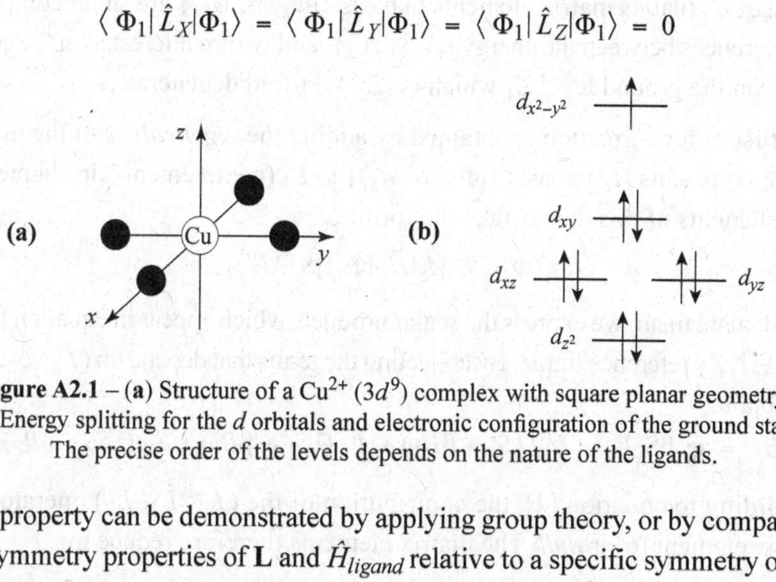

Figure A2.1 – (**a**) Structure of a Cu^{2+} ($3d^9$) complex with square planar geometry. (**b**) Energy splitting for the d orbitals and electronic configuration of the ground state. The precise order of the levels depends on the nature of the ligands.

This property can be demonstrated by applying group theory, or by comparing the symmetry properties of **L** and $\hat{H}_{ligand}$ relative to a specific symmetry operation called "time-reversal" [Abragam and Bleaney, 1986].

Equations [4] are generally interpreted by saying that the motion of the electrons responsible for the orbital angular momentum **L**–which is favoured in atoms and free ions by the spherical symmetry of their distribution–is "quenched" in the complex because of the very anisotropic electrostatic interactions with ligands. We will see that quenching of orbital angular momentum has significant consequences on the *magnetic properties* of the complex.

3 – Magnetic interactions

We now introduce *magnetic interactions*, which are generally weaker than electrostatic interactions with ligands. These interactions remove the spin degeneracy of the ground level, E_1, to produce the levels between which EPR transitions take place. We will see that this splitting can be reproduced by an equivalent Hamiltonian which only contains *spin operators*.

3.1 – Effects of spin-orbit coupling and interaction with a magnetic field

Within an (L, S) term, the sum of these interactions is described by the operator

$$\hat{H}_P = \hat{H}_{SO} + \hat{H}_{Zeeman}; \; \hat{H}_{SO} = \lambda\, \mathbf{L}\cdot\mathbf{S}; \; \hat{H}_{Zeeman} = \beta(\mathbf{L} + g_e\, \mathbf{S})\cdot\mathbf{B} \qquad [5]$$

λ is the spin-orbit coupling constant for the transition ion and we have used expression [2] for the magnetic moment. We assume that $\hat{H}_P$ is a *perturbation* of $\hat{H}_{ligand}$, i.e., that its matrix elements in basis $\{|\Phi_k\rangle|S, M_S\rangle\}$ are small compared to differences between the energy levels $\{E_k\}$, and we are interested in the effect of $\hat{H}_P$ on the ground level E_1 which is $(2S + 1)$-fold degenerate.

The first order correction is obtained by adding the *eigenvalues* of the matrix which represents $\hat{H}_P$ in basis $\{|\Phi_1\rangle|S, M_S\rangle\}$ to E_1 (complement 2 in chapter 3). The elements of this matrix take the form:

$$\langle\Phi_1|\langle S, M_S|\hat{H}_P|\Phi_1\rangle|S, M_S'\rangle \qquad [6]$$

To calculate them, we express the scalar products which appear in equation [5] in the $\{X, Y, Z\}$ reference frame. By collecting the terms that depend on $(\hat{L}_X, \hat{L}_Y, \hat{L}_Z)$, we obtain:

$$\hat{H}_P = g_e\, \beta\mathbf{S}\cdot\mathbf{B} + [\hat{L}_X(\lambda\hat{S}_X + \beta B_X) + \hat{L}_Y(\lambda\hat{S}_Y + \beta B_Y) + \hat{L}_Z(\lambda\hat{S}_Z + \beta B_Z)]$$

According to equation [4], the contribution of the $(\hat{L}_X, \hat{L}_Y, \hat{L}_Z)$ operators to matrix element [6] is *null*. The matrix elements therefore reduce to:

$$\langle S, M_S|\, g_e\beta\mathbf{S}\cdot\mathbf{B}\, |S, M_S'\rangle \qquad [7]$$

The first order calculation therefore leads to the following expression for the magnetic moment of the complex:

$$\boldsymbol{\mu} = -g_e \beta \mathbf{S} \qquad [8]$$

This result does not correlate with the experimental data which indicate that the magnetic moment for transition ion complexes is *anisotropic*. The first order approximation is therefore insufficient, and we must proceed to second order calculations.

These calculations are detailed in works devoted to magnetism, and here we will simply provide the results [Abragam and Bleaney, 1986]. The splitting of the energy level E_1 is found to be identical to that produced by a "spin Hamiltonian" $\hat{H}_S$ with the form:

$$\hat{H}_S = \mathbf{S}\cdot\tilde{\mathbf{D}}\cdot\mathbf{S} + \beta\,\mathbf{B}\cdot\tilde{\mathbf{g}}\cdot\mathbf{S} \qquad [9]$$

The first term, called the "zero-field splitting", contributes to the splitting when the spin S of the paramagnetic centre is greater than ½; its effects on the EPR spectrum are described in chapter 6. The second term is equivalent to the Zeeman term in equation [5] *for the lowest energy level*, E_1. For this level, expression [2] for the magnetic moment can therefore be replaced by:

$$\boldsymbol{\mu} = -\beta\,\tilde{\mathbf{g}}\cdot\mathbf{S}$$

The contribution of the orbital angular momentum of the electrons to the magnetic properties of the complex is included in the $\tilde{\mathbf{D}}$ and $\tilde{\mathbf{g}}$ matrices, which depend on the *orbital states* $\{|\Phi_k\rangle\}$ and their *energies* $\{E_k\}$. In particular, the $\tilde{\mathbf{g}}$ matrix takes the form:

$$\tilde{\mathbf{g}} = g_e\tilde{\mathbf{1}} + \Delta\tilde{\mathbf{g}} \qquad [10]$$

where $\tilde{\mathbf{1}}$ is the unit matrix and $\Delta\tilde{\mathbf{g}}$ is a matrix for which the elements are given by

$$(\Delta g)_{ij} = -\lambda\sum_k \frac{2\langle\Phi_1|\hat{L}_i|\Phi_k\rangle\langle\Phi_k|\hat{L}_j|\Phi_1\rangle}{E_k - E_1} \qquad i,j = X,Y,Z \qquad [11]$$

The summation is made over all excited $|\Phi_k\rangle$ orbital states of energy E_k. In the right-hand side of equation [10], the first term represents the first order approximation (equation [8]) and the second one represents the second order correction.

When the $\tilde{\mathbf{g}}$ matrix is expressed in the reference frame of its principal axes $\{x, y, z\}$, the off-diagonal elements are null and the diagonal elements are the *principal values* which are therefore written:

$$g_i = g_e - \lambda \sum_k \frac{2|\langle \Phi_1 |\hat{L}_i| \Phi_k \rangle|^2}{E_k - E_1} \qquad i = x, y, z \qquad [12]$$

They take the form indicated in equation [4.2].

We now return to the example of the Cu^{2+} complex with square planar geometry. Using equations [1] and [3], we find that the only non-null matrix elements in equation [11] are

$$\langle \Phi_{x^2-y^2} | \hat{L}_x | \Phi_{yz} \rangle = \langle \Phi_{x^2-y^2} | \hat{L}_y | \Phi_{xz} \rangle = \mathrm{i} \; ; \; \langle \Phi_{x^2-y^2} | \hat{L}_z | \Phi_{xy} \rangle = -2\mathrm{i}$$

The $\tilde{\mathbf{g}}$ matrix is therefore diagonal in the $\{x, y, z\}$ reference frame shown in figure A2.1a, and its principal values are given by:

$$\begin{aligned} g_x = g_y = g_\perp &= g_e - 2\lambda/(E_{xz} - E_{x^2-y^2}) \\ g_z = g_{//} &= g_e - 8\lambda/(E_{xy} - E_{x^2-y^2}) \end{aligned} \qquad [13]$$

The energies are those shown in figure A2.2b. The axial symmetry of the $\tilde{\mathbf{g}}$ matrix reflects the axial symmetry of the complex. As the $3d$ subshell for the Cu^{2+} ion is more than half-full, the spin-orbit coupling constant λ is negative and the principal values $g_{//}$ and $g_\perp$ are greater than $g_e = 2.0023$. Because of the factor 8 which appears in the expression for $g_{//}$, we expect $g_{//} > g_\perp$. This can effectively be observed on the spectrum shown in figure 4.11.

We will investigate whether expression [13] adequately describes the experimental data, by considering the case of the $Cu(NH_3)_4{}^{2+}$ complex mentioned in section 4.4.1. For this complex, the spectroscopic data indicate that $E_{xz} - E_{x^2-y^2} = 17{,}500\ \mathrm{cm}^{-1}$ and $E_{xy} - E_{x^2-y^2} = 14{,}000\ \mathrm{cm}^{-1}$ [Neese, 2004]. With the value $\lambda = -830\ \mathrm{cm}^{-1}$ for the *free* Cu^{2+} ion (table 4.4), equation [13] gives $g_{//} = 2.477$ and $g_\perp = 2.097$; these values are quite different from the experimental values $g_{//} = 2.241$ and $g_\perp = 2.047$ [Scholl and Hüttermann, 1992].

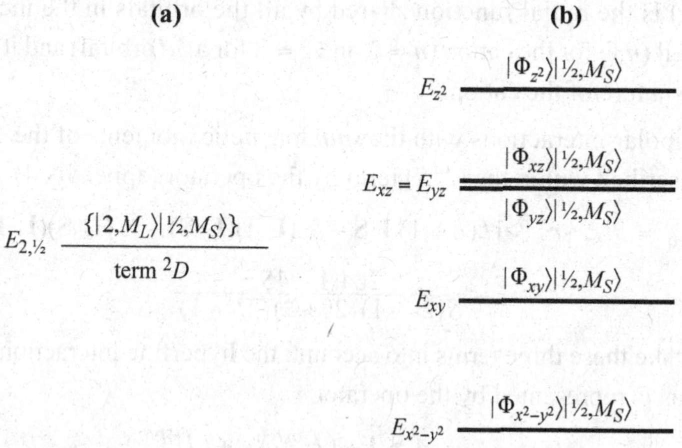

Figure A2.2 – Energy levels and possible states for a Cu^{2+} ion. (**a**) Free ion, ground term. (**b**) Square planar geometry complex. The spin Hamiltonian describes the effect of magnetic interactions on the lowest energy level.

This type of discrepancy is common when using expressions given by ligand field theory, in which delocalisation of the unpaired electrons on the ligands is completely neglected. Some authors take it into account by applying an appropriate "reduction factor" to λ. In the present case, a factor of around 0.5 is necessary to obtain good results.

3.2 – Effects of hyperfine interactions

Interaction between unpaired electrons and the nucleus of the transition ion produces three terms (section 4.4.1):

1. The polarisation of the electrons in the *s* orbitals by the *d* electrons gives a term $A_s\mathbf{S}\cdot\mathbf{I}$.
2. The dipolar interactions between the magnetic moment $g_N\beta_N\mathbf{I}$ of the nucleus and the *orbital* magnetic moments of the electrons are described within the (L,S) term by the equivalent operator (appendix 4):

$$(\hat{H}_{dip}^{orb})_{eq} = P\,<r^{-3}>\,\mathbf{L}\cdot\mathbf{I}$$

$$P = (\mu_0/4\pi)g_e g_N\,\beta\beta_N \quad \text{(writing } g_e = 2)$$

$$<r^{-3}> = \int_0^{+\infty} r^{-3}R_{n,\ell}^2(r)r^2\,dr$$

$R_{n,\ell}(r)$ is the radial function shared by all the orbitals in the incomplete subshell (n,ℓ) for the cation ($n = 3$ and $\ell = 2$ for a $3d$ orbital) and it depends on the nature of the cation.

3. The dipolar interactions with the *spin* magnetic moments of the electrons are described within the (L, S) term by the operator (appendix 4):

$$(\hat{H}_{dip}^{spin})_{eq} = P\,\xi <r^{-3}>[L(L+1)\,\mathbf{I}\cdot\mathbf{S} - \tfrac{3}{2}\,(\mathbf{L}\cdot\mathbf{I})(\mathbf{L}\cdot\mathbf{S}) - \tfrac{3}{2}\,(\mathbf{L}\cdot\mathbf{S})(\mathbf{L}\cdot\mathbf{I})] \quad [14]$$

$$\xi = \frac{2\ell + 1 - 4S}{S(2\ell - 1)(2\ell + 3)(2L - 1)}$$

If we take these three terms into account, the hyperfine interaction with the nucleus is represented by the operator

$$\hat{H}_{hyperfine} = A_s\mathbf{S}\cdot\mathbf{I} + (\hat{H}_{dip}^{orb})_{eq} + (\hat{H}_{dip}^{spin})_{eq} \quad [15]$$

This operator is added to $\hat{H}_P$ (equation [5]). To determine the effects of this interaction on the ground state, we have to re-examine the different steps detailed in section 3.1.

▹ *To the first order*, the operator $(\hat{H}_{dip}^{orb})_{eq}$ does not contribute since the matrix elements of $(\hat{L}_X, \hat{L}_Y, \hat{L}_Z)$ are null (equation [4]). The operators $A_s\mathbf{S}\cdot\mathbf{I}$ and $(\hat{H}_{dip}^{spin})_{eq}$ give matrix elements with the following form:

$$\langle\Phi_1|\langle S, M_S\,|A_s\mathbf{S}\cdot\mathbf{I}\,|\Phi_1\rangle|S, M_S'\rangle + \langle\Phi_1|\langle S, M_S|(\hat{H}_{dip}^{spin})_{eq}\,|\Phi_1\rangle|S, M_S'\rangle$$

The first term is equal to $\langle S, M_S|A_s\mathbf{S}\cdot\mathbf{I}|S, M_S'\rangle$. By replacing $(\hat{H}_{dip}^{spin})_{eq}$ by its expression (equation [14]) and rearranging, we can write the second term in the following compact form:

$$\langle S, M_S|\mathbf{S}\cdot\tilde{\mathbf{T}}\cdot\mathbf{I}|S, M_S'\rangle$$

$\tilde{\mathbf{T}}$ is the "dipolar matrix", the elements of which are given by:

- diagonal elements:

$$T_{ii} = -\tfrac{3}{2}\,P\,\xi <r^{-3}>\,\langle\Phi_1|2\,\hat{L}_i^2 - 2L(L+1)/3\;|\Phi_1\rangle \qquad i = X, Y, Z$$

- off-diagonal elements:

$$T_{ij} = -\tfrac{3}{2}\,P\,\xi <r^{-3}>\,\langle\Phi_1|\,\hat{L}_i\,\hat{L}_j + \hat{L}_j\,\hat{L}_i\,|\Phi_1\rangle \qquad i,j = X, Y, Z$$

This matrix is symmetric and *its trace is null*. Indeed, since ket $|\Phi_1\rangle$ is part of the (L, S) term, it verifies $(\hat{L}_X^2 + \hat{L}_Y^2 + \hat{L}_Z^2)|\Phi_1\rangle = \hat{\mathbf{L}}^2|\Phi_1\rangle = L(L+1)|\Phi_1\rangle$, which leads to $T_{XX} + T_{YY} + T_{ZZ} = 0$.

▹ *Second order calculation* [Abragam and Bleaney, 1986]:

- The operators $\lambda\mathbf{L}\cdot\mathbf{S}$ and $(\hat{H}^{orb}_{dip})_{eq}$ give a term $P{<}r^{-3}{>}\ \mathbf{S}\cdot\Delta\tilde{\mathbf{g}}\cdot\mathbf{I}$, where $\Delta\tilde{\mathbf{g}}$ is the matrix involved in equation [10].
- The operators $\lambda\mathbf{L}\cdot\mathbf{S}$ and $(\hat{H}^{spin}_{dip})_{eq}$ give a term with the form $P\xi\ {<}r^{-3}{>}\ \mathbf{S}\cdot\Delta\tilde{\mathbf{g}}'\cdot\mathbf{I}$. where $\Delta\tilde{\mathbf{g}}'$ is a matrix with elements similar to those of $\Delta\tilde{\mathbf{g}}$ (equation [11]).

The splitting of the ground level E_1 induced by the hyperfine interaction described by $\hat{H}_{hyperfine}$ (equation [15]) is therefore identical to that produced by the equivalent operator $\mathbf{S}\cdot\tilde{\mathbf{A}}\cdot\mathbf{I}$ where:

$$\tilde{\mathbf{A}} = A_s\tilde{\mathbf{1}} + \tilde{\mathbf{T}} + P{<}r^{-3}{>}(\Delta\tilde{\mathbf{g}} + \xi\Delta\tilde{\mathbf{g}}')$$

We now detail these three terms for a Cu^{2+} complex with square planar geometry. To obtain the matrices $\tilde{\mathbf{T}}$ and $\Delta\tilde{\mathbf{g}}'$, we use the expressions for the matrix elements of operators $\hat{L}_i^2$ and $\hat{L}_i\hat{L}_j$ given in appendix 7. As expected, we find that matrix $\tilde{\mathbf{A}}$ is *diagonal* in the $\{x,y,z\}$ reference frame. It is axially symmetric, with:

$$\begin{aligned} A_{//} &= A_s - \tfrac{4}{7}P{<}r^{-3}{>} + P{<}r^{-3}{>}(\Delta g_{//} + \tfrac{3}{7}\Delta g_{\perp}) \\ A_{\perp} &= A_s + \tfrac{2}{7}P{<}r^{-3}{>} + P{<}r^{-3}{>}(\Delta g_{\perp} - \tfrac{3}{14}\Delta g_{\perp}) \end{aligned} \qquad [16]$$

If these expressions are applied to the $Cu(NH_3)_4^{2+}$ complex with the parameters $A_s \approx -300$ MHz and $P{<}r^{-3}{>} = 1200$ MHz calculated for a *free* Cu^{2+} ion [Freeman and Watson, 1965] and the experimental values $g_{//} = 2.241$ and $g_{\perp} = 2.047$, they produce $A_{//} \approx -680$ MHz, $A_{\perp} \approx +85$ MHz. These values should be compared to the experimental values $|A_{//}| = 586$ MHz, $|A_{\perp}| = 68$ MHz [Scholl and Hüttermann, 1992]. Once again, the agreement is mediocre as $A_{\perp}$ is in fact negative [Neese, 2004]. This calculation nevertheless demonstrates that the three contributions of equation [16] are of the same order of magnitude, that $A_{//}$ is negative and that $|A_{//}| \gg |A_{\perp}|$.

3.3 – The case of complexes of cations in an *S* state

For ions with a $3d^5$ configuration, like Mn^{2+} and Fe^{3+}, the ground term is $^6S\ (L = 0, S = \tfrac{5}{2})$ for which the basis is $\{|0,0\rangle|\tfrac{5}{2}, M_S\rangle\}$. As this term is not orbitally degenerate, interaction of the electrons with ligands modifies their energy without causing splitting. The contribution of the ground term to the $\tilde{\mathbf{D}}$ and $\Delta\tilde{\mathbf{g}}$ matrices is therefore null, but the contribution of the excited terms is sometimes significant, in particular in the case of Fe^{3+} heme complexes where the elements of the $\tilde{\mathbf{D}}$ matrix reach tens of cm^{-1} [Scholes, 1971]. With regard to the hyperfine interaction, the $\tilde{\mathbf{T}}$ and $\Delta\tilde{\mathbf{g}}$ matrices are null and the $\tilde{\mathbf{A}}$ matrix should reduce to the isotropic term $A_s\tilde{\mathbf{1}}$. In practice, weak anisotropy is observed due to the contributions of the excited terms.

◇ *Some comments*

Ligand field theory has the advantage of producing simple expressions which can be used to define useful concepts, but the results it provides are disappointing in quantitative terms. To obtain detailed information on the electronic properties of a complex from the principal values of the $\tilde{\mathbf{g}}$ and $\tilde{\mathbf{A}}$ matrices deduced from the EPR spectrum, more realistic models based on the *molecular orbital* concept must be used. DFT-type calculations were performed on the $Cu(NH_3)_4{}^{2+}$ complex [Neese, 2004]. They effectively reproduced the experimentally-determined principal values for the $\tilde{\mathbf{g}}$ and $\tilde{\mathbf{A}}$ matrices for the Cu^{2+} ion, as well as those for the matrix characterising the superhyperfine interactions with the nitrogen atoms for the 4 ligands (section 4.4.3). In particular, they show that the spin population of the $d_{x^2-y^2}$ orbital of the Cu^{2+} ion is equal to around 0.6, and that the remainder of the spin density is mainly delocalised on the nitrogen atoms.

References

ABRAGAM A. & BLEANEY B. (1986) *Electron Paramagnetic Resonance of transition ions*, Dover, New York.

FREEMANN A.J. & WATSON R.E. (1965) "Hyperfine Interactions in Magnetic Materials"
in *Magnetism* vol II A, RADO G.T. & SUHL H. eds, Academic Press, New York.

JEAN Y. (2005) *Molecular Orbitals of Transition Metal Complexes*, Oxford University Press, Oxford.

NEESE F. (2004) *Magnetic Resonance in Chemistry* **42**: S187-S198.

SCHOLES C.P. *et al.* (1971) *Biochimica et Biophysica Acta* **244**: 206-210.

SCHOLL H.J. & HÜTTERMANN J. (1992) *Journal of Physical Chemistry* **96**: 9684-9691.

Appendix 3

Dipolar interactions between a nuclear magnetic moment and electron spin magnetic moments

1 – The dipolar matrix $\tilde{\mathbf{T}}$

According to classical electromagnetic theory, interaction between two macroscopic moments $\boldsymbol{\mu}_A$ and $\boldsymbol{\mu}_B$ located at points O and M can be written:

$$H_{dip} = (\mu_0/4\pi)\,[r^2\,\boldsymbol{\mu}_A\cdot\boldsymbol{\mu}_B - 3(\boldsymbol{\mu}_A\cdot\mathbf{r})(\boldsymbol{\mu}_B\cdot\mathbf{r})]/r^5 \qquad [1]$$

where $\mu_0 = 4\pi \times 10^{-7}$ H m^{-1} is the vacuum permeability, $\mathbf{r}$ is the vector $\overrightarrow{OM}$ and r is its length. The interaction depends on r, on the relative orientation of $\boldsymbol{\mu}_A$ and $\boldsymbol{\mu}_B$ and on their orientation relative to $\mathbf{r}$.

Let us now consider an atom or molecule where a *nucleus* with angular momentum $\mathbf{I}$ located at O interacts with an electron of spin angular momentum $\mathbf{s}$ located at M. This interaction can be described using equation [1] on the condition that $\boldsymbol{\mu}_A$ and $\boldsymbol{\mu}_B$ are replaced by $g_N\beta_N\,\mathbf{I}$ and $-g_e\beta\mathbf{s}$, respectively, and that H_{dip} is considered as an *operator*:

$$\hat{H}_{dip}^{spin} = P\,[3\,(\mathbf{I}\cdot\mathbf{r})(\mathbf{s}\cdot\mathbf{r}) - r^2(\mathbf{I}\cdot\mathbf{s})]/r^5 \qquad [2]$$

where:

$$P = (\mu_0/4\pi)\,g_e g_N \beta\beta_N$$

The components of $\mathbf{s}$ and $\mathbf{I}$ are operators which act, respectively, on the kets which represent the spin states of the electron and the nucleus. If we express $\hat{H}_{dip}^{spin}$ in a Cartesian reference frame $\{x,y,z\}$ centred at O, we obtain a sum of 9 terms which can be written in the following matrix form:

P. Bertrand, *Electron Paramagnetic Resonance Spectroscopy*,
https://doi.org/10.1007/978-3-030-39663-3

$$\hat{H}_{dip}^{spin} = P[\hat{s}_x\ \hat{s}_y\ \hat{s}_z] \begin{bmatrix} \frac{3r_x^2 - r^2}{r^5} & \frac{3r_x r_y}{r^5} & \frac{3r_x r_z}{r^5} \\ \frac{3r_y r_x}{r^5} & \frac{3r_y^2 - r^2}{r^5} & \frac{3r_y r_z}{r^5} \\ \frac{3r_z r_x}{r^5} & \frac{3r_z r_y}{r^5} & \frac{3r_z^2 - r^2}{r^5} \end{bmatrix} \begin{bmatrix} \hat{I}_x \\ \hat{I}_y \\ \hat{I}_z \end{bmatrix}$$

This expression includes (r_x, r_y, r_z), the components of **r** which defines the position of the electron relative to the nucleus (figure A3.1). In an atom or a molecule, the *electron probability density* at point M is given by $|\varphi(\mathbf{r})|^2$, where $\varphi(\mathbf{r})$ is the *orbital* occupied by the electron. We must thus replace each matrix element by its *average value* $<>_\varphi$ obtained by integrating over the position of the electron in $\varphi(\mathbf{r})$. The result can be written in the form:

$$\hat{H}_{hyperfine}^{spin} = \mathbf{s} \cdot \tilde{\mathbf{T}} \cdot \mathbf{I} \qquad [3]$$

where $\tilde{\mathbf{T}}$ is the *dipolar matrix* for which the elements are as follows:

- ▹ diagonal elements: $T_{xx} = P <(3r_x^2 - r^2)/r^5>_\varphi$ and similar expressions for T_{yy} and T_{zz}.
- ▹ off-diagonal elements: $T_{xy} = P <3\ r_x\ r_y/r^5>_\varphi$ and similar expressions for T_{xz} and T_{yz}.

The $\tilde{\mathbf{T}}$ matrix is symmetric and has a null trace. Indeed:

$$T_{xx} + T_{yy} + T_{zz} = P < [3(r_x^2 + r_y^2 + r_z^2) - 3r^2]/r^5 >_\varphi = 0$$

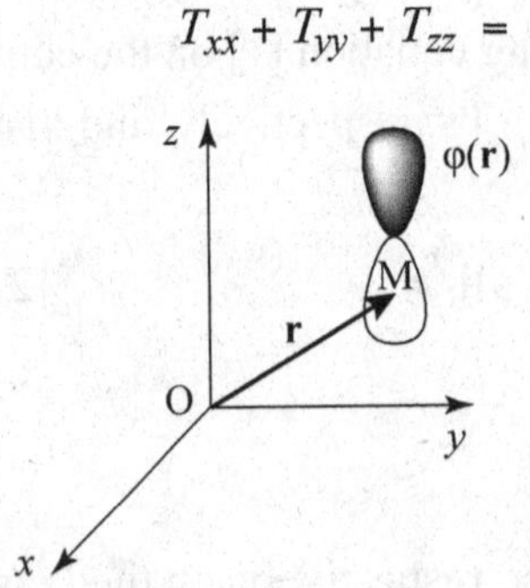

Figure A3.1 – Dipolar interaction between a nucleus located at O and an electron located in an orbital $\varphi(\mathbf{r})$.

Case where the $\varphi(\mathbf{r})$ orbital is spherically symmetric relative to the nucleus

When the orbital is spherically symmetric relative to the nucleus, $|\varphi(\mathbf{r})|^2$ only depends on the *distance r*. It corresponds to the case of a *s* atomic orbital centred at O. In this case, the $\tilde{\mathbf{T}}$ matrix is *null* whatever the $\{x, y, z\}$ reference frame. Indeed:

▷ As the $\{x,y,z\}$ directions are equivalent, the three diagonal elements are equal. Since their sum is null, this leads to $T_{xx} = T_{yy} = T_{zz} = 0$.
▷ Consider any off-diagonal term such as T_{xy}. The points $M(r_x, r_y, r_z)$ and $M'(-r_x, r_y, r_z)$, which are symmetric with respect to the (yOz) plane give opposite contributions to the mean value $<3\, r_x\, r_y/r^5>_\varphi$. As this plane is a symmetry plane for the orbital $\varphi(\mathbf{r})$, the quantity $<3\, r_x\, r_y/r^5>_\varphi$ is null.

2 – Principal axes and principal values of the $\tilde{\mathbf{T}}$ matrix

There exists a system of axes $\{X, Y, Z\}$ such that the matrix is *diagonal*:

$$\begin{bmatrix} T_{XX} & 0 & 0 \\ 0 & T_{YY} & 0 \\ 0 & 0 & T_{ZZ} \end{bmatrix}$$

These are the principal axes of the $\tilde{\mathbf{T}}$ matrix. The principal values are the diagonal elements:

$$\begin{aligned} T_{XX} &= P<(3\, r_X{}^2 - r^2)/r^5>_\varphi \\ T_{YY} &= P<(3\, r_Y{}^2 - r^2)/r^5>_\varphi \\ T_{ZZ} &= P<(3\, r_Z{}^2 - r^2)/r^5>_\varphi \end{aligned} \quad [4]$$

Frequently, the orbital $\varphi(\mathbf{r})$ has a rotation axis z which passes through the nucleus. The directions x and y are then equivalent with regard to the dipolar interactions, and the $\tilde{\mathbf{T}}$ matrix is axial, with $T_{zz} = T_{//} = -2T_\perp$. Some examples are detailed below. They are illustrated in figure A3.2.

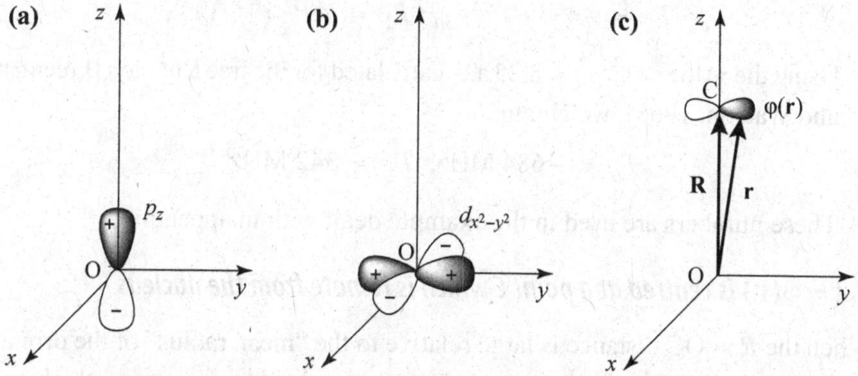

Figure A3.2 – Examples of situations where the dipolar matrix $\tilde{\mathbf{T}}$ is axial. **(a, b)** The electron occupies an orbital centred on the nucleus, which possesses a rotation axis z. **(c)** The electron occupies an orbital centred at C, which is far from the nucleus located at O.

2.1 – $\varphi(\mathbf{r})$ *is an atomic orbital centred at O*

▹ In free radicals, the anisotropic components of the hyperfine interaction with a ^{13}C or ^{14}N nucleus are mainly determined by the dipolar interaction with the electron in the $2p$ orbital of the same atom (figure A3.2a). If we integrate over the angular part of this orbital, equation [4] becomes [Atherton, 1993]:

$$T_{//} = \tfrac{4}{5}\, P <r^{-3}>_{2p},\ T_{\perp} = -\tfrac{2}{5}\, P <r^{-3}>_{2p}$$

The mean value $<r^{-3}>_{2p}$, which only depends on the radial part of the $2p$ orbital, is tabulated [Morton et Preston, 1978]. Its value is generally given in atomic units (a_0^{-3}, where $a_0 = 4\pi\varepsilon_0\hbar^2/m_e e^2 \approx 0.52918$ Å) and we obtain its value in the international system of units by multiplying by 6.748×10^{30}.

Thus, for example, for the nitrogen atom we have $<r^{-3}>_{2p} = 3.599$ a.u. The dipolar matrix which describes the interaction of a $2p_z$ electron from this atom with the ^{14}N nucleus ($g_N = 0.4038$) is axial with:

$$T_{//} = 110 \text{ MHz},\ T_{\perp} = -55 \text{ MHz}$$

In a radical, these quantities must be weighted by the *spin population* for the $2p$ orbital (exercise 4.6).

▹ In a Cu^{2+} complex with square planar geometry, the single unpaired electron occupies a $d_{x^2-y^2}$ orbital. The dipolar interaction between this electron and the ^{63}Cu or ^{65}Cu nucleus (figure A3.2b) is described by a $\tilde{\mathbf{T}}$ matrix which is axial with:

$$T_{//} = -\tfrac{4}{7}\, P <r^{-3}>_{3d},\ T_{\perp} = \tfrac{2}{7} P <r^{-3}>_{3d}$$

Using the value $<r^{-3}>_{3d} = 8.25$ a.u. calculated for the free Cu^{2+} ion [Freeman and Watson, 1965], we obtain:

$$T_{//} = -684 \text{ MHz},\ T_{\perp} = 342 \text{ MHz}$$

These numbers are used in the example dealt with in appendix 2.

2.2 – $\varphi(\mathbf{r})$ *is centred at a point C which is remote from the nucleus*

When the R = OC distance is large relative to the "mean radius" of the orbital $\varphi(\mathbf{r})$, the mean values in equation [4] are determined by $\mathbf{r}$ vectors which are close to $\mathbf{R} = \overrightarrow{OC}$ (figure A3.2c). The "point dipole" approximation consists in assuming that the whole of the spin density is concentrated at point C. The Z direction of $\mathbf{R}$ thus becomes a rotation axis (figure A3.2c):

$$\langle r_X^2 \rangle_\varphi = \langle r_Y^2 \rangle_\varphi = 0, \ \langle r_Z^2 \rangle_\varphi = R^2$$

Which produces:

$$T_{//} = 2P/R^3, \ T_\perp = -P/R^3$$

For example, we get $T_\perp$[MHz] = $-5.7/R^3$ for a ^{14}N nucleus and $T_\perp$[MHz] = $-78/R^3$ for a ^{1}H proton; R is expressed in [Å].

The point dipole approximation is good when the distance R is greater than 2.5 Å for a $3d$ orbital [Atherton and Horsewill, 1979; Atherton and Shackleton, 1984].

3 – Case with several unpaired electrons

All of the above relates to paramagnetic centres with a single unpaired electron, such as free radicals and some transition ion complexes. In complexes where the nucleus interacts with N unpaired electrons, the dipolar Hamiltonian is the sum of N terms similar to that of equation [2]:

$$\hat{H}_{dip}^{spin} = P \sum_{i=1}^{N} \frac{3(\mathbf{I}\cdot\mathbf{r}_i)(\mathbf{s}_i\cdot\mathbf{r}_i) - r_i^2(\mathbf{I}\cdot\mathbf{s}_i)}{r_i^5}$$

where $\mathbf{r}_i(r_{xi}, r_{yi}, r_{zi})$ indicates the position of electron i relative to the nucleus. When the nucleus in question is that of the cation and the electrons are mainly localised in its orbitals, we can use the equivalent operators method to simplify the integration over the position of the electrons (appendix 4). When the ground state is an orbital singlet produced by splitting of an (L, S) term, this integration produces an operator which takes the form (appendix 2):

$$\hat{H}_{hyperfine}^{spin} = \mathbf{S}\cdot\tilde{\mathbf{T}}\cdot\mathbf{I}$$

where $\tilde{\mathbf{T}}$ is a matrix with a null trace.

References

ATHERTON N.M. (1993) *Principles of Electron Spin Resonance*, Ellis Horwood PTR Prentice Hall, New York.

ATHERTON N.M. & HORSEWILL A.J. (1979) *Molecular Physics* **37**: 1349-1361.

ATHERTON N.M. & SHAKLETON J.F. (1984) *Chemical Physics Letters* **103**: 302-304

FREEMANN A.J. & WATSON R.E. (1965) "Hyperfine Interactions in Magnetic Materials" in *Magnetism* vol II A, RADO G.T. & SUHL H. eds, Academic Press, New York.

MORTON J.R. & PRESTON K.F. (1978) *Journal of Magnetic Resonance* **30**: 577-582.

Appendix 4

Some properties of angular momentum operators. Spin coupling coefficients and equivalent operators. Application to Landé's formula and to dipolar hyperfine interactions

The operators defined from angular momenta have numerous properties which considerably simplify the calculation of eigenvalues and eigenvectors for the Hamiltonian operator. In particular, they allow operators to be replaced by "equivalent operators" within some sub-spaces, which are easier to use.

1 – Definition of coupled bases and spin coupling coefficients

1.1 – Product bases and coupled bases

The Hamiltonian through which the EPR spectrum for a paramagnetic centre can be calculated often includes terms involving several angular momenta. Here, we consider the case of two angular momenta, $\mathbf{J}_1$ and $\mathbf{J}_2$. These angular momenta are associated with spaces $\mathcal{E}_{J_1}$ of dimension $(2J_1 + 1)$ and $\mathcal{E}_{J_2}$ of dimension $(2J_2 + 1)$. For any direction z in the Euclidean space, the eigenvectors $\{|J_1, M_1\rangle\}$ common to operators $\hat{\mathbf{J}}_1{}^2$ and $\hat{J}_{1z}$ verify

$$\hat{\mathbf{J}}_1{}^2 |J_1, M_1\rangle = J_1(J_1 + 1)\, |J_1, M_1\rangle$$

$$\hat{J}_{1z}\, |J_1, M_1\rangle = M_1\, |J_1, M_1\rangle \qquad M_1 = -J_1, -J_1 + 1, \ldots, J_1$$

P. Bertrand, *Electron Paramagnetic Resonance Spectroscopy*,
https://doi.org/10.1007/978-3-030-39663-3

and they make up a basis of $\mathcal{E}_{J_1}$. In the same way, we can define a basis $\{|J_2, M_2\rangle\}$ of $\mathcal{E}_{J_2}$. Hereafter, we use *the same quantization axis z* for the bases $\{|J_1, M_1\rangle\}$ and $\{|J_2, M_2\rangle\}$, and we complement it with two axes (x, y) to produce a Cartesian reference frame.

To determine the eigenvalues and eigenvectors for the Hamiltonian, we can construct the matrix which represents it in the "product basis" $\{|J_1, M_1\rangle|J_2, M_2\rangle\}$ for the space $\mathcal{E} = \mathcal{E}_{J_1} \otimes \mathcal{E}_{J_2}$, of dimension $(2J_1 + 1)(2J_2 + 1)$. However, in some cases, which we will discuss below, it is easier to use another basis of $\mathcal{E}$ constructed from the total angular momentum **J** defined by:

$$\mathbf{J} = \mathbf{J}_1 + \mathbf{J}_2 \qquad [1]$$

The components of **J** define the operators $(\hat{J}_x, \hat{J}_y, \hat{J}_z)$ which act on the kets $\{|J_1, M_1\rangle|J_2, M_2\rangle\}$ of $\mathcal{E}$. Thus, for example, we have:

$$\begin{aligned} \hat{J}_z|J_1, M_1\rangle\, |J_2, M_2\rangle &= (\hat{J}_{1z} + \hat{J}_{2z})\, |J_1, M_1\rangle\, |J_2, M_2\rangle \\ &= (M_1 + M_2)\, |J_1, M_1\rangle\, |J_2, M_2\rangle \end{aligned} \qquad [2]$$

By using the general properties of the operators defined from angular momenta, it can be shown that J can take the following values [Ayant and Belorizky, 2000; Cohen-Tannoudji *et al.*, 2015]:

$$J = |J_1 - J_2|, |J_1 - J_2| + 1, \ldots, J_1 + J_2 \qquad [3]$$

If the momenta are numbered such that $J_2 \leq J_1$, there exist $(2J_2 + 1)$ possible values for J. For example J can take the values 0 and 1 for $J_1 = J_2 = ½$, and the values 1, 2, 3 for $J_1 = 2$ and $J_2 = 1$.

For each value of J in list [3] there is a corresponding sub-space $\mathcal{E}_J$ of $\mathcal{E}$, of dimension $(2J + 1)$. A particular basis of $\mathcal{E}_J$ is composed of the kets $\{|J, M\rangle\}$, eigenvectors shared by $\hat{\mathbf{J}}^2$ and $\hat{J}_z$, which verify:

$$\begin{aligned} \hat{\mathbf{J}}^2|J, M\rangle &= J(J + 1)\, |J, M\rangle \\ \hat{J}_z|J, M\rangle &= M\, |J, M\rangle \qquad M = -J, -J + 1, \ldots, J \end{aligned} \qquad [4]$$

It can readily be shown that the sum of dimensions of the $\mathcal{E}_J$ sub-spaces is equal to the dimension $(2J_1 + 1)(2J_2 + 1)$ of $\mathcal{E}$, and $\mathcal{E}$ is said to be the "direct sum" of the $\mathcal{E}_J$ sub-spaces.

All of the kets $\{|J, M\rangle\}$, where J takes all the values in list [3], make up the "coupled basis" of $\mathcal{E}$. We will see that this basis is sometimes easier to use than the product bases, and that it can be used to define "equivalent operators".

The move from the product basis to the coupled basis involves linear relations such as:

$$|J,M\rangle = \sum_{M_1=-J_1}^{J_1} \sum_{M_2=-J_2}^{J_2} C(J_1,M_1,J_2,M_2;J,M)|J_1,M_1\rangle|J_2,M_2\rangle \quad [5]$$

The components $C(J_1,M_1,J_2,M_2;J,M)$, known as the "Clebsch-Gordan coefficients", are tabulated for various values of (J_1,J_2) [Ayant and Belorizky, 2000; Cohen-Tannoudji *et al.*, 2015]. These coefficients have numerous symmetry properties. For example, by applying operator $\hat{J}_z$ to both sides of equation [5] and using equation [2], we obtain the relation $M_1 + M_2 = M$.

To illustrate the above, we will consider the case where $J_1 = J_2 = ½$. The space $\mathcal{E}$ is of dimension 4, and the product basis $\{|J_1,M_1\rangle|J_2,M_2\rangle\}$ is as follows:

$$\{|½,-½\rangle\,|½,-½\rangle, |½,-½\rangle\,|½,½\rangle, |½,½\rangle\,|½,-½\rangle, |½,½\rangle\,|½,½\rangle\}$$

The coupled basis $\{|J,M\rangle\}$ can be written $\{|0,0\rangle,|1,-1\rangle,|1,0\rangle,|1,1\rangle\}$ and the Clebsch-Gordan coefficients are such that equation [5] is written:

$$\begin{aligned} |0,0\rangle &= (|½,½\rangle\,|½,-½\rangle - |½,-½\rangle\,|½,½\rangle)/\sqrt{2} \\ |1,-1\rangle &= |½,-½\rangle\,|½,-½\rangle \\ |1,0\rangle &= (|½,½\rangle\,|½,-½\rangle + |½,-½\rangle\,|½,½\rangle)/\sqrt{2} \\ |1,1\rangle &= |½,½\rangle\,|½,½\rangle \end{aligned} \quad [6]$$

1.2 – Construction of the matrices representing operators defined from $\mathbf{J}_1$ *and* $\mathbf{J}_2$ *in the coupled basis*

We now wish to construct the matrices representing the operators $(\hat{J}_{1x},\hat{J}_{1y},\hat{J}_{1z},\hat{\mathbf{J}}_1^2)$ and similar operators produced by $\mathbf{J}_2$, in the coupled basis. These matrices are particularly simple for $\hat{\mathbf{J}}_1^2$ and $\hat{\mathbf{J}}_2^2$. Indeed, for any ket $|J_1,M_1\rangle$ of $\mathcal{E}_{J_1}$ and $|J_2,M_2\rangle$ of $\mathcal{E}_{J_2}$, we can write (section 3.2.1):

$$\hat{\mathbf{J}}_1^2\,|J_1,M_1\rangle = J_1(J_1+1)\,|J_1,M_1\rangle;\ \hat{\mathbf{J}}_2^2\,|J_2,M_2\rangle = J_2(J_2+1)\,|J_2,M_2\rangle$$

Which gives rise to:

$$\hat{\mathbf{J}}_1^2\,|J_1,M_1\rangle\,|J_2,M_2\rangle = J_1(J_1+1)\,|J_1,M_1\rangle\,|J_2,M_2\rangle$$

$$\hat{\mathbf{J}}_2^2\,|J_1,M_1\rangle\,|J_2,M_2\rangle = J_2(J_2+1)\,|J_1,M_1\rangle\,|J_2,M_2\rangle$$

By applying $\hat{\mathbf{J}}_1^2$ and $\hat{\mathbf{J}}_2^2$ to both sides of equation [5], we therefore obtain:

$$\hat{\mathbf{J}}_1^2\,|J,M\rangle = J_1(J_1+1)\,|J,M\rangle$$

$$\hat{\mathbf{J}}_2^2\,|J, M\rangle = J_2(J_2+1)\,|J, M\rangle$$

Like $\hat{\mathbf{J}}^2$ and $\hat{J}_z$, the $\hat{\mathbf{J}}_1{}^2$ and $\hat{\mathbf{J}}_2{}^2$ operators are represented by *diagonal matrices* in the basis $\{|J,M\rangle\}$. We show how the matrices representing the other operators can be constructed by considering the specific case $J_1 = J_2 = ½$. By applying $\hat{J}_{1x}$ to both sides of the set of equations [6] and by using the following relations (section 3.2.5):

$$\hat{J}_{1x}|½, ½\rangle = ½\,|½, -½\rangle;\ \hat{J}_{1x}\,|½, -½\rangle = ½\,|½, ½\rangle$$

we obtain:

$$\begin{aligned}
\hat{J}_{1x}\,|0, 0\rangle &= (|1, -1\rangle - |1, 1\rangle)\,/(2\sqrt{2})\\
\hat{J}_{1x}\,|1, -1\rangle &= (|1, 0\rangle + |0, 0\rangle)\,/(2\sqrt{2})\\
\hat{J}_{1x}\,|1, 0\rangle &= (|1, -1\rangle + |1, 1\rangle)\,/(2\sqrt{2})\\
\hat{J}_{1x}\,|1, 1\rangle &= (|1, 0\rangle - |0, 0\rangle)\,/(2\sqrt{2})
\end{aligned}$$

By proceeding in the same way for the other operators, we obtain the matrices representing the operators $(\hat{J}_{1x}, \hat{J}_{1y}, \hat{J}_{1z}, \hat{J}_{2x}, \hat{J}_{2y}, \hat{J}_{2z})$ in the basis $\{|0,0\rangle, |1,-1\rangle, |1,0\rangle, |1,1\rangle\}$:

$$\tilde{\mathbf{J}}_{1x} = \frac{1}{2\sqrt{2}}\begin{bmatrix} 0 & 1 & 0 & -1\\ 1 & 0 & 1 & 0\\ 0 & 1 & 0 & 1\\ -1 & 0 & 1 & 0\end{bmatrix},\ \tilde{\mathbf{J}}_{1y} = \frac{1}{2\sqrt{2}}\begin{bmatrix} 0 & -i & 0 & -i\\ i & 0 & i & 0\\ 0 & -i & 0 & i\\ i & 0 & -i & 0\end{bmatrix},\ \tilde{\mathbf{J}}_{1z} = \frac{1}{2}\begin{bmatrix} 0 & 0 & 1 & 0\\ 0 & -1 & 0 & 0\\ 1 & 0 & 0 & 0\\ 0 & 0 & 0 & 1\end{bmatrix}$$

$$\tilde{\mathbf{J}}_{2x} = \frac{1}{2\sqrt{2}}\begin{bmatrix} 0 & -1 & 0 & 1\\ -1 & 0 & 1 & 0\\ 0 & 1 & 0 & 1\\ 1 & 0 & 1 & 0\end{bmatrix},\ \tilde{\mathbf{J}}_{2y} = \frac{1}{2\sqrt{2}}\begin{bmatrix} 0 & i & 0 & i\\ -i & 0 & i & 0\\ 0 & -i & 0 & i\\ -i & 0 & -i & 0\end{bmatrix},\ \tilde{\mathbf{J}}_{2z} = \frac{1}{2}\begin{bmatrix} 0 & 0 & -1 & 0\\ 0 & -1 & 0 & 0\\ -1 & 0 & 0 & 0\\ 0 & 0 & 0 & 1\end{bmatrix}$$

1.3 – Spin coupling coefficients

If we extract the matrices of dimension 3 which represent the $(\hat{J}_{1x}, \hat{J}_{1y}, \hat{J}_{1z})$ operators *in the sub-space* $\mathcal{E}_1$ $\{|1,-1\rangle, |1,0\rangle, |1,1\rangle\}$ from the matrices $(\tilde{\mathbf{J}}_{1x}, \tilde{\mathbf{J}}_{1y}, \tilde{\mathbf{J}}_{1z})$ above, we obtain:

$$\frac{1}{2}\begin{bmatrix} 0 & \frac{1}{\sqrt{2}} & 0\\ \frac{1}{\sqrt{2}} & 0 & \frac{1}{\sqrt{2}}\\ 0 & \frac{1}{\sqrt{2}} & 0\end{bmatrix},\ \frac{1}{2}\begin{bmatrix} 0 & \frac{i}{\sqrt{2}} & 0\\ \frac{-i}{\sqrt{2}} & 0 & \frac{i}{\sqrt{2}}\\ 0 & \frac{-i}{\sqrt{2}} & 0\end{bmatrix},\ \frac{1}{2}\begin{bmatrix} -1 & 0 & 0\\ 0 & 0 & 0\\ 0 & 0 & 1\end{bmatrix}$$

The same result is obtained for the matrices representing the operators $(\hat{J}_{2x}, \hat{J}_{2y}, \hat{J}_{2z})$. We observe that these matrices are identical to those representing the operators $(\hat{J}_x, \hat{J}_y, \hat{J}_z)$ in this sub-space, appart from a proportionality factor of ½ (exercise 3.3). This result is general: the matrices representing the operators defined from $\mathbf{J}_1$ and $\mathbf{J}_2$ in the basis $\{|J, M\rangle\}$ of the $\mathcal{E}_J$ sub-space are *proportional* to those of the corresponding operators defined from $\mathbf{J} = \mathbf{J}_1 + \mathbf{J}_2$, which are much easier to calculate. *Within this sub-space*, we therefore have the equivalence:

$$\mathbf{J}_1 \equiv K_1\,\mathbf{J};\quad \mathbf{J}_2 \equiv K_2\,\mathbf{J} \tag{7}$$

where the "spin coupling coefficients" K_1 and K_2 are linked by $K_1 + K_2 = 1$ according to equation [1]. To determine their expression, we write the matrix element $\langle J, M|\mathbf{J}_1 \cdot \mathbf{J}|\, J, M\rangle$ in two different ways:

▷ We can first write:

$$\langle J, M\,|\mathbf{J}_1 \cdot \mathbf{J}|\, J, M\rangle = \sum_{\substack{i = x,y,z \\ M'}} \langle J, M\,|\hat{J}_{1i}|\, J, M'\rangle\langle J, M'\,|\hat{J}_i|\, J, M\rangle$$

Which, using the equivalence described by equation [7], gives:

$$\begin{aligned}\langle J, M\,|\mathbf{J}_1 \cdot \mathbf{J}|\, J, M\rangle &= \sum_{\substack{i = x,y,z \\ M'}} K_1 \langle J, M\,|\hat{J}_i|\, J, M'\rangle\langle J, M'\,|\hat{J}_i|\, J, M\rangle \\ &= K_1 \langle J, M\,|\mathbf{J}^2|\, J, M\rangle = K_1 J(J+1)\end{aligned}$$

▷ We also have

$$\mathbf{J}_1 \cdot \mathbf{J} = (\mathbf{J} - \mathbf{J}_2)\cdot\mathbf{J} = \mathbf{J}^2 + ½\,[(\mathbf{J} - \mathbf{J}_2)^2 - \mathbf{J}^2 - \mathbf{J}_2{}^2] = ½\,(\mathbf{J}^2 + \mathbf{J}_1{}^2 - \mathbf{J}_2{}^2)$$

which leads to:

$$\begin{aligned}\langle J, M|\mathbf{J}_1 \cdot \mathbf{J}|J, M\rangle &= ½\,\langle J, M|\hat{\mathbf{J}}^2 + \hat{\mathbf{J}}_1{}^2 - \hat{\mathbf{J}}_2{}^2)|\, J, M\rangle \\ &= ½\,[J(J+1) + J_1(J_1+1) - J_2(J_2+1)]\end{aligned}$$

We therefore deduce that:

$$\begin{aligned}K_1 &= \frac{J(J+1) + J_1(J_1+1) - J_2(J_2+1)}{2J(J+1)} \\ K_2 &= \frac{J(J+1) + J_2(J_2+1) - J_1(J_1+1)}{2J(J+1)}\end{aligned} \tag{8}$$

For $J_1 = J_2 = ½$, $J = 1$, we obtain $K_1 = K_2 = ½$, which are the proportionality factors found above.

The equivalence described by equation [7] only applies within a *sub-space* $\mathcal{E}_J$ and it is only useful if these sub-spaces appear naturally in the calculation.

1.4 – Application to the energy of multiplets. Demonstration of the Landé formula

We consider a free atom or ion which has an incomplete subshell. Its electrons are subjected to electrostatic interactions represented by a Hamiltonian $\hat{H}_0$. We note (L, S) the ground term and $E_{L,S}$ its energy (appendix 2). Any Z direction in the Euclidean space defines a basis $\{|L, M_L\rangle\}$ for $\mathcal{E}_L$ and a basis $\{|S, M_S\rangle\}$ for $\mathcal{E}_S$, and therefore a "product basis" $\{|L, M_L\rangle|S, M_S\rangle\}$ for this term. We can also define a "coupled basis" $\{|J, M\rangle\}$ from the total angular momentum:

$$\mathbf{J} = \mathbf{L} + \mathbf{S} \tag{9}$$

The possible values for J are given by:

$$J = |L - S|, |L - S| + 1, \ldots, L + S \tag{10}$$

We will now introduce the spin-orbit interaction represented by the operator $\hat{H}_{SO} = \lambda \mathbf{L} \cdot \mathbf{S}$. The parameter λ is linked to the spin-orbit coupling constant ζ by $\lambda = \pm \zeta/2S$, with the plus sign if the subshell is less than half full. The Hamiltonian thus becomes:

$$\hat{H} = \hat{H}_0 + \hat{H}_{SO}$$

We assume that $\hat{H}_{SO}$ is a perturbation of $\hat{H}_0$, and we examine how the ground term is split by the spin-orbit interaction. As this term is degenerate with an order of $(2L + 1)(2S + 1)$, we obtain the energy levels to the first order of perturbation theory by adding to $E_{L,S}$ the eigenvalues for the matrix representing $\hat{H}_{SO}$. This matrix is not particularly simple in the product basis, but it is in the coupled basis $\{|J, M\rangle\}$. Indeed, from equation [9], we deduce the identity:

$$\lambda \mathbf{L} \cdot \mathbf{S} = (\lambda/2)(\hat{\mathbf{J}}^2 - \hat{\mathbf{L}}^2 - \hat{\mathbf{S}}^2) \tag{11}$$

According to the results obtained in sections 1.1 and 1.2, all the operators that appear in the right-hand side of equation [11] are represented by a diagonal matrix in the *coupled basis* $\{|J, M\rangle\}$. This is therefore also the case for the operator $\lambda \mathbf{L} \cdot \mathbf{S}$. For each value of J in list [10] there is a corresponding *multiplet* (L, S, J) for which the energy and eigenvectors are given by:

$$E_{L,S,J} = E_{L,S} + (\lambda/2)\,[J(J+1) - L(L+1) - S(S+1)] \quad \{|J, M\rangle, M = -J, -J+1, \ldots J\}$$

If the incomplete subshell is less than half full, λ is positive and the ground multiplet is characterised by $J_{min} = |L - S|$. Otherwise, it is $J_{max} = L + S$.

We will now focus on the total magnetic moment of the electrons, which is written as follows:

$$\boldsymbol{\mu} = -\beta\,(\mathbf{L} + g_e\mathbf{S})$$

Within the multiplet (L, S, J), we can replace $\mathbf{L}$ by $K_L\mathbf{J}$ and $\mathbf{S}$ by $K_S\mathbf{J}$, where K_L and K_S are the spin coupling coefficients obtained by substituting (L,S,J) for (J_1,J_2,J) in equations [8]. We can therefore write:

$$\boldsymbol{\mu} = -g_{Landé}\,\beta\,\mathbf{J}$$

with $g_{Landé} = K_L + g_e\,K_S$.

Using the coupled basis and the spin coupling coefficients, we have readily obtain the results presented in appendix 1, and in particular equation [7] therein.

2 – Calculation of the dipolar components of the hyperfine interaction within an (L, S) term

Equations [7] express the *equivalence* between $K_1\mathbf{J}$ and $\mathbf{J}_1$ on the one hand, and $K_2\mathbf{J}$ and $\mathbf{J}_2$ on the other, within the $\mathcal{E}_J$ sub-space. From $\hat{J}_x^{\,2}, \hat{J}_y^{\,2}, \hat{J}_z^{\,2}$, we can also define operators which are equivalent to $\hat{J}_{1x}^{\,2}, \hat{J}_{1y}^{\,2}, \hat{J}_{1z}^{\,2}, \hat{J}_{2x}^{\,2}, \hat{J}_{2y}^{\,2}, \hat{J}_{2z}^{\,2}$ [Bencini and Gatteschi, 1990]. This equivalence is the result of the Wigner-Eckart theorem, which relies on the symmetry properties for some operators with respect to the rotation group [Ayant and Belorizky, 2000; Cohen-Tannoudji *et al.*, 2015]. This theorem can be used to generalise the notion of spin coupling coefficients when more than two angular momenta exist, and to define equivalent operators for operators other than those defined by angular momenta. In particular, it considerably simplifies calculation of quantities involving integration over the *coordinates* of the unpaired electrons.

To illustrate this latter point, consider a transition ion or rare earth ion for which the nucleus, with a magnetic moment $g_N\,\beta_N\,\mathbf{I}$, interacts with N unpaired electrons. The interactions involve a term $A_s\mathbf{S}\cdot\mathbf{I}$ due to the core polarisation mechanism, and a term due to the dipolar interactions with the orbital and spin magnetic moments of the electrons (section 4.4.1). We will see that calculation of the dipolar terms is considerably simplified by the use of equivalent operators.

First, consider the dipolar interaction between the nucleus and the *orbital magnetic moments* of the electrons. It can be written [Abragam and Bleaney, 1986]:

$$H_{dip}^{orb} = 2\frac{\mu_0}{4\pi} g_N \beta\beta_N \sum_{i=1}^{N} \frac{\mathbf{I}\cdot\mathbf{l}_i}{r_i^3} \qquad [12]$$

where $\mu_0 = 4\pi \times 10^{-7}$ H m^{-1} is the vacuum permeability, r_i is the distance between the nucleus and electron i with orbital angular momentum $\mathbf{l}_i$. The electron *probability density* at a given point is determined by the wave function. We must therefore calculate the *mean value* for $\hat{H}_{dip}^{orb}$ by integrating the second part of equation [12] over the coordinates of the electrons, taking the orbitals they occupy into account. We can avoid this complex calculation by replacing $\hat{H}_{dip}^{orb}$ by the following equivalent operator, which is valid within an (L, S) term [Abragam and Bleaney, 1986]:

$$\left(\hat{H}_{dip}^{orb}\right)_{eq} = 2\frac{\mu_0}{4\pi} g_N \beta\beta_N < r^{-3} > \mathbf{L}\cdot\mathbf{I} \qquad [13]$$

$$< r^{-3} > = \int_0^{\infty} r^{-3} R_{n,\ell}^2(r) r^2 \mathrm{d}r$$

$R_{n,\ell}(r)$ is the radial function shared by all the orbitals in the incomplete subshell (n, ℓ) of the cation ($n = 3, \ell = 2$ for $3d$, $n = 4$, $\ell = 3$ for $4f$); it depends on the nature of the cation.

In expression [13], the prefactor $<r^{-3}>$, which comes from integration over the radial part of the orbitals, is tabulated for free ions [Freeman and Watson, 1965], and integration over the angular part amounts to calculating matrix elements of type $\langle L, M_L | \hat{L}_j | L, M_L' \rangle$, where $j = x, y, z$.

The operator representing the dipolar interactions between the nucleus and the *spin magnetic moments* of the electrons can be written (appendix 3):

$$\hat{H}_{dip}^{spin} = P \sum_{i=1}^{N} \frac{3(\mathbf{I}\cdot\mathbf{r}_i)(\mathbf{s}_i\cdot\mathbf{r}_i) - r_i^2(\mathbf{I}\cdot\mathbf{s}_i)}{r_i^5} \qquad [14]$$

where $\mathbf{s}_i$ is the spin angular momentum of electron i, and $P = \frac{\mu_0}{4\pi} g_e g_N \beta\beta_N$.

Within an (L,S) term, we can replace $\hat{H}_{dip}^{spin}$ by the equivalent operator [Abragam and Bleaney, 1986]:

$$(\hat{H}_{dip}^{spin})_{eq} = P\xi <r^{-3}>[L(L+1)\,\mathbf{I}\cdot\mathbf{S} - \tfrac{3}{2}(\mathbf{L}\cdot\mathbf{I})(\mathbf{L}\cdot\mathbf{S}) - \tfrac{3}{2}(\mathbf{L}\cdot\mathbf{S})(\mathbf{L}\cdot\mathbf{I})] \qquad [15]$$

where:

$$\xi = \frac{2\ell + 1 - 4S}{S(2\ell - 1)(2\ell + 3)(2L - 1)}$$

Integration over the angular part of the wave function amounts to calculating matrix elements of type $\langle L, M_L | \hat{L}_j \hat{L}_k | L, M_L' \rangle$, where $j, k = x, y, z$.

The equivalent operators defined by equations [13] and [15] are used in appendix 2 to obtain the expression of the hyperfine matrix $\tilde{\mathbf{A}}$ given by ligand field theory.

References

ABRAGAM A. & BLEANEY B. (1986) *Electron Paramagnetic Resonance of transition ions*, Dover, New York.

AYANT Y. & BELORIZKY E. (2000) *Cours de mécanique quantique*, Dunod, Paris.

BENCINI A. & GATTESCHI D. (1990) *EPR of Exchange Coupled Systems*, Springer-Verlag, Berlin.

COHEN-TANNOUDJI C., DIU B. & LALOË F. (2015) *Quantum Mechanics*, Wiley-VcH, New York.

FREEMANN A.J. & WATSON R.E. (1965) "Hyperfine Interactions in Magnetic Materials" in *Magnetism* vol II A, RADO G.T. & SUHL H. eds, Academic Press, New York.

Appendix 5

The notion of spin density

Spin density describes the spatial distribution of unpaired electrons in a paramagnetic molecule. It is a very useful link between the spin Hamiltonian parameters deduced by analysing the spectrum, and the electronic structure of the molecules.

- ▹ In the isotropic regime, the EPR spectrum for *free radicals* depends directly on the distribution of spin density over the different atoms in the molecule. Indeed, the *shape* of the spectrum is determined by the hyperfine interactions involving the nuclei of the atoms with a spin population (section 2.2.2). In addition, the mean g_{iso} for the principal values of the $\tilde{\mathbf{g}}$ matrix, which determines the *position* of the spectrum, increases when the spin density is delocalised onto heavy atoms with a large spin-orbit coupling constant (section 4.2.2).
- ▹ In *transition ion complexes*, delocalisation of the spin density onto the ligands reduces the orbital moment's contribution to the principal values of the $\tilde{\mathbf{g}}$ matrix, it weakens the hyperfine interactions with the nucleus of the transition ion (appendix 2), and creates superhyperfine interactions with the ligand nuclei (section 4.4.1).
- ▹ In *polynuclear complexes*, delocalisation of the electrons onto the bridging ligands favours overlap of the orbitals and, consequently, exchange interactions (section 7.2.1). More importantly, we will see that the concept of spin density gives physical meaning to the "spin coupling coefficients" K_i which play a key role in the interpretation of EPR spectra for these complexes (section 7.4).

P. Bertrand, *Electron Paramagnetic Resonance Spectroscopy*,
https://doi.org/10.1007/978-3-030-39663-3

1 – Definition

Consider an electron i of spin angular momentum $\mathbf{s}_i$ which occupies the orbital $\varphi_i(\mathbf{r}_i)$, for which the spin state is defined by $\sigma(i)$. $\sigma(i)$ is one of the two functions $\{\alpha(i), \beta(i)\}$ which verify:

$$\hat{\mathbf{s}}_i^2\,\alpha(i) = \tfrac{3}{4}\,\alpha(i) \qquad \hat{\mathbf{s}}_i^2\,\beta(i) = \tfrac{3}{4}\,\beta(i)$$

$$\hat{s}_{iz}\,\alpha(i) = \tfrac{1}{2}\,\alpha(i) \qquad \hat{s}_{iz}\,\beta(i) = -\tfrac{1}{2}\,\beta(i)$$

where z is any direction in the Euclidean space. By definition, at point $\mathbf{r}$ this electron creates a *spin density* $\rho_i(\mathbf{r})$ equal to $|\varphi_i(\mathbf{r})|^2$ if $\sigma(i) = \alpha(i)$ or to $-|\varphi_i(\mathbf{r})|^2$ if $\sigma(i) = \beta(i)$. It is therefore an *algebraic* quantity which can be written:

$$\rho_i(\mathbf{r}) = |\varphi_i(\mathbf{r})|^2\,\langle\sigma(i)|2\,\hat{s}_{iz}\,|\sigma(i)\rangle$$

The dimension of $\rho_i(\mathbf{r})$ is that of $|\varphi_i(\mathbf{r})|^2$, i.e., $[\text{length}]^{-3}$. The spin density created by all of the electrons in a molecule is obtained by adding their contributions as follows:

$$\rho(\mathbf{r}) = \sum_i |\varphi_i(\mathbf{r})|^2 \langle\sigma(i)\,|2\hat{s}_{iz}|\,\sigma(i)\rangle \qquad [1]$$

When two electrons occupy the same orbital with different spin functions, their contributions to the spin density *cancel each other out*: as expected only *unpaired* electrons contribute to the spin density.

The spin density created by the electrons in a molecule depends on its *spin state*. For a molecule with n unpaired electrons of spin angular momenta $\{\mathbf{s}_i, i = 1,2,\ldots,n\}$, there exist $(2)^n$ possible spin states. In the absence of spin-orbit interactions, these states are described by functions $\chi^S_{M_S}(1,2,\ldots,n)$ which are linear combinations of products of $\sigma(i)$ functions verifying:

$$\hat{\mathbf{S}}^2\,\chi^S_{M_S}(1,2,\ldots,n) = S(S+1)\,\chi^S_{M_S}(1,2,\ldots,n)$$

$$\hat{S}_z\,\chi^S_{M_S}(1,2,\ldots,n) = M_S\,\chi^S_{M_S}(1,2,\ldots,n), \quad M_S = -S, -S+1, \ldots, S$$

where $\mathbf{S} = \mathbf{s}_1 + \mathbf{s}_2 + \ldots + \mathbf{s}_n$ is the total spin angular momentum of the molecule. The number S can take values between S_{min}, which is 0 if n is even and ½ if n is odd, and $S_{max} = n/2$. Construction of the $\chi^S_{M_S}$ functions is described for example in [Davidson, 1969]. For $n = 2$, these functions are given by equations [7.6]. If we note $\{\varphi_i(\mathbf{r}_i), i = 1,2,\ldots,n\}$ the orbitals occupied by the electrons, the possible electronic states for the molecule are described by wave functions with the following form:

$$\Psi(1,2,\ldots,n) = (1/n!)^{\frac{1}{2}} A\,[\varphi_1(\mathbf{r}_1)\varphi_2(\mathbf{r}_2)\ldots\,\varphi_n(\mathbf{r}_n)\,\chi^S_{M_S}(1,2,\ldots,n)] \qquad [2]$$

In this expression, $(1/n!)^{\frac{1}{2}}$ is a normalisation factor and A is the "antisymmetrisation operator". From the quantity in brackets, this operator generates a linear combination of the $(n!)$ possible permutations of the electrons coordinates such that $\Psi(1,2,\ldots,n)$ is *antisymmetric*, i.e., changes sign when the coordinates of two electrons are exchanged (section 7.2.1). If the orbitals are *normalised* and *orthogonal*, it can be shown that for the state defined by equation [2], the spin density defined by equation [1] can be written:

$$\rho_{S,M_S}(\mathbf{r}) = \sum_{i=1}^{n} |\varphi_i(\mathbf{r})|^2 \left\langle \chi^S_{M_S} \middle| 2\hat{s}_{iz} \middle| \chi^S_{M_S} \right\rangle \qquad [3]$$

It is therefore determined by the *nature of the orbitals* and by the molecule's *spin state*.

Example

To illustrate the above, we will calculate the spin densities produced by two simplified models of the electronic structure of the allyl radical $H_2C^{\bullet} - CH = CH_2$.

▹ In the "molecular orbital" description, we construct 3 linear combinations $\{\Phi_1, \Phi_2, \Phi_3\}$ of the $2p_z$ orbitals for the 3 carbon atoms, which comply with the symmetry properties of the molecule. Calculation reveals that the wave function minimising the energy is the following:

$$\Psi(1,2,3) = (1/\sqrt{6})\,A\,[\Phi_1(\mathbf{r}_1)\Phi_1(\mathbf{r}_2)\Phi_2(\mathbf{r}_3)\,\chi^{1/2}_{1/2}(1,2,3)]$$

where:

$$\Phi_1(\mathbf{r}) = (\tfrac{1}{2})\,2p_{z1}(\mathbf{r}) + (1/\sqrt{2})\,2p_{z2}(\mathbf{r}) + (\tfrac{1}{2})\,2p_{z3}(\mathbf{r})$$

$$\Phi_2(\mathbf{r}) = (1/\sqrt{2})\,2p_{z1}(\mathbf{r}) - (1/\sqrt{2})\,2p_{z3}(\mathbf{r})$$

$$\chi^{1/2}_{1/2}(1,2,3) = 1/\sqrt{2}\,[\alpha(1)\beta(2) - \beta(1)\alpha(2)]\,\alpha(3)$$

For this wave function, the spin density calculated from equation [3] is equal to:

$$\rho(\mathbf{r}) = |\Phi_2(\mathbf{r})|^2 \langle \chi^{1/2}_{1/2} | 2\hat{s}_{3z} | \chi^{1/2}_{1/2} \rangle$$

$$= |\Phi_2(\mathbf{r})|^2$$

This model leads to a *spin population* equal to zero on carbon C_2, and equal to 0.5 on carbons C_1 and C_3. This distribution differs from the spin populations deduced from the hyperfine constants measured by EPR, which are equal to 0.63 for C_1 and C_3 and to -0.18 for C_2 (exercise 2.2).

▹ We will now use a "valence bond" function with the following form:

$$\Psi'(1,2,3) = (1/\sqrt{6})A\,[2p_{z1}(\mathbf{r}_1)\,2p_{z2}(\mathbf{r}_2)\,2p_{z3}(\mathbf{r}_3)\,\chi^S_{M_S}(1,2,3)]$$

The spin function which minimises the energy is as follows:

$$\chi^{1/2}_{1/2}{}'(1,2,3) = (1/\sqrt{6})\,[\alpha(1)\alpha(2)\beta(3) + \beta(1)\alpha(2)\alpha(3) - 2\alpha(1)\beta(2)\alpha(3)]$$

The ground state is characterised by $S = ½$, and equation [3] gives the following spin density:

$$\begin{aligned}\rho(\mathbf{r}) &= \sum_{i=1}^{3} |2p_{zi}(\mathbf{r})|^2 \langle \chi^{1/2'}_{1/2} | 2\hat{s}_{iz} | \chi^{1/2'}_{1/2} \rangle \\ &= ⅔\,|2p_{z1}(\mathbf{r})|^2 - ⅓|2p_{z2}(\mathbf{r})|^2 + ⅔\,|2p_{z3}(\mathbf{r})|^2\end{aligned}$$

The spin populations of the $2p_z$ orbitals for carbon C_1 and C_3 are equal to 0.66 and that of the $2p_z$ orbital of carbon C_2 is equal to – 0.33. This model accounts for the existence of a *negative* spin density on carbon C_2. Agreement with the experimental data can be improved by using more elaborate wave functions obtained by variational methods. This simple example shows how the distribution of the spin density obtained by analysing the EPR spectrum can be used to test a model of the electronic structure.

2 – Spin density in a mononuclear complex

In this section, we are interested in a mononuclear complex with a transition ion in the *high spin* situation, with $S = S_{max} = n/2$, where n is the number of unpaired electrons. For this value of S, we can simplify expression [3] using the Wigner-Eckart theorem [Ayant and Belorizky, 2000; Cohen-Tannoudji *et al.*, 2015]. Thus, we can write:

$$\begin{aligned}\langle \chi^S_{M_S} | 2\hat{s}_{iz} | \chi^S_{M_S} \rangle &= \langle \chi^S_{M_S} | 2\hat{S}_z | \chi^S_{M_S} \rangle\,(S||s_i||S) \\ &= 2M_S\,(S||s_i||S) \qquad [4]\end{aligned}$$

where $(S||s_i||S)$ is a "reduced matrix element" which depends neither on z, nor on M_S. To determine its value, we consider the specific case $M_S = S$. The spin function then takes the very simple form:

$$\chi^S_S\,(1,2,\ldots,n) = \alpha(1)\alpha(2)\ldots\,\alpha(n)$$

The left-hand side of equation [4] is equal to 1, which shows that $(S||s_i||S)$ is equal to $\frac{1}{2S}$. We can therefore write equation [3] as:

$$\rho_{S,M_S}(\mathbf{r}) = 2M_S\, d(\mathbf{r}) \qquad [5]$$

where:

$$d(\mathbf{r}) = \frac{1}{2S}\sum_{i=1}^{n} |\varphi_i(\mathbf{r})|^2 \qquad [6]$$

We can verify that integration of $d(\mathbf{r})$ over the whole space gives:

$$\int d(\mathbf{r})\mathrm{d}^3\mathbf{r} = 1$$

$d(\mathbf{r})$ is the *normalised spin density* which determines the value of some of the spin Hamiltonian parameters.

Example

Noodleman and collaborators calculated the distribution of the spin density in the model complexes $Fe(SCH_3)_4{}^{1-}$ and $Fe(SCH_3)_4{}^{2-}$ where ions Fe^{3+} and Fe^{2+} of spin ⁵⁄₂ and 2 are coordinated to 4 sulfur atoms in tetrahedral symmetry [Noodleman *et al.*, 1985]. In table 1, we have indicated the spin populations deduced from the density $\rho_{5/2,5/2}(\mathbf{r})$ calculated for the spin state ($S = 5/2, M_S = 5/2$) of complex $Fe(SCH_3)_4{}^{1-}$. The populations are given for two sulfur atoms to facilitate subsequent comparison.

Table 1 – Spin populations in the $Fe(SCH_3)_4{}^{1-}$ complex, deduced from the spin density $\rho_{5/2,5/2}(\mathbf{r})$ and the normalised density $d(\mathbf{r})$ (equation [5]).

	2S	Fe^{3+}	2S
$\rho_{5/2,5/2}$	0.48	3.64	0.48
d	0.096	0.728	0.096

We obtain the populations corresponding to the normalised density $d(\mathbf{r})$ by dividing those of $\rho_{5/2,5/2}(\mathbf{r})$ by 5 (equation [5]). The sum of the populations is less than 1 due to a weak spin density on the carbon atoms (table 1).

Similarly, Table 2 lists the populations deduced from the spin density $\rho_{2,2}(\mathbf{r})$ calculated for the state ($S = 2, M_S = 2$) of the $Fe(SCH_3)_4{}^{2-}$ complex, along with those corresponding to the normalised density $d(\mathbf{r})$, calculated by dividing by 4 (equation [5]).

Table 2 – Spin populations in the $Fe(SCH_3)_4^{2-}$ complex, deduced from the spin density $\rho_{2,2}(\mathbf{r})$ and the normalised density $d(\mathbf{r})$ (equation [5]).

	2S	Fe^{2+}	2S
$\rho_{2,2}$	0.20	3.30	0.20
d	0.05	0.825	0.05

Delocalisation of the unpaired electrons onto the sulfur atoms increases when the ion is ferric. This difference is manifested in the magnetic characteristics of the two complexes:

- ▹ In the case of Fe^{3+} complexes, the principal values for the $\tilde{\mathbf{g}}$ matrix are close to 2.020 [Schneider *et al.*, 1968; Sweeney and Coffman, 1972]. They differ significantly from the $g_e = 2.0023$ value predicted by ligand field theory (section 4.2.2).
- ▹ In contrast, we can describe the magnetic properties of Fe^{2+} complexes with this model by using the spin-orbit coupling constant for the free ion [Bertrand and Gayda, 1988].

3 – Spin density in a dinuclear complex

Let us now consider two centres A and B in which the transition ions are in the *high-spin* situation with $S_A = n_A/2$ and $S_B = n_B/2$: n_A electrons occupy the $(\varphi_1, \varphi_2, \ldots, \varphi_{n_A})$ orbitals centred on A and n_B electrons occupy the $(\varphi_{n_A+1}, \varphi_{n_A+2}, \ldots, \varphi_n)$ orbitals centred on B, with $n = n_A + n_B$. When these two centres are coupled by a strong exchange interaction, we can describe the states of the (A, B) system by "valence bond" type wave functions of the following form:

$$\Psi(1,2,\ldots,n) = [1/(n)!]^{1/2} A\,[\varphi_1(\mathbf{r}_1)\varphi_2(\mathbf{r}_2)\ldots\,\varphi_n(\mathbf{r}_n)\,\chi^S_{M_S}(1,2,\ldots,n)]$$

The functions $\chi^S_{M_S}$ are the eigenfunctions shared by $(\hat{\mathbf{S}}^2, \hat{S}_z)$, where $\mathbf{S} = \mathbf{S}_A + \mathbf{S}_B$ is the total spin angular momentum and S can take the values $\{|S_A - S_B|, |S_A - S_B| + 1, \ldots, S_A + S_B\}$ (appendix 4). Expression [3] for the spin density can be written in the form:

$$\rho_{S,M_S}(\mathbf{r}) = \rho^A_{S,M_S}(\mathbf{r}) + \rho^B_{S,M_S}(\mathbf{r})$$

$$\rho^A_{S,M_S}(\mathbf{r}) = \sum_{i=1}^{n_A} |\varphi_i(\mathbf{r})|^2 \langle \chi^S_{M_S} | 2\hat{s}_{iz} | \chi^S_{M_S} \rangle$$

$$\rho^B_{S,M_S}(\mathbf{r}) = \sum_{i=n_A+1}^{n} |\varphi_i(\mathbf{r})|^2 \langle \chi^S_{M_S} | 2\hat{s}_{iz} | \chi^S_{M_S} \rangle$$

The densities $\rho^A_{S,M_S}(\mathbf{r})$ and $\rho^B_{S,M_S}(\mathbf{r})$ are the contributions from the orbitals centred on A and B, respectively. They can be simplified by expressing the $\chi^S_{M_S}$ functions in the "product basis" $\{\chi^{S_A}_{M_{S_A}}\chi^{S_B}_{M_{S_B}}\}$, where $\chi^{S_A}_{M_{S_A}}$ and $\chi^{S_B}_{M_{S_B}}$ are the spin functions associated with the angular momenta $\mathbf{S}_\mathrm{A}$ and $\mathbf{S}_\mathrm{B}$ (appendix 4):

$$\chi^S_{M_S}(1,2,\ldots n) = \sum_{M_A=-S_A}^{S_A}\sum_{M_B=-S_B}^{S_B} C(S_A,M_{S_A},S_B,M_{S_B},S,M_S)\chi^{S_A}_{M_{S_A}}(1,\ldots n_A)\chi^{S_B}_{M_{S_B}}(n_A+1,\ldots n)$$

In this relation, the quantities $C(S_\mathrm{A},M_{S_{A}},S_\mathrm{B},M_{S_{B}},S,M_S)$ are the Clebsch-Gordan coefficients. When this expression is substituted into $\rho^A_{S,M_S}(\mathbf{r})$, the quantities $\langle\chi^{S_A}_{M_{S_A}}|\,2\,\hat{s}_{iz}|\chi^{S_A}_{M_{S_A}}\rangle$ emerge. According to equation [4], they are equal to $\frac{1}{2S_A}\langle\chi^{S_A}_{M_{S_A}}|2\hat{S}_{Az}|\chi^{S_A}_{M_{S_A}}\rangle$. We therefore obtain:

$$\rho^A_{S,M_S}(\mathbf{r}) = \frac{1}{2S_A}\sum_{i=1}^{n_A}|\varphi_i(\mathbf{r})|^2\sum_{M_A=-S_A}^{S_A}\sum_{M_B=-S_B}^{S_B}|C(S_A,M_{S_A},S_B,M_{S_B},S,M_S)|^2\left\langle\chi^{S_A}_{M_{S_A}}\middle|2\hat{S}_{Az}\middle|\chi^{S_A}_{M_{S_A}}\right\rangle$$

The double sum is simply the matrix element $\langle\chi^S_{M_S}|2\hat{S}_{Az}|\chi^S_{M_S}\rangle$. We can therefore write:

$$\rho^A_{S,M_S}(\mathbf{r}) = \left\langle\chi^S_{M_S}\middle|2\hat{S}_{Az}\middle|\chi^S_{M_S}\right\rangle\frac{1}{2S_A}\sum_{i=1}^{n_A}|\varphi_i(\mathbf{r})|^2$$

Or

$$\rho^A_{S,M_S}(\mathbf{r}) = \langle\chi^S_{M_S}|2\hat{S}_{Az}|\chi^S_{M_S}\rangle\, d_\mathrm{A}(\mathbf{r})$$

where $d_\mathrm{A}(\mathbf{r})$ is the local *normalised spin density* for centre A (equation [6]). The matrix element included in this expression is linked to the *spin coupling coefficient* K_A by (appendix 4):

$$\langle\chi^S_{M_S}|2\hat{S}_{Az}|\chi^S_{M_S}\rangle = 2M_S K_\mathrm{A}$$

Finally, the spin density can be written in a compact form similar to that of equation [5]:

$$\rho_{S,M_S}(\mathbf{r}) = 2M_S\, d_S(\mathbf{r}) \qquad [7]$$

where:

$$d_S(\mathbf{r}) = K_\mathrm{A}\, d_\mathrm{A}(\mathbf{r}) + K_\mathrm{B}\, d_\mathrm{B}(\mathbf{r}) \qquad [8]$$

This expression gives physical meaning to the spin coupling coefficients. Indeed, as their sum is equal to 1, $d_S(\mathbf{r})$ is a *normalised spin density* which plays the same role as $d(\mathbf{r})$ in equation [5], and K_A, K_B are the *total spin populations* on centres A and B.

Example

We consider the model complex $Fe_2S_2(SH)_4^{2-}$, where two Fe^{3+} ions coordinated to 4 sulfur atoms in distorted tetrahedral symmetry are bridged by two sulfur ions, which we will denote S* to distinguish them from the terminal sulfurs (figure 7.11). This double bridge promotes strong antiferromagnetic exchange interactions between ions with spin $S_A = S_B = 5/2$, producing energy levels identified by $S = 0, 1, 2, 3, 4, 5$ (section 7.4). The spin populations deduced from the density $\rho_{5,5}(\mathbf{r})$ calculated for state ($S = 5, M_S = 5$) are listed in Table 3 [Noodleman *et al.*, 1985].

Table 3 – Spin populations in the $Fe_2S_2(SH)_4^{2-}$ complex, deduced from the spin densities $\rho_{5,5}(\mathbf{r})$, $d_5(\mathbf{r})$ (equation [7]) and the local densities $d_A(\mathbf{r})$ and $d_B(\mathbf{r})$ (equation [8]).

	2S	Fe^{3+}	2S*	Fe^{3+}	2S
$\rho_{5,5}$	0.66	3.29	1.16	3.29	0.66
d_5	0.066	0.329	0.116	0.329	0.066
d_A	0.132	0.658	0.116	0	0
d_B	0	0	0.116	0.658	0.132

We obtain the populations corresponding to the normalised density $d_5(\mathbf{r})$ by dividing those of $\rho_{5,5}$ by 10 (equation [7]). To extract the spin populations corresponding to local densities $d_A(\mathbf{r})$ and $d_B(\mathbf{r})$ (equation [8]), the spin coupling coefficients K_A and K_B must be known. For $S_A = 5/2$, $S_B = 5/2$, $S = 5$, equation [8] in appendix 4 gives $K_A = K_B = 1/2$. We can therefore write:

$$d_5(\mathbf{r}) = \tfrac{1}{2}\, d_A(\mathbf{r}) + \tfrac{1}{2}\, d_B(\mathbf{r})$$

If we assume that the local densities only overlap on the bridging sulfur atoms, we obtain the populations reported in table 3.

We will now consider the $Fe_2S_2(SCH_4)^{3-}$ complex where one of the iron ions is in the Fe^{2+} state. In the calculations, the extra electron is "trapped" on the B site thanks to a dissymetry created by extending the Fe – S bonds for this site by 0.07 Å [Noodleman *et al.*, 1985]. The antiferromagnetic interactions between ions of spin $S_A = 5/2$ and $S_B = 2$ lead to $S = 1/2, 3/2, 5/2, 7/2, 9/2$. The EPR spectrum given by the ground state of spin $S = 1/2$ is described in section 7.4.3 (figure 7.11). The spin populations deduced from the density $\rho_{9/2,9/2}(\mathbf{r})$ calculated for the state $(S = 9/2, M_S = 9/2)$ are listed in table 4.

Table 4 – Spin populations in the $Fe_2S_2(SH)_4^{3-}$ complex, deduced from the spin densities $\rho_{9/2,9/2}(\mathbf{r})$, $d_{9/2}(\mathbf{r})$ (equation [7]), and the local densities $d_A(\mathbf{r})$ and $d_B(\mathbf{r})$ (equation [8]).

	2S	Fe^{3+}	2S*	Fe^{2+}	2S
$\rho_{9/2,9/2}$	0.48	3.19	0.90	3.15	0.44
$d_{9/2}$	0.053	0.354	0.100	0.350	0.049
d_A	0.095	0.637	0.100	0	0
d_B	0	0	0.100	0.788	0.110

The populations corresponding to the normalised density $d_{9/2}(\mathbf{r})$ are obtained by dividing by 9 (equation [7]). For $S_A = 5/2$, $S_B = 2$, $S = 9/2$, we have $K_A = 5/9$ and $K_B = 4/9$. We can therefore write (equation [8]):

$$d_{9/2}(\mathbf{r}) = 5/9\, d_A(\mathbf{r}) + 4/9\, d_B(\mathbf{r})$$

By applying the same hypothesis as above, "local" spin populations can be obtained by multiplying those of the Fe^{3+} site by 9/5 and those of the Fe^{2+} site by 9/4 (Table 4). The bridging sulfur atom populations are indeterminate and we assume them to be identical for the two sites.

Tables 3 and 4 show that delocalisation of the unpaired electrons on the Fe^{3+} sites is similar in the two dinuclear complexes. If we compare the local densities obtained to those of mononuclear complexes $Fe(SCH_3)_4^{1-}$ (table 1) and $Fe(SCH_3)_4^{2-}$ (table 2), we observe that electron delocalisation is greater in dinuclear complexes. A more complete analysis of these data and their implications in the field of spectroscopies can be found in [Bertrand, 1993].

Equations [7] and [8] can be modified to take the phenomenon of valence delocalisation into account, and they can be extended to complexes with nuclearity greater than two.

References

AYANT Y. & BELORIZKY E. (2000) *Cours de mécanique quantique*, Dunod, Paris.

BERTRAND P. (1993) *Inorganic Chemistry* **32**: 741-745.

Bertrand P. & Gayda J.-P. (1988) *Biochimica et Biophysica Acta* **954**: 347-350.

Cohen-Tannoudji C., Diu B. & Laloë F. (2015) *Quantum Mechanics*, Wiley-VcH, New York.

Davidson E.R. (1969) "Valence Bond and Molecular Orbital Methods" in *Physical Chemistry: an Advanced Treatise* vol.III, Henderson D. ed., Academic Press, New York.

Noodleman L. *et al.* (1985) *Journal of the American Chemical Society* **107**: 3418-3426.

Schneider J. *et al.* (1968) *Journal of Physics and Chemistry of Solids* **29**: 451-462.

Sweeney W.V. & Coffman R.E. (1972) *Biochimica et Biophysica Acta* **286**: 26-35.

Appendix 6

Example of calculation of the spin-lattice relaxation time T_1: the direct process

Here, we describe the calculation which demonstrates the $T_1 \propto 1/T$ temperature dependence of the direct process. As a real example, we will consider a mononuclear transition ion complex. We denote $V(\{\mathbf{r}^p\}, \{\mathbf{X}^L\})$ the electrostatic interaction between the electrons of the cation, the positions of which are referenced by $\{\mathbf{r}^p\}$, and the ligands for which the position is referenced by $\{\mathbf{X}^L\}$. Vibrations of the medium (the "lattice") cause the distance between the cation and the ligands to vary; this effect produces a time-dependent interaction which can induce transitions between the ($|a\rangle, |b\rangle$) states of the cation (figure 5.2). For small *relative* ligand displacements $\{\mathbf{x}^L\}$ relative to the cation, we can write

$$V = V_0 + V_1(t) + \ldots \qquad [1]$$

$$V_1(t) = \sum_L \sum_{i=1}^{3} \left(\frac{\partial V}{\partial X_i^L}\right)_0 x_i^L(t) \qquad [2]$$

where the zero subscript refers to the equilibrium geometry. (x_1^L, x_2^L, x_3^L) are the components of the relative displacement $\mathbf{x}^L$ in any reference frame, and we have omitted the $(\{\mathbf{r}^p\}, \{\mathbf{X}^L\})$ variables to simplify the expression.

The direct process is the most efficient process at very low temperatures (section 5.4.1) where only low-frequency vibrational modes are populated. These modes can be assimilated to *normal modes*, each of which is characterised by its angular frequency ω_k and its normal (dimensionless) coordinate Q_k. The (x_i^L) components are linear combinations of the normal modes

$$x_i^L(t) = \sum_k u_i^L(k) Q_k \cos \omega_k t$$

such that the quantity $V_1(t)$ defined by equation [2] becomes:

$$V_1(t) = \sum_k U_k Q_k \cos \omega_k t \qquad [3]$$

P. Bertrand, *Electron Paramagnetic Resonance Spectroscopy*,
https://doi.org/10.1007/978-3-030-39663-3

The U_k quantities are expressed in units of energy. In quantum mechanics, the $\hat{U}_k$ operators act on the *states* ($|a\rangle, |b\rangle$) of the cation and each normal coordinate Q_k is an operator $\hat{Q}_k$ which acts on the kets which represent the *vibrational states* of mode $\{\omega_k\}$ [Ayant and Belorizky, 2000; Cohen-Tannoudji *et al.*, 2015]. Indeed:

▹ A set ot energy levels $E(N_k) = (N_k + ½)\hbar\omega_k$ corresponds to the $\{\omega_k\}$ mode, with $N_k = 0, 1, 2, \ldots$, and a *vibrational state* represented by a ket $|N_k\rangle$ is associated with each level.

▹ The Hermitian operator $\hat{Q}_k$ acts on the kets $\{|N_k\rangle\}$, and its only non-null matrix elements are:

$$\langle N_k|\hat{Q}_k|N_k - 1\rangle = (1/\sqrt{2})\,(N_k)^{½}$$
$$\langle N_k|\hat{Q}_k|N_k + 1\rangle = (1/\sqrt{2})\,(N_k + 1)^{½} \qquad [4]$$

We will now consider two sets of kets $\{|a\rangle|N_1\rangle|N_2\rangle\ldots|N_k\rangle\ldots\}$ and $\{|b\rangle|N_1\rangle|N_2\rangle \ldots |N_k\rangle\ldots\}$ which represent the states of the system formed by the union between the cation and all the normal modes. In order for the $V_1(t)$ interaction (equation [3]) to induce transitions between $|a\rangle|N_k\rangle$ and $|b\rangle|N_k'\rangle$, and for the states of the other modes to remain unchanged, the following conditions must be verified (complement 1 in chapter 5):

▹ The matrix element $\langle a|\langle N_k|\hat{U}_k\,\hat{Q}_k|b\rangle|N_k'\rangle$, which is equal to $\langle a|\hat{U}_k|b\rangle\langle N_k|\hat{Q}_k|N_k'\rangle$, must be non-null. According to equations [4], this is only possible if $|N_k'\rangle = |N_k \pm 1\rangle$: the transitions take place between *adjacent* levels of mode $\{k\}$.

▹ The energy conservation condition implies that $\hbar\omega_k \approx \Delta E$: a transition between $|a\rangle$ and $|b\rangle$ is accompanied by a transition in the opposite direction between two states of a "resonant" mode $\{\omega_r\}$ of angular frequency $\omega_r \approx \Delta E/\hbar$ (figure 5.10a).

We will calculate the probability per second for the allowed transitions:

▹ In the case of the transition $|b\rangle|N_r\rangle \rightarrow |a\rangle|N_r + 1\rangle$, the matrix element is $(1/\sqrt{2})\langle a|\hat{U}_r|b\rangle(N_r + 1)^{½}$ and the transition probability per second can be written (complement 1, chapter 5):

$$P_{ba} = (\pi/4\hbar)|\langle b|\hat{U}_r|a\rangle|^2\,(N_r + 1)\,d(E)$$

where $d(E)$ is the density of the energy E for states $|a\rangle|N_r + 1\rangle$. As this energy is equal to $E_a + E(N_r + 1)$, its density is equal to the *convolution product* of the density of E_a by that of the energy $\hbar\omega_r$ of the resonant vibrational modes. The density of vibrational modes is much greater than that of the levels of the paramagnetic centre (figure 5.9), and thus $d(E)$ can be identified with $g(\omega_r)/\hbar$. The probability per second w_{ba} for the transition from $|b\rangle$ to $|a\rangle$ is the averaged value of P_{ba} for all the resonant mode's energy levels; the contribution of each level is weighted by its population at temperature T:

$$w_{ba} = (\pi/4\hbar^2)\,|\langle b|\hat{U}_r|a\rangle|^2\,(\overline{N_r}+1)\,g(\omega_r)$$

$\overline{N_r}$ is the mean value of N_r at temperature T, which is equal to

$$\frac{1}{\exp(\hbar\omega_r/k_B T)-1}$$

▹ In the case of the transition $|a\rangle|N_r\rangle \rightarrow |b\rangle|N_r-1\rangle$, the matrix element is $(1/\sqrt{2})\langle b|\hat{U}_r|a\rangle(N_r)^{½}$ and the transition probability per second is $P_{ab} = (\pi/4\hbar)|\langle a|\hat{U}_r|b\rangle|^2\,N_r d(E)$. The transition probability per second from $|a\rangle$ to $|b\rangle$ is equal to:

$$w_{ab} = (\pi/4\hbar^2)|\langle a|\hat{U}_r|b\rangle|^2\,\overline{N_r}\,g(\omega_r)$$

We can verify that $w_{ab}/w_{ba} = \exp(-\hbar\omega_r/k_B T)$, as indicated in section 5.3.1. From this result we deduce that:

$$1/T_1 = w_{ba} + w_{ab} = (\pi/4\hbar^2)|\langle a|\hat{U}_r|b\rangle|^2\,g(\omega_r)/\tanh(\hbar\omega_r/2k_B T) \qquad [5]$$

At X-band, we can replace $[\tanh(\hbar\omega_r/2k_B T)]^{-1}$ by $2k_B T/\hbar\omega_r$ provided the temperature exceeds a few kelvins, so that $1/T_1$ is found to be proportional to T.

This temperature dependance was obtained simply by assuming *normal* vibrational modes. It is much more difficult to determine the *value* of T_1. Indeed, to do so we must calculate the matrix elements and the density $g(\omega_r)$ involved in equation [5]. This requires a detailed description of the states of the paramagnetic centre, of the vibrational modes for the medium and of the interaction $V(\{\mathbf{r}^P\}, \{\mathbf{X}^L\})$ [Orbach and Stapleton, 1972]. We highlight the fact that the operator $\hat{U}_r$ acts only on the *orbital states* of the paramagnetic centre. In the frequent situation where the orbital momentum is quenched, the slight mixing provided by the spin-orbit coupling and by the magnetic field **B** allows the matrix element $\langle a|\hat{U}_r|b\rangle$ to be different from zero.

References

AYANT Y. & BELORIZKY E. (2000) *Cours de mécanique quantique*, Dunod, Paris.

COHEN-TANNOUDJI C., DIU B. & LALOË F. (2015) *Quantum Mechanics*, Wiley-VcH, New York.

ORBACH R. & STAPLETON H.J (1972) "Electron Spin-Lattice Relaxation" in *Electron Paramagnetic Resonance*, GESCHWIND S. ed. Plenum Press, New York.

Appendix 7

Matrix elements of operators defined from components of an angular momentum

We note $R(M_S, M_S') = [S(S+1) - M_S M_S']^{1/2}$. The non-null matrix elements of operators $(\hat{S}_x, \hat{S}_y, \hat{S}_z)$ in the $\{|M_S\rangle\}$ basis, made of simultaneous eigenvectors of $(\hat{S}^2, \hat{S}_z)$, are as follows:

$$\langle M_S - 1|\hat{S}_x|M_S\rangle = ½\, R(M_S - 1, M_S);\ \langle M_S - 1|\hat{S}_y|M_S\rangle = \tfrac{i}{2}\, R(M_S - 1, M_S)$$

$$\langle M_S + 1|\hat{S}_x|M_S\rangle = ½\, R(M_S + 1, M_S);\ \langle M_S + 1|\hat{S}_y|M_S\rangle = -\tfrac{i}{2}\, R(M_S + 1, M_S)$$

$$\langle M_S|\hat{S}_z|M_S\rangle = M_S$$

$$\langle M_S - 2|\hat{S}_x^2|M_S\rangle = ¼\, R(M_S - 2, M_S - 1)\, R(M_S - 1, M_S)$$

$$\langle M_S|\hat{S}_x^2|M_S\rangle = ½\, [S(S+1) - M_S^2]$$

$$\langle M_S + 2|\hat{S}_x^2|M_S\rangle = ¼\, R(M_S + 2, M_S + 1)\, R(M_S + 1, M_S)$$

$$\langle M_S - 2|\hat{S}_y^2|M_S\rangle = -¼\, R(M_S - 2, M_S - 1)\, R(M_S - 1, M_S)$$

$$\langle M_S|\hat{S}_y^2|M_S\rangle = ½\, [S(S+1) - M_S^2]$$

$$\langle M_S + 2|\hat{S}_y^2|M_S\rangle = -¼\, R(M_S + 2, M_S + 1)\, R(M_S + 1, M_S)$$

$$\langle M_S|\hat{S}_z^2|M_S\rangle = M_S^2$$

$$\langle M_S - 2|\hat{S}_x\, \hat{S}_y|M_S\rangle = \tfrac{i}{4}\, R(M_S - 2, M_S - 1)\, R(M_S - 1, M_S)$$

$$\langle M_S|\hat{S}_x\, \hat{S}_y|M_S\rangle = \tfrac{i}{2}\, M_S$$

$$\langle M_S + 2|\hat{S}_x\, \hat{S}_y|M_S\rangle = -\tfrac{i}{4}\, R(M_S + 2, M_S + 1)\, R(M_S + 1, M_S)$$

$$\langle M_S - 2|\hat{S}_y\, \hat{S}_x|M_S\rangle = \tfrac{i}{4}\, R(M_S - 2, M_S - 1)\, R(M_S - 1, M_S)$$

P. Bertrand, *Electron Paramagnetic Resonance Spectroscopy*,
https://doi.org/10.1007/978-3-030-39663-3

$$\langle M_S\,|\hat{S}_y\,\hat{S}_x|M_S\rangle = -\tfrac{i}{2}\,M_S$$

$$\langle M_S+2|\hat{S}_y\,\hat{S}_x|M_S\rangle = -\tfrac{i}{4}\,R(M_S+2, M_S+1)\,R(M_S+1, M_S)$$

$$\langle M_S-1|\hat{S}_x\,\hat{S}_z|M_S\rangle = \tfrac{1}{2}\,M_S\,R(M_S-1, M_S)$$

$$\langle M_S+1|\hat{S}_x\,\hat{S}_z|M_S\rangle = \tfrac{1}{2}\,M_S\,R(M_S+1, M_S)$$

$$\langle M_S-1|\hat{S}_z\,\hat{S}_x|M_S\rangle = \tfrac{1}{2}\,(M_S-1)\,R(M_S-1, M_S)$$

$$\langle M_S+1|\hat{S}_z\,\hat{S}_x|M_S\rangle = \tfrac{1}{2}\,(M_S+1)\,R(M_S+1, M_S)$$

$$\langle M_S-1|\hat{S}_y\,\hat{S}_z|M_S\rangle = \tfrac{i}{2}\,M_S\,R(M_S-1, M_S)$$

$$\langle M_S+1|\hat{S}_y\,\hat{S}_z|M_S\rangle = -\tfrac{i}{2}\,M_S\,R(M_S+1, M_S)$$

$$\langle M_S-1|\hat{S}_z\,\hat{S}_y|M_S\rangle = \tfrac{i}{2}\,(M_S-1)\,R(M_S-1, M_S)$$

$$\langle M_S+1|\hat{S}_z\,\hat{S}_y|M_S\rangle = -\tfrac{i}{2}\,(M_S+1)\,R(M_S-1, M_S)$$

Comment – It is often convenient to use the operators $\hat{S}_+$ and $\hat{S}_-$ in calculations, they are defined by:

$$\hat{S}_+ = \hat{S}_x + i\,\hat{S}_y;\quad \hat{S}_- = \hat{S}_x - i\,\hat{S}_y$$

and verify:

$$\hat{S}_+|M_S\rangle = R(M_S, M_S+1)\,|M_S+1\rangle$$

$$\hat{S}_-|M_S\rangle = R(M_S, M_S-1)\,|M_S-1\rangle$$

However, it should be noted that these operators are *not Hermitian* [Ayant and Belorisky, 2000; Cohen-Tannoudji, 2015].

References

AYANT Y. & BELORIZKY E. (2000) *Cours de mécanique quantique*, Dunod, Paris.

COHEN-TANNOUDJI C., DIU B. & LALOË F. (2015) *Quantum Mechanics*, Wiley-VcH, New York.

Glossary

Degenerate: An energy level is n-fold degenerate ($n > 1$) if it corresponds to n different states.

Effective spin $S = \frac{1}{2}$: When a paramagnetic centre has a doubly-degenerate energy level which is well separated from the other levels, the effect of magnetic interactions on this level can be described with the help of a spin Hamiltonian for which the operators are the components of an angular momentum **S** with $S = \frac{1}{2}$. This notion is applied to Kramers doublets in sections 6.6 and 8.3.

Electronic orbital: Real or complex function defined at each point in the space, for which the square of the modulus represents the electron probability density (probability that an electron is present) at this point in a given stationary state.

Equivalent operator: An operator that acts on the kets of a vector space can often be replaced within a given sub-space by an equivalent operator which is simpler to use. This equivalence is due to the properties of angular momenta.

Exchange interactions: Effect of electrostatic interactions which appears when the orbitals occupied by unpaired electrons overlap. In atoms, exchange interactions are due to electrostatic interactions between electrons. These interactions preferentially stabilise the highest spin state (Hund's rule). In molecules, exchange interactions create several contributions with opposing signs.

Fine structure: Structure of the spectrum due to the zero-field splitting terms for the Hamiltonian. This structure only exists when the spin of the paramagnetic centres is greater than ½.

Homogeneous line: Resonance line, the width of which is determined by the lifetime of the energy levels between which the transition takes place. Sometimes referred to as "natural" width.

Hyperfine structure: Structure of the spectrum due to interaction between unpaired electrons and one or several paramagnetic nuclei. In complexes containing

P. Bertrand, *Electron Paramagnetic Resonance Spectroscopy*,
https://doi.org/10.1007/978-3-030-39663-3

transition, rare earth or actinide ions, *hyperfine* is often reserved for the nucleus of the cation, and the term *superhyperfine* refers to the nuclei of ligands.

Inhomogeneous line: Resonance line which results from the superposition of a large number of homogeneous lines.

Isotropic interaction: An interaction between two material entities is isotropic when its effect on the energy levels is independent of their orientation relative to the axis linking them. Interaction between a paramagnetic centre and a magnetic field is isotropic if its effect on energy levels is independent of the direction of the field relative to the centre.

Kramers doublet: sub-space of dimension 2 associated with a two-fold degenerate energy level in the absence of a magnetic field. The two kets of a basis are conjugated by "time reversal". If one is written as $| u \rangle = \sum_{M_J} c(M_J) \,|\, J, M_J \rangle$, the other is $| \bar{u} \rangle = \sum_{M_J} (-1)^{J-M_J} c^*(M_J) \,|\, J, -M_J \rangle$.

Line intensity: Area under the absorption signal which describes the line.

Magnetic dipolar interaction: Interaction between the magnetic moments of two paramagnetic entities; varies with the distance according to $1/r^3$.

- ▹ Dipolar interactions between nuclei and unpaired electrons in a molecule are weak and cause anisotropy in hyperfine interactions.
- ▹ Dipolar interactions between unpaired electrons can be much stronger. In organic molecules in a triplet state and some complexes containing transition ions, these interactions determine the zero-field splitting terms. These terms can lead to spreading of the spectrum over a large field range. Dipolar interactions between two paramagnetic centres are generally weaker and cause splitting of the resonance lines.

Magnetic moment: Vector quantity characteristic of the ground state of a paramagnetic entity. The magnetic moment is used to express the interaction between its unpaired electrons and a magnetic field or another paramagnetic entity.

Paramagnetic centre: Microscopic entity (atom, molecule) with unpaired electrons.

Paramagnetic entity: nucleus, atom or molecule which has a magnetic moment. Nuclear paramagnetism is due to protons and neutrons, whereas that of paramagnetic centres (atoms, molecules) is due to unpaired electrons.

Powder spectrum: Spectrum resulting from the superposition of resonance lines produced by a large number of paramagnetic centres randomly oriented relative to the applied magnetic field. Samples made up of a polycrystalline powder or a frozen solution produce a powder spectrum.

Resonance line: Signal due to transitions between two energy levels for a paramagnetic centre, induced by its interaction with the radiation. As a rule, the position and intensity of the line depend on the orientation of the centre relative to the magnetic field.

Saturation: A transition, resonance line or spectrum are said to be saturated when the energy absorbed at resonance by the paramagnetic centres is not transferred quickly enough to the lattice through relaxation processes. In this case, the centres are not at thermal equilibrium. This phenomenon generally occurs at low temperatures where the spin-lattice relaxation time T_1 is long. Saturation can be recognised by the fact that the intensity increases more slowly than the square root of the power of the radiation.

Spectroscopy: Technique to study the elementary constituents of matter based on their interaction with a particular type of radiation, generally electromagnetic radiation.

- ▹ The *position* of the lines of a spectrum is determined by the splitting between energy levels of the elementary constituents.
- ▹ The *intensity* of the lines is determined by interaction of the elementary constituents with the radiation.

Spectroscopy must be distinguished from imaging techniques, which aim to *visualise* the spatial organisation of certain entities at a microscopic or macroscopic scale.

Spectrum intensity: Sum of intensities of all the lines making up the spectrum. This intensity can be calculated by two successive integration of the experimental spectrum. The number of paramagnetic centres contained in a sample can be determined by comparing its value to that of the spectrum given by a reference sample.

Spin density: Function which describes how unpaired electrons are spatially distributed in a molecule. Its value at any point is equal to the difference between probability densities of electrons with ($m_S = ½$) and ($m_S = -½$). It can therefore be positive, negative or null and its dimension is [length]$^{-3}$. See **spin population**.

Spin Hamiltonian: Operator associated with a level of a paramagnetic centre (in general the ground level) which is well isolated from other levels. This operator makes it possible to reproduce the splitting created by magnetic interactions to which unpaired electrons are subjected. It consists of:

- ▹ Spin operators built from the angular momenta of the centre's ground state and of the paramagnetic entities with which it interacts.
- ▹ Phenomenological parameters; their values can be calculated using molecular models of magnetic interactions.

Spin-lattice relaxation: All the processes which ensure that paramagnetic centres return to thermal equilibrium ("thermalisation"). These processes generate energy exchanges between paramagnetic centres and the "lattice", which is composed of the molecules' translation, rotation and vibration modes. These modes determine the sample temperature. The efficiency of relaxation processes increases with temperature.

Spin-lattice relaxation time: In some cases, the efficiency of a relaxation process can be assessed using a characteristic time, the spin-lattice relaxation time (T_1): the more efficient the process, the shorter T_1.

Spin operator: Operator built from components of a spin angular momentum **S**, which acts on the kets of the vector space $\mathcal{E}_S$ associated with **S**.

Spin population: Positive or negative dimensionless number, which represents the fraction of a molecule's spin density present in one orbital of an atom.

Trace: The *trace of a square matrix* is the sum of its diagonal elements. When a matrix represents an *operator*, its trace is the sum of the operator's eigenvalues. In that case, it is independent of the chosen representation, and is also called the operator trace. When the trace is null, the *mean* of eigenvalues is null.

Transition: This term is used to mean two slightly different things:

- ▹ A pair of energy levels defines "a transition" characterised by the splitting between the levels (the transition energy), their associated states and populations. A resonance line is recorded for each allowed transition.
- ▹ The interaction between the radiation and a paramagnetic centre is also said to induce "transitions" between the states associated with these energy levels; their rate is characterised by a "transition probability per second".

Index

P. Bertrand, *Electron Paramagnetic Resonance Spectroscopy*,
https://doi.org/10.1007/978-3-030-39663-3

Printed in the United States
By Bookmasters